MW00835390

FUNDAMENTALS OF HEALTH PHYSICS & RADIATION PROTECTION

FUNDAMENTALS OF HEALTH PHYSICS & RADIATION PROTECTION

Philip C. Fulmer, PhD, CHP
Professor of Physics & Health Physics
Francis Marion University
Florence SC

ISBN: 979-8-9895979-0-1
First printing: January 2024

Cover design: Daniel Flagel
Cover and title page artwork used under license from stock.adobe.com

Printed in the United States of America

TABLE OF CONTENTS

DEDICATION

Any good that comes from this book should not be attributed to me but to many people whom I had the great privilege to learn from during my career (however, I will accept blame for any shortcomings). The list below is not all-inclusive; I worked in the DOE complex and in the nuclear power industry with outstanding individuals who taught me more than they could realize. Though several in this list have passed away, their influence continues. I am fortunate to call them friends.

Dr. Lynn D. "Skip" Hendrick and **Dr. David M. Peterson** were my two primary undergraduate professors, and Dr. Peterson was later my department chair and colleague. Both men drove a sense of determination and rigor into my thought processes.

Dr. John W. Poston, Sr., was my graduate advisor and mentor at Texas A&M University and instilled in me to focus on the concepts first and then let the math take care of itself. I model my classroom approach largely on what I learned from John. Though he has retired, his impact in health physics remains immense.

Dr. Wesley Bolch was a professor at Texas A&M University early in his career for some of my graduate classes, and I co-taught several undergraduate classes with him. He is a consummate professional, and it was a joy to learn from him and work with him.

Dr. Gerald Schlapper and **Dr. Milton McLain** were my other graduate professors at Texas A&M University. Both men had practical experience and brought that to the classroom and showed me that a textbook was only the beginning of learning about health physics.

Dr. Dade W. Moeller was a true gentleman in addition to being a superb health physicist and took a chance on hiring me early in my career to work for his then-young company with world-class professionals. He was an inspiration and encouragement, and it was a pleasure to work for him.

My colleagues in the Physics & Engineering Department at Francis Marion University are simply the best. We enjoy each other's company, and they make it a delight for me to do what I enjoy doing. In addition to being my department chair, **Dr. Derek Jokisch** is a valued friend. He encouraged me to write this book and arranged for my classes to be covered during my sabbatical to do the bulk of the rough draft. He was also a sounding board for many ideas I had for presenting complex information in an understandable fashion.

Students in my classes have been a huge source of inspiration. I have enjoyed experimenting to find what seems to work best in the classroom; this book is a result of those years of trial and error.

Finally, my wife **Judy Fulmer** (also a health physicist) deserves a special thanks. She has encouraged me for years to write this book and has cheered me on through this monumental effort.

Introduction

WARNING: THIS BOOK IS A LITTLE DIFFERENT.

As a reader of many books, I know that it is tempting to skip any introduction or foreword, especially in a textbook. However, I hope that you will recognize the value of the information contained here to get the most out of this resource.

Readers may rightly question "Why another health physics textbook? Aren't there enough good ones out there? My school used ________ or ________ all these years, and it was the standard book to use." There are a couple of answers to those good questions.

The development of this book should in no way be taken as an insult against the existing books. They are excellent, have a lot of information, and are wonderful tools and references. However, I also remember what it was like as a student using those books for the first time and feeling overwhelmed in many cases by the sheer volume of information. In addition, sometimes the books assumed knowledge that I did not possess on practical topics of how things operated in the workplace. Therefore, I have tried to include some basic information that I wish I had known when I entered my first health physics course.

The field of health physics and radiation protection has changed tremendously in the modern era with the advent of the internet and computational tools. Calculations that were done by hand and with estimates (rules of thumb) in the past can be performed with high precision and greater documentation (very important for quality assurance) today than at any point in the past. I have embraced those changes in daily life as well as in authoring this text.

What is going to make this book different from other health physics/radiation protection textbooks? Here is a quick list of just some of the features:

1. In contrast to some books, I assume that the reader has already had some coursework in basic physics, radioactive decay, and radiation interactions. Ideally, you should have already had a dedicated class on radioactive decay and interactions before attempting the material in this book. There will be a very brief review of some radioactive decay and interaction concepts, but the review is by no means sufficient to learn it for the first time.

2. This book is about fundamentals. There is no single book that can cover every health physics topic. That is one reason why there are so many books. This book gives a solid technical underpinning so that mastery of the topics in this book will make the reader more adaptable in all the other areas. Dose calculations, radiological measurements, and movement of radioactivity in the workplace and environment form the basis for practical aspects such as waste management, transportation of radioactive materials, environmental remediation, radiological engineering, emergency management, medical use of radionuclides, decommissioning, and many other specialties. There is no effort to teach regulatory aspects in this book. Those are very detailed and can be learned later at one's own pace.
3. Extensive use of online resources is mandatory in today's world, and this book will show how to do that. There are not many reference tables in the chapters. That is by design. The most versatile professional should be able to use online resources and computational methods to research and get the most recent data available and import it electronically so that transcription errors are at a minimum. Phone apps and online web resources are discussed and used extensively in this textbook.
4. Computational tools are a huge part of being a professional. While we cannot possibly cover all the computer programs and other tools that are available, it is imperative that you at least know the names of some of the common software that is used. In addition, I have included sufficient information for you to be conversant about what the software does, and in some cases have provided links for free software.
5. To the extent practical, I have tried to bring in real-world examples and applications as part of the discussion. Discussion of concepts can thus be reinforced when you see how those concepts are used and why they matter.
6. I have included some examples within the text, although not in every section. However, the main resource will be the companion website that goes with this book (www.hpfundamentals.com). It is impossible to show all the details in a printed book of what you should know and how to solve problems. Therefore, the companion website contains real-world problems (with included solutions for some of them) and computational tools (usually Excel© spreadsheets) that can be used to help in the learning process.
7. While I have tried to be as technically correct as I can be, some purists may find that I am not overly rigorous in my discussions. In particular, I am not inclined to do extensive mathematical derivations. Yes, there are a couple here and there that are of practical interest. However, I am primarily interested in getting the concepts nailed down and have tried to ensure that I have used the technical terminology correctly and

consistently so that you can be precise in your understanding of the information.

This book is appropriate for either a senior undergraduate class or a first-year graduate level class. Logical breaks in the chapters are included so that if only a single course is available, certain chapters could be skipped without too much harm. If two courses are available, then essentially the entire book can be covered.

I must acknowledge that despite my best efforts, it is probable that some errors have crept into the text. I will update the website at hpfundamentals.com with errata as I discover oversights or mistakes. I will also do my best to make corrections for future printings and future editions. If you find an egregious error, please contact me at pfulmer@fmarion.edu.

Finally, if you are reading this book and are not a member of the Health Physics Society, you should join. If you are going to be a professional, you need to conduct yourself as a professional and be part of a professional society. I joined HPS in 1987 as a student member and was honored to be designated as a Fellow of the Society in 2022. It is a close-knit community, there are a lot of technical and professional benefits, and I have many friends that I have met through Society activities. You can join online at www.hps.org; it is a good decision to make.

Hopefully, you've made it this far and are excited about using a 21st century textbook. I will do my best to keep it interesting.

Philip C. Fulmer, PhD, CHP
December 2023

Chapter 1 Quantities and Units

In any field, it is necessary to have a proper understanding of the terms that are used. In health physics (like many fields), there are terms in use that are not always intuitive in their meanings; in addition, there are shades of meaning between some terms. Worse yet, sometimes there are identical units that are used for different quantities, and this can lead to confusion and outright error in interpreting the result of a measurement or a calculation.

We will begin by introducing quantities in three broad areas: characterization of the source of radiation, characterization of the radiation field from these sources, and characterization of the results of interactions of the radiation field in the material of interest. As we get further into the book, we will introduce additional terms when we encounter them, but by starting with a baseline set of terms, we will have a common frame of reference from this point forward.

Our approach in this book is to attempt to understand and define the conceptual idea behind the term first before engaging in the mathematics behind the term. Often, an equation for a term may be memorized, but without a proper understanding of the concept, it is easy to confuse the term and misapply equations. Further, the equations presented in this text may vary in symbology from equations presented in other references. This is a common occurrence in any scientific field. By understanding the underlying basis of the equations, a change in symbology will not create confusion.

RADIONUCLIDE INFORMATION RESOURCES

Throughout this book, we will make frequent use of radionuclide information with regard to half-life, decay mode, and decay energy. In the modern era, it is easiest to obtain this information from electronic sources rather than printed publications. There are many resources available online, but three of the easiest to use are the following:

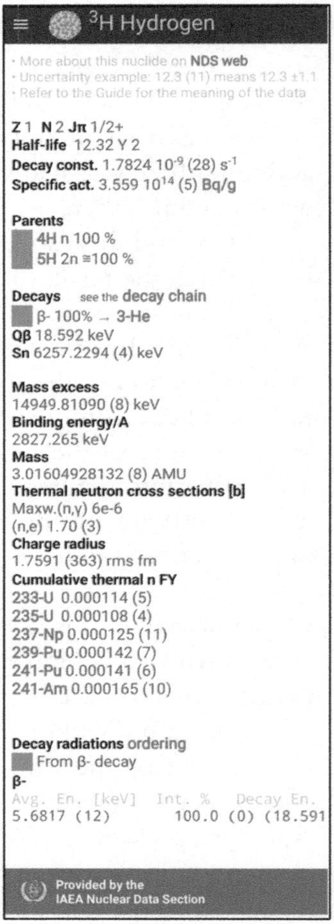

1. **IAEA Isotope Browser App** – The International Atomic Energy Agency (IAEA) has published a phone app that is free to download for both Apple and Android devices. The interface between the two device types is slightly different, but still easy to navigate. To the right is a screenshot of the app (image courtesy of IAEA) and the type of information it contains.

^{3}H Hydrogen

· More about this nuclide on **NDS web**
· Uncertainty example: 12.3 (11) means 12.3 ±1.1
· Refer to the Guide for the meaning of the data

Z 1 **N** 2 **Jπ** 1/2+
Half-life 12.32 Y 2
Decay const. 1.7824 10^{-9} (28) s^{-1}
Specific act. 3.559 10^{14} (5) Bq/g

Parents
4H n 100 %
5H 2n ≅100 %

Decays see the decay chain
β- 100% → 3-He
Qβ 18.592 keV
Sn 6257.2294 (4) keV

Mass excess
14949.81090 (8) keV
Binding energy/A
2827.265 keV
Mass
3.01604928132 (8) AMU
Thermal neutron cross sections [b]
Maxw.(n,γ) 6e-6
(n,e) 1.70 (3)
Charge radius
1.7591 (363) rms fm
Cumulative thermal n FY
233-U 0.000114 (5)
235-U 0.000108 (4)
237-Np 0.000125 (11)
239-Pu 0.000142 (7)
241-Pu 0.000141 (6)
241-Am 0.000165 (10)

Decay radiations ordering
From β- decay
β-
Avg. En. [keV] Int. % Decay En.
5.6817 (12) 100.0 (0) (18.591

Provided by the IAEA Nuclear Data Section

2. **ICRP 107** – The International Commission on Radiological Protection (ICRP) has published its Report 107, which is available for free download at www.icrp.org. The ICRP has also made available a database that is downloadable with radionuclide information.
3. **Live Chart of the Nuclides** – For those who are familiar with the print-style version of the "Chart of the Nuclides," there are several online versions of it published by scientific entities. The IAEA has its version at https: //www-nds.iaea.org/relnsd/vcharthtml/VChartHTML.html, which gives a graphical interface to see all available nuclides on one screen to observe the trends of nuclide properties, such as seeing the "line of stability" and the trends of decay modes based on neutron to proton ratio.

1.1 ICRU Report 85

While several different organizations define radiological units, many of the terms are defined by the International Commission on Radiation Units and Measurements (ICRU). The ICRU attempts to define terms precisely, sometimes using rigorous mathematics, to give the best possible opportunity for health physicists to use terms consistently. ICRU Report 85 *(Revised) Fundamental Quantities and Units for Ionizing Radiation* is available through the ICRU website (www.icru.org) and defines many (but not necessarily all) of the terms we will use in this book.

1.2 Activity

In many calculations or measurements that we make in the health physics/radiation protection field, the activity of a radioactive source is used. While a formal definition of activity helps us to quantify it, we first need to understand the conceptual definition of activity. Simply put, the activity of a radionuclide is the instantaneous number of radioactive transformations or decays that occur per unit time. Although we can measure/calculate activity, it is important to understand that activity by itself does not give us any insight on what radiation is emitted by the radionuclide or any direct indication of the radiological impact from the radionuclide. Each decay produces one or more radiations (unique to that specific radionuclide) that are emitted; those radiations then interact with matter. Our goal ultimately will be to measure/calculate the results of those interactions.

In keeping with the prerequisites identified in the introduction of this book, we assume that the reader is already familiar with the concept of activity, although a few minor notes are in order. It is important to note that activity is the instantaneous number of **decays** that occur per unit time for a given radionuclide. It is determined based on the number of radioactive atoms/nuclei and the radionuclide-specific radioactive decay constant λ. The decay constant has units of reciprocal time and gives the probability of a nucleus decaying per unit time. A large decay constant indicates a large probability of decay, which means that the overall number of radioactive nuclei would decline quickly; similarly, a small decay constant indicates a small probability of decay, which means that the overall number of radioactive nuclei would decline at a slower rate.

For easy mental reference, we tend to document the half-life of a radionuclide instead of its decay constant. The half-life has units of time and is the amount of time necessary for one-half of the radioactive nuclei to decay. It is inversely proportional to the decay constant. In fact, the half-life is calculated as shown below in Equation 1.1.

$$T_{1/2} = \frac{\ln(2)}{\lambda} \quad (1.1)$$

Where

- $T_{1/2}$ = the half life
- ln(2) is the natural logarithm of 2, approximately 0.693
- λ is the decay constant

Activity is a time-dependent quantity; as such, the ICRU defines it in terms of the rate of change of radioactive nuclei (N) with respect to time as shown in Equation 1.2:

$$A = -\frac{dN}{dt} \quad (1.2)$$

Where

- A is the activity
- N is the number of radioactive nuclei of a particular energy state
- t is time

If the number of radioactive nuclei at a particular time is known, the instantaneous activity can also be calculated in terms of the decay constant as shown in Equation 1.3.

$$A = \lambda N \quad (1.3)$$

As time elapses, the amount of activity of a particular radionuclide decreases such that the time-dependent activity can be determined using the radioactive decay equation as shown in Equation 1.4.

$$A(t) = A_0 e^{-\lambda t} \qquad (1.4)$$

Where

- $A(t)$ is the activity at a particular time t during a time interval
- A_0 is the activity at the beginning of the time interval (wherever time t=0 is defined)
- and the other terms are as defined before.

Since there is a direct correlation between activity and the number of radioactive atoms, this means that the number of radioactive atoms also follows the same exponential decay as shown in Equation 1.5:

$$N(t) = N_0 e^{-\lambda t} \qquad (1.5)$$

Where

- $N(t)$ is the number of radioactive atoms at a particular time t during a time interval
- N_0 is the number of radioactive atoms at the beginning of the time interval
- and the other terms are as defined before.

The SI unit for activity is the bequerel (Bq) and is equivalent to 1 decay/second. While useful for our calculations, it is nonetheless a small unit. Historically, a unit called the curie (Ci) was used. It was originally defined as the activity of 1 gram of pure Ra-226 and is equal to 3.7×10^{10} Bq. Of course, the standard metric prefixes can be applied to either unit to accommodate very large or very small amounts of activity. Thus, we can use units of kBq, MBq, mCi, nCi, etc.

1.3 Integrated Activity/Cumulative Activity/Cumulated Activity

In introductory kinematic physics, we do many calculations with velocity and distance. Sometimes we want to know the rate at which objects are moving (velocity) and sometimes we want to know the total distance traveled by the objects.

Similarly, when doing our radiation protection calculations, the value of activity gives us a sense of the rate at which events are happening (e.g., radiations being emitted, interactions occurring), but we are often interested in the total impact over some time interval. The impact might be damage to human tissue, damage to equipment, or some other impact that depends on the total amount of radiation that strikes a particular target of interest. The total amount of emitted radiation is directly proportional to the number of decays that occur during a time interval. We call this the integrated activity or cumulative activity or cumulated activity because it tells us the total number of decays over time due to some amount of activity. In this book, we will call it the integrated activity, but the reader should be aware of the other terms' usage.

If the activity does not change drastically over the time interval of interest, then it stands to reason that the total number of decays during the interval should be equal simply to the product of the activity (decays/second) and time (seconds). This situation occurs quite frequently if the time interval of interest is very short compared to the half-life of the radionuclide.

However, there is also the possibility that the radionuclide may decay appreciably during the time interval. This may occur because the half-life is simply so short that any reasonable time interval of interest is too long. Similarly, it can occur because the time interval of interest is so long that the half-life of many radionuclides is too short by comparison. Either way, the time interval of interest is long compared to the half-life. Therefore, we must determine the number of decays while considering the changing activity. This is illustrated below in Figure 1-1.

For this particular example, the decay constant λ has a value of 0.046 sec^{-1} based on the curve fit equation in the figure. Likewise, the initial activity is given as 265 decays per second (or Bq). If we wanted to know the number of decays that occur in the first 40 seconds, it would be incorrect to simply multiply 265 Bq by 40 seconds, as that would assume that the activity <u>remained</u> 265 Bq for that entire interval when it is obvious from the graph that it is decaying.

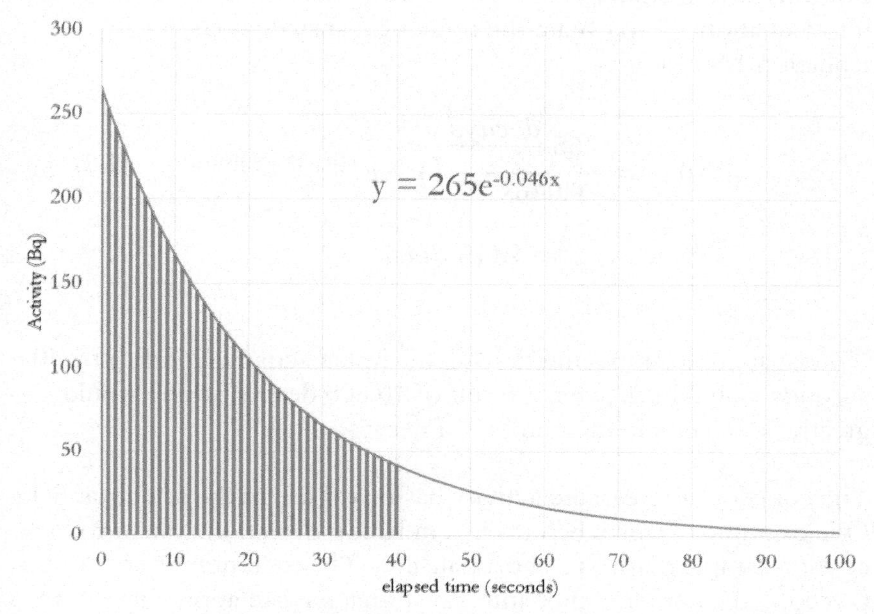

Figure 1-1. The total number of decays (integrated activity) that occur for a radionuclide in the time interval between 0 seconds and 40 seconds is the area under the curve evaluated for that time interval.

Rather, we should use some calculus to determine the area under the curve as shown in Equation 1.6.

$$\int_0^T A_0 e^{-\lambda t} dt = \tilde{A}(T) \quad (1.6)$$

The quantity $\tilde{A}(T)$ is thus the integrated activity. The exact value of T (the time interval of interest) depends on the problem at hand. In Figure 1-1, we are interested in the integrated activity for the first 40 seconds. Therefore, we use integral calculus to solve this integral using a u-substitution and evaluating it for the definite interval, thus obtaining the solution shown in Equation 1.7.

$$\tilde{A}(T) = \frac{A_0}{\lambda}(1 - e^{-\lambda T}) \quad (1.7)$$

For our current situation in Figure 1-1, we substitute T=40 seconds into Equation 1.7 to obtain the numerical answer as shown in Equation 1.8 below.

$$\tilde{A}(40\ s) = \frac{265\ \frac{decays}{s}}{0.046s^{-1}}\left(1 - e^{-(0.046s^{-1})(40\ s)}\right) \quad (1.8)$$

$$= 4845\ decays$$

Note again that simply multiplying the initial activity (265 Bq) by 40 seconds would have given a result of 10,600 decays, which would greatly overestimate the number of decays.

The concept of integrated activity has broad use in the scientific field. One example occurs when using a radiation detection system to count a sample that has a short half-life. The total number of counts is recorded over some time interval so that we can apply the detection efficiency (how many counts are recorded for each radiation emitted by the source) to determine the number of decays that occurred during the interval. However, we may wish to calculate the activity at the beginning of the counting interval. In that case, we would know the integrated activity and the time interval T based on our measurements, and we would use that information to calculate A_0.

Example: How many decays occur in a 10 μCi sample of (a) P-32 and (b) Na-24 over a time period of 10 days?

Solution: This example will give us a little practice in unit conversion as well as calculating the integrated activity. From the IAEA Browser app, we can learn the following information about the two radionuclides:

P-32: $T_{1/2}$ = 14.268 days → $\lambda = 5.62\times10^{-7}$ sec^{-1} (from Eq. 1.1)

Na-24: $T_{1/2}$ = 14.997 hours → $\lambda = 1.28\times10^{-5}$ sec^{-1} (from Eq. 1.1)

Next, we need to convert the activity into units of Bq as follows:

$$10\ \mu Ci \times \frac{10^{-6} Ci}{1\ \mu Ci} \times \frac{3.7 \times 10^{10}\ Bq}{1\ Ci} = 370{,}000\ Bq$$

$$= 370{,}000\ \frac{decays}{sec}$$

This value is the initial activity A_0.

Our time interval of interest is 10 days, which is 864,000 seconds.

Part a (P-32) using Equation 1.7:

$$\tilde{A}(864{,}000\ s) = \frac{370{,}000\ \frac{decays}{s}}{5.62 \times 10^{-7} s^{-1}}\left(1 - e^{-(5.62\times10^{-7}s^{-1})(864{,}000\ s)}\right)$$

$$= 2.53 \times 10^{11}\ decays$$

Again, as we expect, this is less than if we had simply multiplied the initial activity by the time interval of interest.

Part b (Na-24) using Equation 1.7:

$$\tilde{A}(864{,}000\ s) = \frac{370{,}000\ \frac{decays}{s}}{1.28 \times 10^{-5} s^{-1}}\left(1 - e^{-(1.28\times10^{-5}s^{-1})(864{,}000\ s)}\right)$$

$$= 2.88 \times 10^{10}\ decays$$

One additional point to Part b is that the half-life of Na-24 is so short compared to the time interval of interest that the nuclide essentially decays away completely during the time interval of interest. This forces the $(1-e^{-\lambda T})$ term in Equation 1.7 to become essentially unity so that the integrated activity is A_0/λ, a totally logical result since it means that the entire number of radioactive atoms decayed during the time interval (remember that $A=\lambda N$ and thus $N=A/\lambda$).

This example emphasizes that even though both radionuclides had the same initial activity, they have different integrated activities over the 10-day time interval due to the difference in their respective half-lives.

Another case where the integrated activity is important is in the assessment of internal dose: that is, the dose to the human body from internal emitters that remain in the body but decay as time elapses. The dose is proportional to the total number of decays that occur in a given organ or tissue, which we will cover further in Chapter 7. As we will discuss later, the general assumption in internal dose calculations for occupational exposure (not medical exposures) is for the integrated activity to be evaluated for a 50-year time period, which often exceeds the radiological half-life of the radionuclide.

An obvious question to ask at this point is whether Equation 1.7 can ever reduce mathematically to the solution where we can simply multiply the activity by the time. If we hold T at a constant value and allow the half-life to extend to longer and longer values, then λ will get increasingly smaller. Thus, we can attempt to take the limit of Equation 1.7 as λ approaches zero as follows in Equation 1.9, **but** we run into a mathematical impossibility of division by zero.

$$\lim_{\lambda \to 0} \frac{A_0}{\lambda}(1 - e^{-\lambda T}) = \frac{0}{0} \tag{1.9}$$

From calculus, we know that in this case, we can apply L'Hospital's Rule and take the first derivative of the function (shown below in Equation 1.10). Then we can take the limit of <u>that derivative</u> with respect to λ (not T, as might be our first inclination), and that will give us the same limit as the limit of the original function. Doing this indeed results in the integrated activity being approximated as the more intuitive solution A_0T as shown in Equation 1.11 for instances where the half-life is long compared to the evaluation time.

$$\frac{\frac{d}{d\lambda}[A_0 - A_0 e^{-\lambda T}]}{\frac{d}{d\lambda}[\lambda]} = \frac{A_0 T e^{-\lambda T}}{1} \tag{1.10}$$

$$\lim_{\lambda \to 0} A_0 T e^{-\lambda T} = A_0 T \tag{1.11}$$

1.4 Radiation Fluence and Radiation Fluence Rate

The concepts of fluence and fluence rate can be confusing when entering the field of health physics, but again a conceptual understanding of these terms will help us in developing the more mathematical approach to calculating numerical values.

Fluence is related to the intensity of radiation that strikes an area. It can be visualized easily by making a comparison to radiant heat from a fire or other heat source as shown below in Figure 1-2.

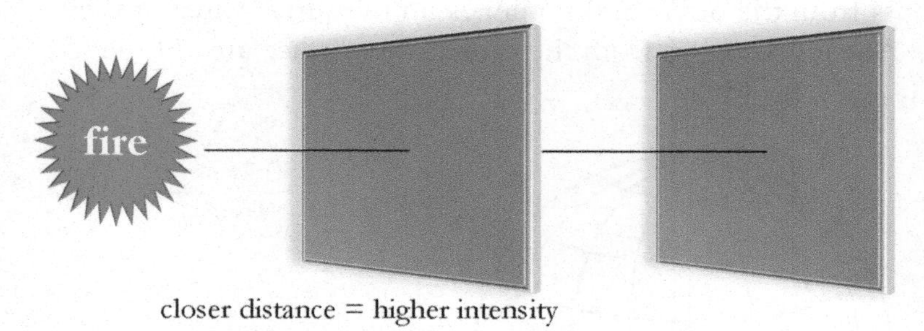

Figure 1-2. Illustration of variation in intensity of heat from a fire as distance is increased.

Intuitively, we recognize that we would feel a greater amount of heat from the fire at closer distances. The reason for this is that the concentration of radiant energy (i.e., the amount of radiant energy per cm^2) on our bodies is greater at closer distances.

We adopt these same ideas when discussing ionizing radiation. The amount of radiation damage or interactions in material is related to the concentration or intensity of the radiation as it strikes a material.

In a general sense, the radiation fluence **Φ** is defined as the number of radiations that strike a unit area as shown in Equation 1.12.

$$\Phi = \frac{number\ of\ radiations}{unit\ area} \tag{1.12}$$

There are some subtleties to the definition, but at the introductory level, it is sufficient to visualize radiation fluence as the number of radiations crossing a 1 cm^2 area, since the usual units for radiation fluence are radiations/cm^2, or cm^{-2}.

For a point source of radiation with no interactions between the source and the receptor, the calculation is relatively straightforward based on the illustration shown below in Figure 1-3.

In this figure, the source of radiation is at the origin of a virtual sphere. The radiation emanates isotropically (uniformly in all directions) from the origin and impinges on the surface area of the sphere. At a given radius from the center, the surface area of the sphere is given by $4\pi r^2$.

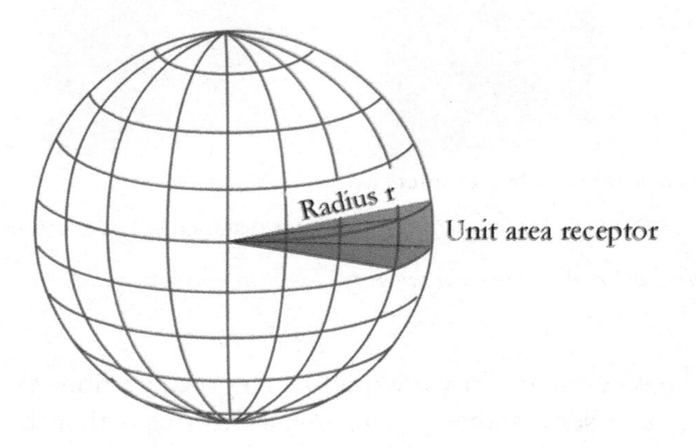

Figure 1-3. Illustration of a radiation source at the center of a sphere where the radiations emanate isotropically.

Hence, the radiation fluence for this particular geometry at a given radius is the total number of radiations divided by the surface area of the sphere as shown below in Equation 1.13.

$$\Phi = \frac{number\ of\ radiations}{4\pi r^2} \tag{1.13}$$

The term $4\pi r^2$ in the denominator gives rise to a common law for point sources called the inverse square law. This law states that the intensity or fluence of the radiation field from a point source varies inversely with the square of the distance. Thus, doubling the distance between the source and the receptor reduces the fluence to a value of ¼ its initial value. This can be useful when trying to reduce the dose to a person, as we will discuss in Chapter 6.

More commonly, we calculate the radiation fluence **rate** ϕ (sometimes written as $\phi(t)$) which is the time-dependent rate of the radiation fluence as shown in Equation 1.14 and has common units of radiations/cm²/second.

$$\phi = \frac{d\Phi}{dt} \quad (1.14)$$

In radiation physics, we typically must calculate a fluence rate for each energy of each radiation type emitted by a radiation source; this is because that there are different probabilities for interactions at different energies. In the case of a point source of a radionuclide, we do this as shown in Equation 1.15.

$$\phi_{point} = \frac{A \cdot Y_i}{4\pi r^2} \quad (1.15)$$

Where

- A = the activity of the radionuclide (Bq)
- Y_i = the yield of the radiation of interest
- r = the radius (distance) from the source

The product of the activity and the yield gives the number of radiations emitted per second, and then dividing by $4\pi r^2$ gives the proper fluence rate.

Example: What is the unshielded fluence rate of 662 keV photons from a 250,000 Bq point source of Cs-137 at a distance of 47 cm?

Solution: Using either the IAEA Isotope Browser App available for mobile devices or the Live Chart of the Nuclides available through the IAEA, the yield for the 662 keV photons is 85.1% or 0.851 γ/decay

Using Equation 1.15, we can substitute in the following values for the variables:

A= 250,000 Bq (250,000 decays/second)

$Y_{662\ keV}$ = 0.851 γ/decay

r= 47 cm

$$\phi = \frac{(250{,}000\ \frac{decays}{second})(0.851\ \frac{\gamma}{decay})}{4\pi(47cm)^2} = 7.66\ \frac{\gamma}{cm^2 \cdot sec}$$

For situations where a receptor location may be struck by radiations from a variety of trajectories, the total fluence is the sum of the radiations striking from all directions. The ICRU accounts for this by defining the fluence as the number of radiations that intersect a sphere with a cross sectional area of 1 cm^2. In essence, this can be visualized by imagining a disc of area 1 cm^2 that is able to rotate in three-dimensional space so that radiation from any direction is always impinging perpendicularly to the disc. This allows the calculation of fluence using integral calculus for geometries other than point sources as well as summing the fluence from multiple discrete sources. Figure 1-4 below shows how using this definition of fluence allows the fluence to be tallied regardless of the direction of the incoming radiation. In the figure, any radiation that strikes any portion of the sphere is counted as part of the fluence even if it does not go through the center of the sphere.

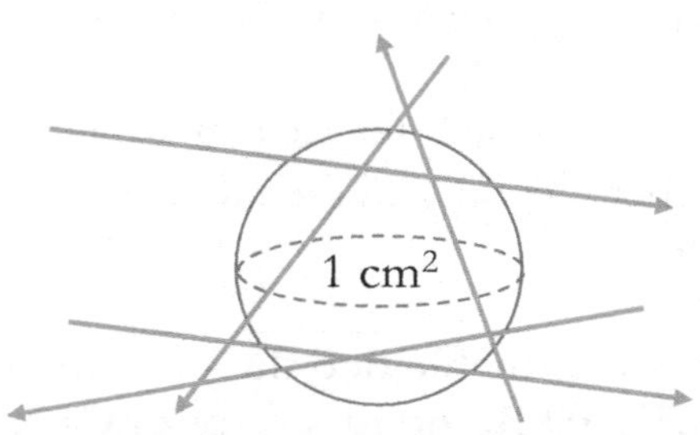

Figure 1-4. The ICRU defines the fluence as the number of radiations that strike a sphere of cross-sectional area 1 cm^2, regardless of the direction of the radiation and regardless of how much of the sphere is intersected by the radiation.

One other source geometry for which the radiation fluence can be readily calculated by hand is for a radiation source uniformly distributed in a line. Again, we assume that there are no interactions of the radiation between the source and the receptor. Unlike the point source, though, there are three different common configurations for how the receptor can be arranged relative to the line source as shown below in Figure 1-5. At first glance, this looks more complicated than it really is. The equation for the calculation of fluence rate for a line source is the same regardless of the configuration and is shown below in Equation 1.16. This equation can be derived using calculus by integrating the fluence rate along the line source, but the derivation itself is not a requirement to be able to use the resulting equation.

$$\phi_{line} = \frac{S_L}{4\pi h}(\theta_{subtended}) \qquad (1.16)$$

Where

- S_L = the radiation emission rate from the line source (radiations/cm along the line) as shown below in Equation 1.17
- h = the perpendicular distance between the line source and the receptor
- $\theta_{subtended}$ = the angle subtended between the receptor and the ends of the line source. This angle **must** be in radians. As shown in Figure 1-5, the main challenge is using trigonometry to determine the angle. The most difficult is the third scenario where a virtual extension of the line source is used to obtain the angle as shown in the figure.

Unlike a point source, a line source does not follow the inverse square law. Increasing the distance between source and receptor does reduce the fluence, but the reduction is not as drastic as it is for the point source.

For a radioactive line source, S_L can be calculated using Equation 1.17.

$$S_L = \frac{A \cdot Y_i}{L} \qquad (1.17)$$

Where

- A = the activity of the source
- Y_i= the yield of the radiation of interest
- L = the physical length of the line source

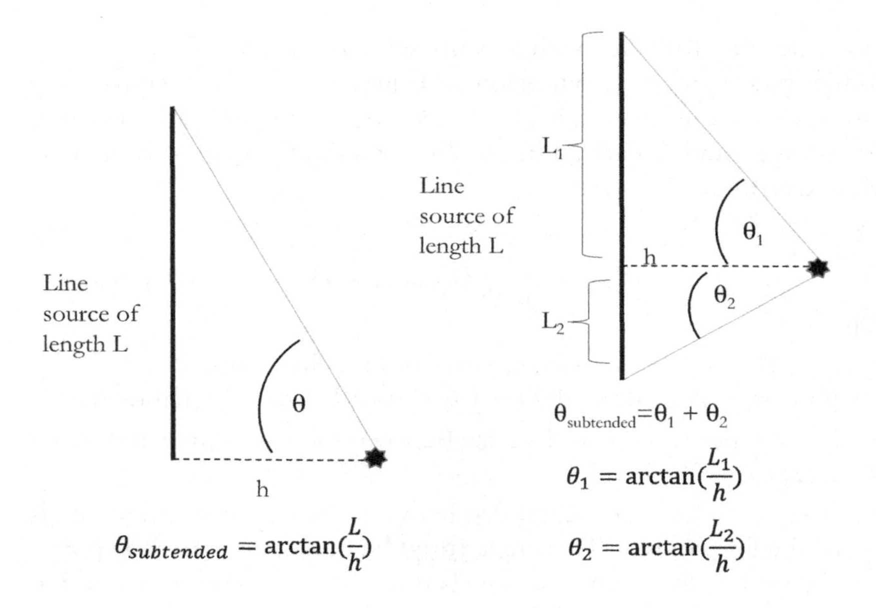

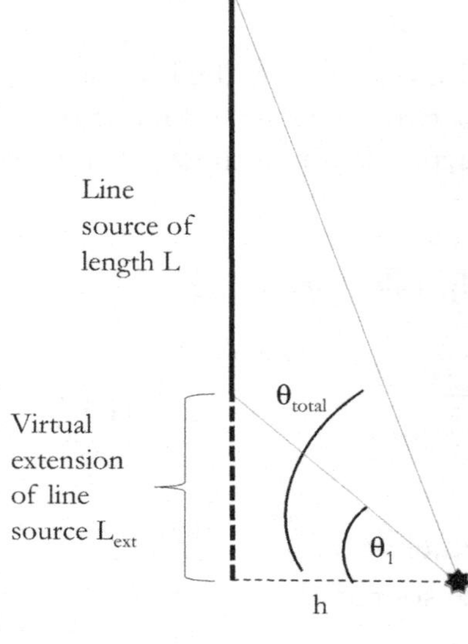

Figure 1-5. Common arrangements of the receptor with respect to a line source. In the first configuration, the receptor is at one end of the line. In the second configuration, the receptor is between the ends of the line source. In the third configuration, the receptor is beyond the end of the physical line source, and a virtual line extension is mathematically employed to solve for the total subtended angle. The virtual line extension is whatever length is necessary so that h is perpendicular to the line. NOTE: the arctangent of an angle may also be called the inverse tangent (tan^{-1}) of the angle.

Example: What is the unshielded fluence rate of 662 keV photons from a 250,000 Bq source of Cs-137 uniformly distributed in a 200 cm line source if the receptor is at a distance of 47 cm from the line as shown below?

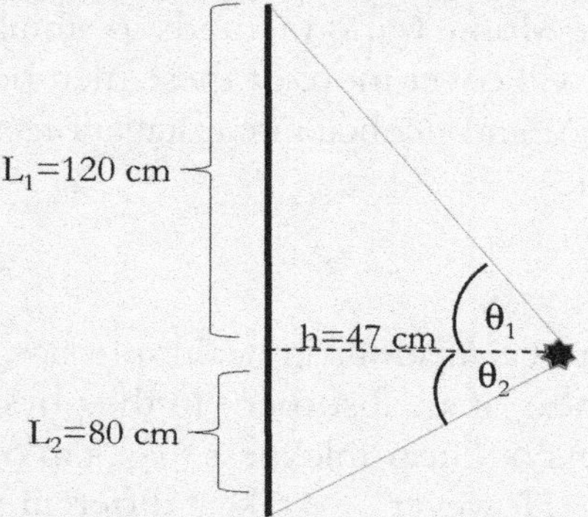

Solution: As we saw in our previous example, the yield of the 662 keV photon in Cs-137 is 85.1% (0.851 γ/decay). It remains only to identify the appropriate parameters and apply them to the second configuration in Figure 1-5.
A= 250,000 Bq (250,000 decays per second)
$Y_{662\ keV}$ = 0.851 γ/decay
L= 200 cm, with L_1 = 120 cm and L_2 = 80 cm

$$\text{Therefore, } S_L = \frac{A \cdot Y_i}{L} = \frac{(250{,}000\,\frac{decays}{second})(0.851\,\frac{\gamma}{decay})}{200\ cm} = 1064\ \frac{\gamma}{cm \cdot sec}$$

θ_1= arctan (L_1/h) = arctan (120 cm/47 cm) = 1.20 (radians)
θ_2= arctan (L_2/h) = arctan (80 cm/47 cm) = 1.04
$\theta_{subtended} = \theta_1 + \theta_2$ = 1.20+1.04 = 2.24

$$\phi_{line} = \frac{S_L}{4\pi h}(\theta_{subtended}) = \frac{1064\ \frac{\gamma}{cm \cdot sec}}{4\pi(47\ cm)}(2.24) = 4.04\ \frac{\gamma}{cm^2 \cdot sec}$$

Note that even though the source activity in this example and in the previous example is the same and the perpendicular distance from the source to receptor is the same, the fluence rate in this example is lower than in the previous example. This is because most of the activity in the line source is at a distance greater than 47 cm due to the geometry of a line source relative to the receptor location.

While it is possible to do hand calculations for other more complex geometries of fluence rate (for example, planar sources or volumetric sources), in the modern era, we typically use spreadsheets or computer programs to expedite the process of approximating integrals, or we use Monte Carlo methods to simulate the tracks of the radiation. We will cover more of these methods in later chapters when we use fluence and fluence rate calculations to calculate dosimetric quantities.

1.5 Density Thickness

In the course of our calculations in health physics, we will need to quantify the thickness of an absorber. In the physical world, we make measurements of linear thickness (e.g., cm of a material, inches of a material, etc.). However, we take a different approach often in our health physics calculations where we use the concept of density thickness. As its name implies, density thickness is the product of the material's linear thickness and its density. If we express the density in units of g/cm^3 and the linear thickness in units of cm, then the density thickness of a material will have units of g/cm^2 as we can see below in Equation 1.18.

$$density\ thickness = \rho t \qquad (1.18)$$

Where

- ρ = density of the material (g/cm^3)
- t = linear thickness of the material (cm)

There are several reasons for using density thickness in health physics. One reason is that it removes the density dependence when calculating an absorber thickness. For example, suppose that a particular shielding calculation shows that a density thickness of 16 g/cm^2 of concrete is needed to shield a source. A quick online search reveals that concrete can be mixed with differing densities depending on the job specifications. Low-density concrete may be used in some instances where the concrete does not have to support excessive weight loads; however, high-density concrete may be mixed to support a lot of weight. By calculating our initial results in density thickness, we can determine the linear thickness at the end of the problem. That way, if the density of the material should change due to design modifications, we can adjust the solution quickly to achieve the correct linear thickness.

Another reason for using density thickness in calculations is that it is easier to tabulate values of interaction probabilities as a probability per density thickness of material. Thus, we can have a fixed set of probabilities for a given material that is not dependent on density. We can adjust the probability when needed for given densities; thus, it makes the tabulation of the fundamental probabilities easier.

As we delve into dosimetric calculations later in the text, the use of density thickness will become more apparent.

1.6 Dosimetric Quantities

Throughout this book, we will refer to several different quantities that are used to express the impact of radiation on materials. In this chapter we simply define the terms without regard for how to calculate them at this point. In later chapters, we will delve into the intricacies of how to calculate these quantities and (in some cases) how to measure these quantities.

We call these "dosimetric" quantities as we borrow a concept from the medical field where medical personnel may dispense a "dose" of medicine to a patient. The dose that is dispensed is not the impact itself, but based on the dose, the impact of the medicine can be assessed. Similarly, in the radiation protection/health physics field, the dose of radiation to a person does not give the impact, but the purpose of assessing the dose is to be able to predict what impact it may have.

It is difficult to understand many of these dosimetric quantities without following some of the history behind the development of them. Therefore, we will proceed through the units in a somewhat historical manner while explaining along the way the shortcomings of each unit and why additional units became necessary. There are other dosimetric quantities that are beyond the introductory level, but we will discuss the most common ones for purposes of radiation protection.

1.6.1 Exposure

In the late 1800s and early 1900s, the use of radiation and radioactive material was not as carefully controlled as it is today. In part, this was because there was still much to learn about the properties of radiation and its potential harm to people. As an example, one of the earliest methods used to determine when people had received enough radiation was to observe whether they had received any skin reddening (erythema). Much like exposure to the sun, exposure to radiation is capable of causing burns to the skin. However, also like exposure to the sun, there is a significant amount of variability from person to person as to what amount of radiation would cause the effect. Thus, this method of controlling the amount of radiation to people was discontinued in favor of methods that could be physically measured and would therefore be more objective.

The first dosimetric quantity that could be measured was termed "exposure." This can be confusing because in health physics/radiation protection, we use the word "exposure" in two contexts. Commonly, we talk about a person being exposed to radiation, and we might refer to this as exposure to the radiation (just as we did in the paragraph above). However, the dosimetric quantity "exposure" is something received from the radiation. So in conversation, health physicists must be careful when using the word "exposure" to discern whether it is referring to the act of a person being exposed to the radiation or if it is referring to the dosimetric quantity.

The idea behind using exposure as a dosimetric quantity was based on the fact that people were dealing with ionizing radiation, which by definition is able to create ions in material. Early radiation scientists recognized the ability to measure the amount of electric charge produced by the radiation in a volume of air, since it is a convenient material to use. Basically, a capacitor could be constructed with air as the dielectric material between the capacitor electrodes. When a high voltage is applied to the capacitor, current does not flow under normal circumstances; however, when radiation produces ions in the air between the capacitor electrodes, a measurable current can be detected, the magnitude of which is proportional to the amount of ionization produced in the capacitor. Practical aspects of this measurement will be discussed in Chapter 5.

Quoting from ICRU 85,

> *The exposure, X, is the quotient of dq by dm, where dq is the absolute value of the mean total charge of the ions of one sign produced when all the electrons and positrons liberated or created by photons incident on a mass dm of dry air are completely stopped in dry air.*

The symbol for the quantity exposure is X, and the mathematical definition is thus given by Equation 1.19

$$X = \frac{dq}{dm} \tag{1.19}$$

Where

- X is the exposure
- dq is the amount of charge produced in the mass of air
- dm is the mass of air in which the charge is produced

The SI unit for exposure is the coulomb/kilogram (C/kg). However, this is a relatively large unit. In the United States, we historically used a unit called the roentgen (R), named after Wilhelm Roentgen, who received the Nobel Prize in physics in 1901 for his discovery of X-rays. Most of our measuring devices in the United States will display exposure measurements in R or a multiple thereof (mR, μR); however, our calculations in later chapters will be in C/kg due to the ease of using SI units in equations. If needed, we can then convert the exposure to roentgens, since 1 R=2.58×10^{-4} C/kg.

There is also the time-dependent quantity called exposure rate. The SI unit for exposure rate is C/kg/sec; the common unit in the United States is the roentgen/hour (R/hr).

The ICRU definition, however, points to some limitations of the definition of exposure. First, it applies only to photon radiation. This makes sense because in the early days of radiation research, photons made up a lot of the radiations that were in use, particularly given the use of x-ray tubes. Secondly, the definition of exposure applies only to air but not to other materials. Again, this makes it possible to make consistent physical measurements with air as the detector material, but it leaves us with questions about how to predict biological effects in humans who are made up of elements other than the constituents in air.

While exposure is not an efficient predictor of biological effect in humans, it is still very useful for characterizing photon radiation in the workplace. Therefore, it is still prolific today in radiation protection, but it is apparent that another dosimetric quantity is needed if we are to have a quantity that helps us in predicting biological effects.

1.6.2 Absorbed Dose

With the shortcomings of the definition of exposure, the next dosimetric quantity to be defined was absorbed dose, which could be applied to all radiation types and to all materials.

The idea behind absorbed dose is that when radiation impinges on a material, it deposits energy due to ionizing the atoms in the material. Absorbed dose has a very rigorous definition in ICRU 85; at the introductory level, it is usually sufficient to understand the concept of absorbed dose as the energy deposited per unit mass of material as shown in Equation 1.20.

$$Absorbed\ Dose = \frac{energy\ deposited}{mass\ of\ material} \qquad (1.20)$$

Dimensional analysis of Equation 1.20 shows that the SI unit for absorbed dose is the joule/kilogram (J/kg). When applied to absorbed dose, this unit has the special name gray (Gy), named after Louis Harold Gray, a pioneer in identifying biological effects of radiation. As with exposure, the SI unit of absorbed dose is relatively large. In the United States, we use the more common unit of the "rad," which was defined in the metric system as 100 ergs/gram. As with exposure, we will always calculate absorbed dose in SI units; however, we may find it convenient to convert those results to rads for workplace application (1 Gy= 100 rads). Absorbed dose rates in units of Gy/second or rads/hour are also commonly used.

The absorbed dose to some (but not all) materials can be measured directly. However, there are some constraints on the interpretation of those measurements and on relating those measurements to other materials for which no direct measurement is possible. For example, in the workplace, we use a dosimeter that fundamentally measures the absorbed dose to the dosimeter material; we then apply some assumptions to relate that measurement to the absorbed dose to human tissue, as we will discuss in Chapter 5.

As defined, absorbed dose can be calculated at a point in a material; however, it can also be calculated as an average over a tissue or an organ, which is usually of interest when trying to anticipate biological effects. For example, a highly localized dose in a small section of the kidney typically would be expected to have smaller effect than if that same dose were received uniformly throughout the kidney.

The advent of the absorbed dose concept led to hopes that it could be used as an effective predictor of biological effects, but a fundamental issue arose with increasing research into radiation biology. It was discovered that equal amounts of absorbed dose in tissue due to different radiation types did not produce the same biological effect. For example, 1 Gy of alpha radiation produced noticeably greater biological effects than 1 Gy of gamma radiation. That effect was also different than that due to 1 Gy of neutron radiation. In essence, the quantitative impact from each radiation type was different so that absorbed dose by itself did not adequately help assess potential radiation impact on people. Clearly, another dosimetric quantity would be needed that could be used as a predictor of biological effects.

1.6.3 Dose Equivalent/Equivalent Dose

The concepts of dose equivalent and equivalent dose are very similar to each other. In both cases, the absorbed dose to human tissue is multiplied by a factor to adjust for the fact that different radiation types cause differing amounts of biological effects. Dose equivalent is a historical term that used a Quality Factor (Q) to adjust the absorbed dose to account for the difference in impact on biological systems. The International Commission on Radiological Protection (ICRP) last recommended the use of this term in 1977 with Publication 26.

Beginning with its 1990 recommendations in ICRP Publication 60, the ICRP defined equivalent dose mathematically as shown in Equation 1.21. This continued into the most recent recommendations in 2007 with ICRP Publication 103.

$$H_T = \sum_R w_R D_{T,R} \tag{1.21}$$

Where

- H_T = the equivalent dose to tissue T
- w_R = the radiation weighting factor (dimensionless) for radiation type R; this factor is used to account for the difference in relative biological effectiveness between photons and other radiation types
- $D_{T,R}$ = the average absorbed dose received in tissue T due to radiation type R

Note that in this definition, the absorbed dose to the tissue is an average dose, not the dose at a given point in the tissue. For many situations, the dose across a tissue may be uniform and thus would be equal to the dose at any given point in the tissue. However, for cases where an organ/tissue is nonuniformly irradiated, the average absorbed dose must be used in calculating equivalent dose.

We also come to a point that is easily confusing. Since the radiation weighting factor is dimensionless as specified by the ICRP, absorbed dose and equivalent dose have the same SI physical units of J/kg. However, the equivalent dose is conceptually different than the absorbed dose. Therefore, the ICRP has designated that equivalent dose has the special name of the sievert (Sv), named after Rolf Maximilian Sievert, who was a renowned researcher into the biological effects of radiation.

In the United States, we use the unit of "rem" as the unit for equivalent dose. Like the mathematical relationship between Gy and rads, 1 Sv=100 rem.

As mentioned above, the radiation weighting factor w_R is used to account for the difference in what we call the relative biological effectiveness between photons and other radiation types. Photons are assigned a value of 1 for w_R; other radiation types are assigned a non-zero value based on their impact compared to photons as shown in Table 1-1 below.

Table 1-1. Values of w_R as defined by the ICRP in Publication 103.

Radiation type	**Assigned value of w_R**
Photons	1
Electrons[a] and muons	1
Protons and charged pions	2
Alpha particles, fission fragments, heavy ions	20
Neutrons	Dependent upon neutron energy

a. An exception is made for Auger electrons emitters that are incorporated into the nucleus of a cell. These should be evaluated on a case-by-case basis to determine the appropriate value of w_R.

For radiation types other than neutrons, it is easy to see that a single value of w_R is assigned for the radiation type regardless of energy. Also, the values of w_R are integer values. Obviously, there is some rounding and variation in these values, and thus the radiation weighting factors are not rigorous scientific quantities. However, for purposes of radiation protection, it is sufficient to use a single rounded integer value for w_R for each radiation type.

For neutrons, the ICRP recommends a value of w_R that is dependent on the neutron energy as shown in the equations below. Figure 1-6 shows the relationship between w_R and energy based on these equations where E_n is the neutron energy in MeV.

$$w_R = 2.5 + 18.2e^{-\frac{1}{6}(\ln(E_n))^2}, \text{ for } E_n<1 \text{ MeV}$$

$$w_R = 5.0 + 17.0e^{-\frac{1}{6}(\ln(2E_n))^2}, \text{ for } 1 \text{ MeV} \leq E_n \leq 50 \text{ MeV}$$

$$w_R = 2.5 + 3.25e^{-\frac{1}{6}(\ln(0.04E_n))^2}, \text{ for } E_n>50 \text{ MeV}$$

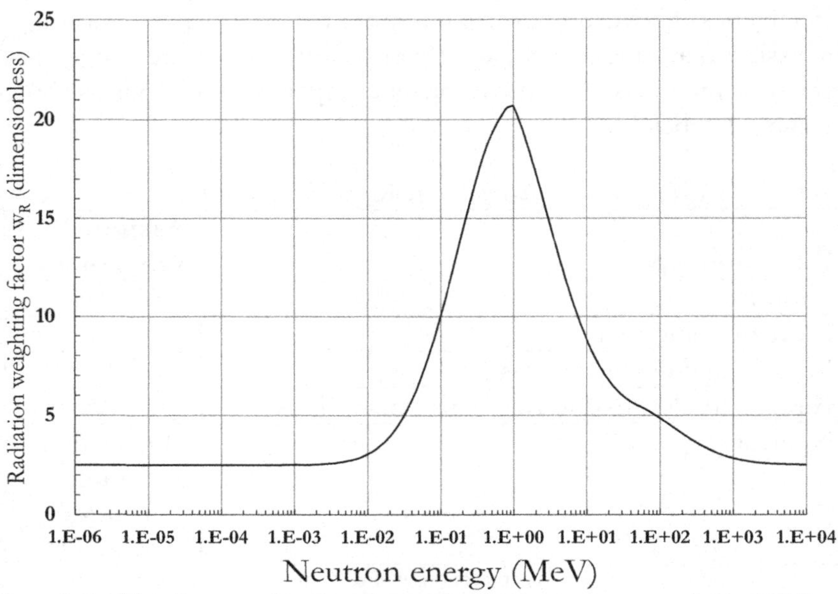

Figure 1-6. Plot of w_R as a function of energy for neutrons as recommended in ICRP Publication 103.

When considering the concept of equivalent dose, it should be apparent that we have now developed a dosimetric quantity that cannot be measured physically. We will discuss in Chapter 5 how we make measurements and then determine the equivalent dose.

With the advent of the equivalent dose/dose equivalent concept, the differences in radiation quality could finally be addressed. However, further research in radiation biology uncovered another conundrum. Different tissues and organs in the body were not equally radiosensitive; some tissues were known to be highly sensitive to certain radiation effects compared to others.

In addition, questions arose as to how to address nonuniform irradiation of the human body. For example, sources of radiation external to the body might irradiate only a small part of the body (one instance would be a worker who handles radioactive material such that the hands receive a much larger radiation dose than other parts of the body). Also, for radionuclides that might be incorporated in the body via inhalation or ingestion, it was known that most radionuclides did not distribute uniformly throughout the body but would have higher concentrations in certain organs and tissues. In these situations, the concept of equivalent dose/dose equivalent is unable to provide a meaningful number that can be used to predict biological effects. For instance, a 0.01 Sv equivalent dose solely to the hands would not be expected to provide the same biological impact as a uniform 0.01 Sv equivalent dose to the entire body; however, the numerical values of equivalent dose give no means to distinguish the potential biological impact.

Due to differing radiosensitivities of tissues and the potential for nonuniform irradiation of the body, another dosimetric quantity was needed to address these issues.

1.6.4 Effective Dose Equivalent/Effective Dose

In its 1977 recommendations in Publication 26, the ICRP introduced the concept of effective dose equivalent, a concept that could be used to address nonuniform exposures and also address differing radiosensitivities of tissues. The effective dose equivalent was defined mathematically as shown in Equation 1.22.

$$H_E = \sum_T w_T H_T \tag{1.22}$$

Where

- H_E = the effective dose equivalent
- w_T = the tissue weighting factor for tissue T
- H_T = the dose equivalent to tissue T

In Publication 60 and Publication 103, the ICRP renamed this concept to be effective dose rather than effective dose equivalent and defined it mathematically as shown in Equation 1.23, which is similar to Equation 1.22.

$$E = \sum_T w_T H_T \tag{1.23}$$

Where

- E = the effective dose
- w_T = the tissue weighting factor for tissue T
- H_T = the equivalent dose to tissue T

As both equations indicate, the assessment of effective dose equivalent or effective dose is a summation over the tissues of the body with a tissue weighting factor w_T applied to each tissue dose equivalent. Conceptually, w_T represents the fractional risk attributable to the tissue of interest as shown in Equation 1.24 below.

$$w_T = \frac{\text{risk to tissue } T}{\text{total risk to whole body}} \qquad (1.24)$$

In the case of a uniform whole-body exposure, w_T would have a value of 1, since the tissue T of interest is the whole body, and the risk to tissue T and the total risk are the same. Logically, the effective dose equivalent and the dose equivalent to the whole body would be the same (a parallel statement can be made for effective dose and equivalent dose). For nonuniform exposures, each tissue would have a w_T value less than 1, but the sum of all the w_T values would be equal to 1.

Effective dose equivalent and effective dose both have SI units of sieverts (Sv), adding potential confusion when doses are reported. Since the sievert is used as the unit for dose equivalent/equivalent dose, it is evident that simply reporting a dose in sieverts is incomplete. Therefore, it is always critical when a dose is reported that the type of dose quantity be clearly understood so that valid conclusions can be drawn.

Equation 1.22 gives a mathematical definition of effective dose equivalent, but we need to understand the conceptual definition of this dosimetric quantity for the equation to make sense. Fundamentally, the effective dose equivalent is a fictitious (but still science-based) dose equivalent that carries the same risk as a uniform whole-body dose equivalent of the same amount. For example, a nonuniform radiation exposure that results in a calculated effective dose equivalent of 0.02 Sv carries the same risk as a uniform whole-body dose equivalent of 0.02 Sv, even if no tissue of the body actually received a dose equivalent of 0.02 Sv. A similar parallel can be drawn for the conceptual definition of effective dose compared to its mathematical definition in Equation 1.23.

The concept of risk as considered by the ICRP has progressed as the ICRP has made newer recommendations. The risk considered in ICRP 26 included the risk of fatal cancer and hereditary effects only. The subsequent recommendations of the ICRP in ICRP 60 and ICRP 103 included the concept of detriment, which was defined as the total harm to health and included principally:

- Probability of fatal cancer due to the radiation exposure
- Weighted probability of non-fatal cancer
- Weighted probability of severe heritable effects
- Length of life lost if the harm occurs

Thus, in ICRP 60 and ICRP 103, the ICRP used the detriment-adjusted risk in determining the values of w_T while ICRP 26 considered only fatal cancer and hereditary effects. Hence, although effective dose equivalent (ICRP 26) and the effective dose (ICRP 60 and 103) are similar in concept, the effective dose is the newer concept and is based on detriment-adjusted risk.

Table 1-2 below shows the evolution of the tissue weighting factors from ICRP 26 to ICRP 60 to ICRP 103. These are included in this text because when evaluating doses from various time periods, it is critical to know the assumptions that were used in the calculation of the dose. As shown in the table, additional tissues were included with specific values of w_T after ICRP 26 rather than assigning them as remainder tissues; also, the values of w_T changed as additional data on biological effects came to light. Finally, the implementation of the remainder w_T changed with each subsequent set of recommendations.

In the ICRP 103 recommendations, it is easy to see based on the w_T values that the individual tissues/organs with the highest risk are the breast, red bone marrow, lung, colon, and stomach. This is in contrast to the earlier recommendations; in ICRP 26 and ICRP 60, for instance, the gonads had the single highest value of w_T. This change in values is the result of both radiobiological research and the change in how risk was defined, as discussed earlier.

Table 1-2. Tissue weighting factors w_T from ICRP Publications 26, 60, and 103

Organ/Tissue of Interest	ICRP 26 w_T	ICRP 60 w_T	ICRP 103 w_T
Gonads	0.25	0.20	0.08
Breast	0.15	0.05	0.12
Red Bone Marrow (RBM)	0.12	0.12	0.12
Lung	0.12	0.12	0.12
Thyroid	0.03	0.05	0.04
Bone Surfaces	0.03	0.01	0.01
Colon	-	0.12	0.12
Stomach	-	0.12	0.12
Bladder	-	0.05	0.04
Liver	-	0.05	0.04
Esophagus	-	0.05	0.04
Skin	-	0.01	0.01
Brain	-	-	0.01
Salivary glands	-	-	0.01
Remainder	0.30[a]	0.05[b]	0.12[c]
Total	1.0	1.0	1.0

a. In ICRP 26, the remainder category is intended to include any five tissues with the highest dose equivalents that were not assigned an explicit w_T; each tissue would be assigned a w_T equal to 0.06.

b. In ICRP 60, the remainder tissues include the following: adrenals, brain, upper large intestine, small intestine, kidney, muscle, pancreas, spleen, thymus, and uterus. The value of 0.05 should be applied to the average dose to these tissues. If the dose to one remainder tissue exceeds the dose to the tissues for which an explicit w_T is provided, then the remainder tissue with the highest dose should have a w_T of 0.025 assigned to it; a w_T of 0.025 would then be applied to the average dose of the other remainder tissues.

c. In ICRP 103, the remainder tissues include the following: adrenals, extrathoracic (ET) region, gall bladder, heart, kidneys, lymphatic nodes, muscle, oral mucosa, pancreas, prostate, small intestine, spleen, thymus, uterus/cervix. The value of $w_T = 0.12$ applies to the arithmetic mean dose of these organs/tissues for each sex (i.e., no splitting of w_T is done as it had been done previously in ICRP 26 and ICRP 60).

As with the concepts of dose equivalent/equivalent dose, the effective dose equivalent/effective dose cannot be measured. We will address this shortcoming in later chapters as we look at what measurements are possible and how those measurements can be used to estimate the effective dose equivalent/effective dose.

1.6.5 Summary of Dose Quantities

The development of dosimetric quantities follows a historical path. As newer quantities were developed, however, the previous quantities remained in use for a variety of reasons, including the ability for physical measurements that are not possible with the newer quantities. Thus, there is a dichotomy that exists in that the quantities that are easiest to measure do not necessarily give a realistic indication of impact from the radiation while the quantities that are designed to give better indications of impact are not measurable directly. It is the purview of the health physicist/radiation protection specialist to understand the distinction in these quantities so that when doses are reported, they can be interpreted meaningfully.

Table 1-3 below summarizes the dosimetric quantities discussed in this chapter. Additional quantities will be introduced later in the text as we address the challenges with measuring dose quantities.

Table 1-3. Summary of the common dosimetric quantities and units.

Quantity	**Unit(s)**	**Directly measurable?**
Exposure	C/kg (SI) R (USA)	Yes
Absorbed dose	J/kg (SI) = Gy rad (USA)	In some materials
Dose Equivalent or Equivalent Dose	J/kg (SI) = Sv rem (USA)	No
Effective Dose Equivalent or Effective Dose	J/kg (SI) = Sv rem (USA)	No

Free Resources for Chapter 1

- Questions and problems for this chapter can be found at www.hpfundamentals.com
- IAEA Isotope Browser app for Android and iPhone devices can be found at the respective app store for the device.
- The Live Chart of the Nuclides website is maintained by the International Atomic Energy Agency and can be located online.
- ICRP Publication 26, *Recommendations of the ICRP*, Annals of the ICRP 1(3), 1977. Available at www.icrp.org
- ICRP Publication 60, *1990 Recommendations of the International Commission on Radiological Protection*, Annals of the ICRP 21(1-3), 1991. Available at www.icrp.org
- ICRP Publication 103, *The 2007 Recommendations of the International Commission on Radiological Protection*, Annals of the ICRP 37(2-4), 2007. Available at www.icrp.org
- ICRP Publication 107, *Nuclear Decay Data for Dosimetric Calculations*, Annals of the ICRP 38(3), 2008. Available at www.icrp.org
- ICRU Report 85 (Revised) *Fundamental Quantities and Units for Ionizing Radiation*, Journal of the ICRU Volume 11, No. 1, 2011. Available at www.icru.org

Additional References

Fluence definitions and the derivation of the fluence from a point source and from a line source are explained in a variety of references. Three common references are the following:

- Attix, F.H. Introduction to Radiological Physics and Radiation Dosimetry, John Wiley & Sons, Hoboken, NJ, 1986, 1991.
- Chilton, A.B., Shultis, J.K. and Faw, R.E. Principles of Radiation Shielding. Prentice Hall, Englewood Cliffs, NJ 1984.
- Shultis, J.K. and Faw, R.E Radiation Shielding, American Nuclear Society, La Grange Park IL, 2000.

In the modern age, numerous scientific publications are available for free download through international agencies. These have the advantage of having been reviewed and are suitable as technical references when justifying assumptions in the workplace. Among the most prolific sources for such documents are:

- International Atomic Energy Agency (IAEA) – www.iaea.org
- International Commission on Radiological Protection (ICRP) – www.icrp.org
- International Commission on Radiological Units (ICRU) – www.icru.org
- Health Physics Society (HPS) – www.hps.org – membership is required but is very affordable for students and gives access to the Health Physics Journal and health physics-related reports of the American National Standards Institute

Chapter 2 Summary of Biological Effects of Radiation

As discussed in Chapter 1 with regard to quantities and units, observations and measurements of the biological impact of radiation on humans have been a driver in the development of standards, dose quantities, and work practices. While this book cannot present an exhaustive treatise on the state of knowledge in radiation biology, there are nevertheless some principles that should be reviewed so that the motivation and reasoning behind certain practices and recommendations in the field are better understood.

More detailed information on the radiobiological effects can be obtained through the resources listed at the end of this chapter.

2.1 Mechanisms and Types of Damage from Radiation

Before delving into the mechanics of radiation damage in biological tissue, it is helpful to consider a few physics principles coupled with our knowledge of units from the previous chapter.

Our dosimetric quantities primarily have at their core a physical component that quantifies the energy deposited per unit mass of material. It would be tempting to conclude that it is solely the deposited energy, then, that leads to biological effects. However, it is not that straightforward.

For instance, it is known that if a person were to receive an absorbed dose uniformly across the entire body of 10 Gy, it is generally not survivable. For a person of mass 70 kg, this would be a total energy deposited in the body of 700 joules (remember that 1 Gy = 1 J/kg). Since 4.184 J=1 calorie, the 700 joules of energy represent 167 calories. Distributed over the entire human body, this amount of energy is insufficient to cause any noticeable increase in body temperature; therefore, the biological impact of radiation is not thermal in nature. In fact, drinking approximately six ounces of water at 38 degrees Celsius (1 degree above body temperature) would give a human the same energy deposition as that lethal dose of radiation.

Clearly, it is not simply the energy deposition at work in causing biological impacts from radiation. Rather, there must be a spatial aspect to the energy deposition that causes the impact. While it may appear that the deposition is uniform at the macroscopic scale, it is nonetheless more grainy at the microscopic scale. This is similar to how we construct grayscale in print format using dots as shown below in Figure 2-1. Much like the grayscale image, we may think that a tissue is uniformly irradiated at all levels, but at the microscopic level, the tissue has some regions that receive greater energy deposition.

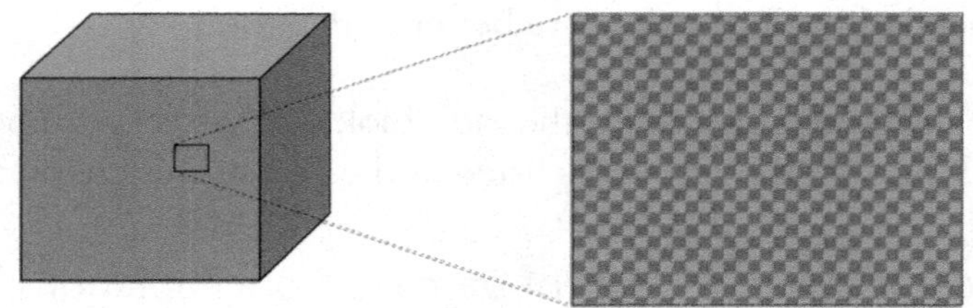

Figure 2-1. Illustration of the illusion of uniformity. The box on the left appears uniformly gray, but zooming in on the surface shows pixelation that indicates that all areas of the box are not exactly the same color.

The hazard of radiation to biological tissue involves a multi-step process. First, the radiation interacts and produces ions; this causes chemical changes to occur, which then can result in biological damage as shown in Figure 2-2 below.

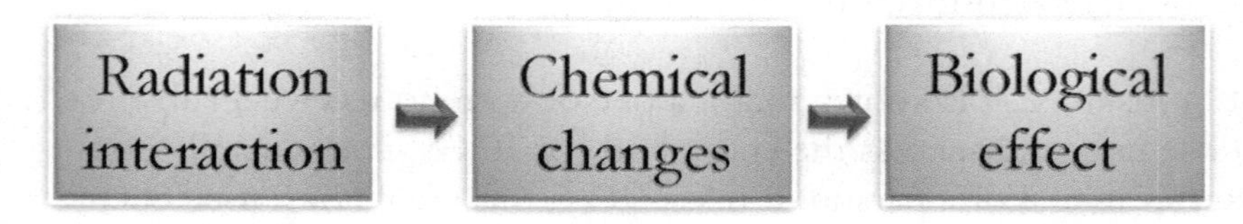

Figure 2-2. The steps leading to possible biological effects from radiation.

In looking at the radiation interaction piece of the chain of events, we realize that the radiation can cause damage ultimately through one of two mechanisms:

1. Direct interaction mechanism – the radiation ionizes biological material (DNA, organelles, cell structures, etc.), which directly damages the material via the chemical changes.
2. Indirect interaction mechanism – the radiation ionizes other molecules (usually water) to form chemical reactants that then proceed to damage biological material.

For the indirect interaction mechanism, it is easy to see how the radiolysis of water (a major component of biological systems) can lead to biological damage. There are a multitude of possible chemical reactions that can occur, but by looking at just a few, it helps us understand some of the mechanisms. Figure 2-3 shows some chemical reactions that can occur, resulting in the production of hydrogen peroxide (an oxidizer) that can then attack biological material. Other chemical species can also be produced as well; the free radicals $OH^{\cdot}$ and $H^{\cdot}$ are very reactive and could do damage (remember that free radicals have an unpaired electron and by nature are quite reactive). By contrast, the H^+ and OH^- ions are readily found in the body already as part of our normal body chemistry; so these are usually not as much of a concern.

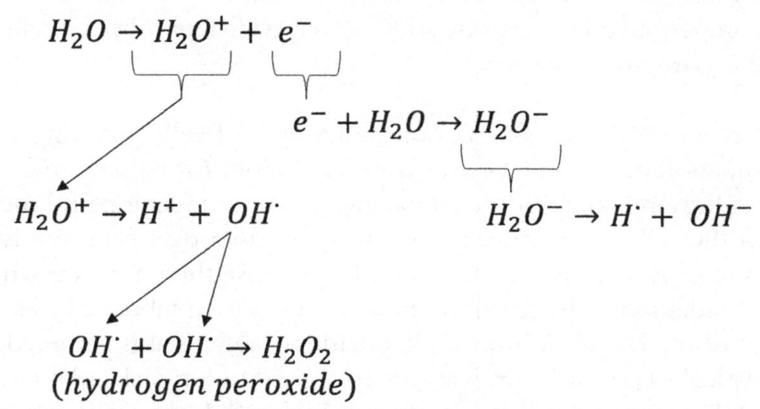

Figure 2-3. Examples of some chemical reactions following the radiolysis of water.

It is a natural question to wonder how many of these reactions must take place to cause significant biological damage. That is the subject of much research. It is known already that the cells of the body have a tremendous repair ability because of damage that occurs independent of any radiation injury. In fact, there are estimates that it is normal for a single cell to have as many as a million DNA changes occur per day (see for example, nature.com/scitable/topicpage/dna-damage-repair-mechanisms-for-maintaining-dna-344/). Clearly, the human body is capable of repairing much damage that is inflicted upon it.

At the cellular level, the damage from radiation could result in one of the following:

1. The cell repairs itself with no long-lasting impact. As noted above, this is very common since the body has error-correcting software code in the DNA.
2. The cell could be damaged to such a degree that the cell dies.
3. The cell is damaged but may be able to reproduce; however, the progeny are lacking in full function and eventually die.
4. The cell's DNA is mutated, and the progeny are likewise mutated; the cells may reproduce out of control and invade other tissues of the body (this uncontrolled growth is called malignancy).

Contrary to first instincts, cell death may be a desirable outcome in some instances rather than accepting the risk of a mutation that could be malignant. In many instances, the death of cells can be compensated by the body so long as a critical number of cells do not die, particularly stem cells.

Cell survival curves (a plot of the fraction of cells surviving as a function of absorbed dose) are a useful tool for helping to understand some of the damaging effects of radiation and the ability of the cell to repair itself. One way to gather data for a survival curve is to grow cells in a culture and then expose them to a known amount of radiation. The fraction that survives is then tabulated as a function of dose. Because of the wide breadth of data that is gathered, it is typical to plot cell survival curves on a semi-log scale. An example of such a curve is presented below in Figure 2-4.

Data such as that represented in Figure 2-4 were responsible for the realization that not all radiation types were equal in their impact. Some radiation types such as alpha particles and neutrons required less absorbed dose to kill cells than did gamma radiation or electron radiation. These types of observations formed the basis for what became the radiation weighting factor w_R discussed in Chapter 1. Sparsely ionizing radiation like photons or electrons produce less concentrated damage and apparently allow the cell more opportunity to repair, while densely ionizing radiation produces more concentrated damage that makes it less likely that the cells can survive. The concentration of ionization is often referred to as the linear energy transfer (LET) of the radiation. Thus, photons and electrons are known as low-LET radiation while alpha particles, protons, and neutrons are known as high-LET radiation. LET has units commonly of keV/μm and represents the amount of energy transferred to electrons in the medium of interest per unit length.

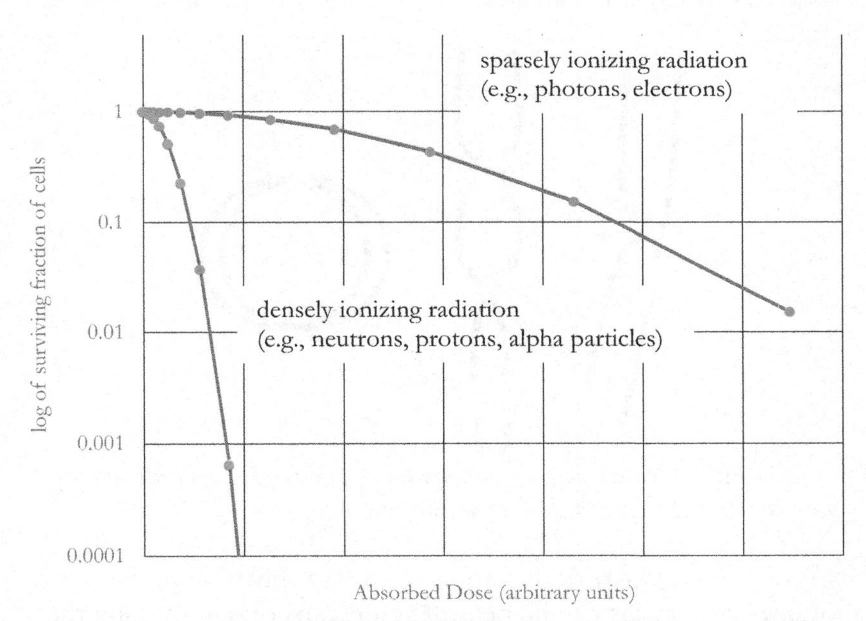

Figure 2-4. A cell survival curve showing the difference in survival between densely ionizing radiation and sparsely ionizing radiation.

The damage to DNA in the cell (termed chromosomal aberrations), which may or may not be lethal, can take on many forms, most of which are beyond the scope of this book. However, there are a couple that are worth mentioning. In these examples, the normal structure of the chromosome with two chromatids and a single centromere after replication (but prior to cell division) is altered, thus giving rise to the aberration.

1. Dicentric – In this aberration, two chromosomes both experience a strand break. Following replication, there are two centromeres in the main chromosome and fragments that are left over.
2. Ring – In this aberration, the two arms of the same chromosome experience a break with the ends of the chromosome joining in on itself to form a ring. Following replication, several different configurations can arise, including an overlapping ring. As with the dicentric, fragments can be left over as well.

Figure 2-5 shows a rudimentary rendering showing the difference between a normal chromosome compared to the dicentric and the ring.

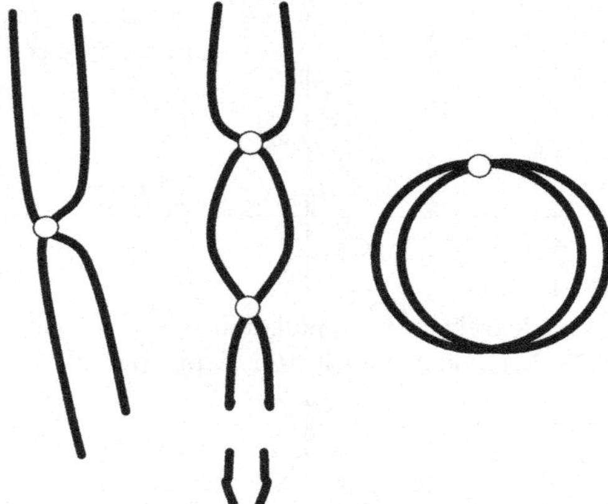

Figure 2-5. Artist rendering of a normal chromosome after replication (left), a dicentric (upper center), a ring (right), and fragments (lower center).

Measuring the number of dicentrics in a cell population under laboratory conditions can be helpful as a means of determining the radiation dose that was received, since the number of dicentrics can be related directly to radiation dose. Further information on this highly specialized dosimetry technique is available through the Radiation Emergency Assistance Center/Training Site (REAC/TS) website at (https://orise.orau.gov/reacts/cytogenetic-biodosimetry-laboratory.html).

2.2 Biological Response to Radiation

Whether at the cellular level or at the whole-body level, there are some general principles gleaned from radiobiological research to explain what factors cause greater or less radiosensitivity. However, it should be noted that the individual factors are not necessarily independent of each other. It is the combination of factors that provides the best explanation for radiobiological observations regarding radiosensitivity.

a. **Total dose received** – As might be expected, the total amount of radiation received by biological material will determine the amount of impact. Higher doses are expected to provide greater impacts. However, this number by itself does not address all variables. The time over which the dose is received plays a part as well. For instance, 1 Gy received uniformly over the course of 1 year does not carry the same impact as 1 Gy over a lifetime. Just as with chemical toxins, cells and the human body seem able to repair damage, and the dose over a longer period of time gives more opportunity for repair to occur.
b. **Dose Rate** – As just mentioned, the rate at which the dose is received also affects the degree of damage to biological material. Higher dose rates cause damage so quickly that biological repair mechanisms may be overwhelmed and unable to repair some or all of the damage. In medical radiation therapy, oncologists take advantage of this principle when exposing tumors to radiation. Multiple doses of radiation are delivered to the tumor but are spread out in time, in part to allow healthy surrounding tissue to recover somewhat in between the doses (eradicating the tumor is of little value if the person's healthy tissue is destroyed in the process). We call this treatment method "dose fractionation."
c. **Quality/Effectiveness of the Radiation** – As discussed in Chapter 1 and illustrated in Figure 2-4, different radiation types produce differing amounts of damage owing to the concentration of chemical species produced because of the linear energy transfer (LET) of the radiation. Alpha particles produce a very dense particle track, thus causing more chemical species and more damage in a smaller volume, while gamma rays produce more sparse ion tracks over a larger volume that gives opportunity for cells to repair. An important note, however, is that at high doses where cell death can occur, the difference in radiation type becomes less significant. In essence, if a cell is dead, then increased ion production in the cell achieves no further impact. For instance, 10 Gy of neutron radiation produces no greater impact in human tissue cells than 10 Gy of

gamma radiation because both absorbed doses are sufficient to kill the cell. This principle will be expounded further when considering the impacts of acute whole-body irradiation at high doses.

d. **Division rate of the cell** – The mitotic division rate of the cell affects the biological impact. Cells that divide at a faster rate seem to be more radiosensitive. Thus, when looking at cancer induction, we tend to be more concerned about tissues that have a higher division rate, such as bone marrow and stem cells in general.
e. **Oxygen content of the cell** – In general, cells that are oxygenated are more radiosensitive than cells with lower molecular oxygen in them. The exact mechanism for this observation is still not fully understood; in addition, this effect is negligible for densely ionizing radiation such as alpha radiation. Therefore, this effect seems to be most noticeable for sparsely ionizing radiation such as gamma rays. One application of this effect is in the treatment of cancerous tumors, where the inner core of the tumor is often hypoxic (low in oxygen). Initial treatment of the tumor via external radiation is used to kill the outer layers of the tumor more so than the inner core. As the outer layers slough off, though, the inner core becomes more oxygenated and can continue to grow. Fractionated radiation therapy, therefore, is used to continue to treat the tumor to maximize the kill rate of the initially poorly oxygenated core in addition to allowing healthy surrounding tissue to recover, as discussed earlier.
f. **Presence of radiosensitizing agents** – The presence of certain chemicals in the cells makes them more radiosensitive, just as certain medications can make people more sensitive to sunburn outdoors. Again, for cancer treatment, this could be a positive aspect if a radiosensitizing agent can be incorporated into a tumor prior to treatment with external radiation. This continues to be an area of research (see, for example, https://www.ncbi.nlm.nih.gov/pmc/articles/PMC7886779/#:~:text=Radiosensitizers%20are%20chemicals%20or%20pharmaceutical,less%20effect%20on%20normal%20tissues) where some Chinese herbs are known to make tumors more radiosensitive. The challenge is to cause the tumor to be more radiosensitive

while allowing the surrounding healthy tissue to be less radiosensitive.

g. **Presence of radioprotector agents** – Just as some chemicals can make cells more radiosensitive, there are some that can make cells less radiosensitive. As with the radiosensitizers, these chemicals must be present before the radiation exposure. The general idea is that the radioprotector scavenges the free radicals that are produced in the early moments following the radiation exposure, thus reducing the chance for these free radicals to do damage to healthy biological tissue. Usually, antioxidants are viewed as helpful, although the degree to which a particular antioxidant works depends on many factors. Specially designed chemicals have been researched for radioprotection; unfortunately, many of them have some chemical toxicity and thus are not practical. If an effective radioprotector can be developed, it would have widespread application, both in civil defense as well as cancer treatment. For cancer treatment, being able to administer radioprotector agents selectively to healthy tissues would allow high doses to be applied to tumors while minimizing side effects to the healthy tissues.

A search of the literature will reveal that there are other factors that affect radiosensitivity as well; this is not intended to be a complete list but hopefully gives a broad understanding of the factors that cause some of the biological variability in biological impacts of radiation.

2.3 Biological Effects of Radiation

Thus far, the discussion of biological impacts/effects has been rather generic without really discussing what those impacts could be. In this section, we will attempt to give an overview of common impacts.

Radiation absorption is a statistical process. No two radiation interactions produce exactly the same damage at the same location in the cell; therefore, radiation effects are largely grouped into two broad categories based on statistical considerations as follows:

Stochastic effects – those effects for which the probability of occurrence (not severity) is a function of dose. The word stochastic relates to statistics and thus conveys the statistical nature of some radiation effects.

Deterministic effects – those effects for which the severity of the effect is a function of dose, and a threshold may exist. In the radiation protection literature, these are sometimes referred to as nonstochastic effects.

2.3.1 Stochastic Effects of Radiation

As noted above, stochastic effects are those for which the probability is a function of dose. If the effect occurs, its severity has no relationship to the radiation dose. The two main stochastic effects are genetic effects and cancer. In both these cases, it is easy to see that the amount of radiation received does not predict the severity of the effect because the effect is binary: it either happens or it does not. Thus, the dose affects only the chance of the effect occurring while the severity of the impact is independent of the dose.

A confounding consideration is that genetic effects and cancer can be caused by things other than radiation. Indeed, there are chemical toxins and biological agents that can cause these same effects. Therefore, when a genetic effect or cancer occurs, it is almost impossible to say conclusively that it was caused by radiation. This is especially true in the case of cancer, where radiobiological data indicate that radiation is not a very efficient carcinogen compared to some other toxins.

Another confounding factor is that, in the case of cancer, there is a latency period between exposure to radiation and the development of cancer. The latency period depends on the type of cancer; for leukemia, it is typically just a few years. For other solid tumors, it could be 10 or more years. Therefore, an individual's cancer diagnosis may not be tied directly to radiation exposure because of the elapsed time between exposure and the cancer. If the person has been exposed to other toxins (such as smoking), trying to separate the risk due to the radiation exposure compared to the risk from the other toxin adds further complication. Worse yet, there are genetic links that make some people predisposed to some cancers; this makes distinguishing the radiation risk more difficult. The best we can do is to calculate a probability of causation (PC) for a stochastic effect: that is, we evaluate the probability that the cancer was caused by the radiation exposure while recognizing that the radiation exposure may not have caused the cancer at all. For high radiation doses, the probability of causation increases so that we may have more confidence that the cancer could have been caused by the radiation.

In 2019, the total death rate in the United States was 715.2 per 100,000. In that same year, cancer deaths represented 146.2 per 100,000; so cancer deaths represented roughly 20% of all deaths in the United States. (Data on death rates and causes are maintained by the Centers for Disease Control and can be found at https://www.cdc.gov/nchs/products/databriefs/db427.htm#Key_finding) Clearly, there is a significant baseline amount of fatal cancer risk that makes it more difficult to distinguish the risk from radiation-induced cancer. However, epidemiologists have reviewed data from several populations of people to try to identify trends and cancer risks from radiation. These populations include, among others:

- Atomic bomb survivors from Hiroshima and Nagasaki
- Medical patients exposed to therapeutic amounts of radiation
- Atomic weapons workers

As might be understood from looking at just these three populations, the data that we have are from people who have had significant radiation dose that is far above what the general population receives (more on this in the next chapter) and more than what present-day nuclear workers receive.

Although there continues to be research in trying to see what low-level doses do in humans, the present-day recommendations of the ICRP for dose limitation still rely in large part on epidemiology based on higher doses. The quandary with that approach can be understood better by referring to Figure 2-6 below.

The data points in Figure 2-6 are not intended to represent actual measured data points but are illustrative of the sorts of observations that may be made by epidemiologists with regard to cancer (similar graphs/figures can be found in some of the references at the end of this chapter). The data that we have are at higher doses, but most people today are not exposed at these levels. Therefore, it creates a dilemma to discern what the calculated risk should be at lower doses to which most people are exposed. The shape of the data is not conclusive to suggest a particular mathematical fit (logarithmic, polynomial, etc.). Even if a linear fit is assumed, there are several different mathematical functions that are a topic of discussion. It is possible that a threshold exists for stochastic effects; however, the data are not clear.

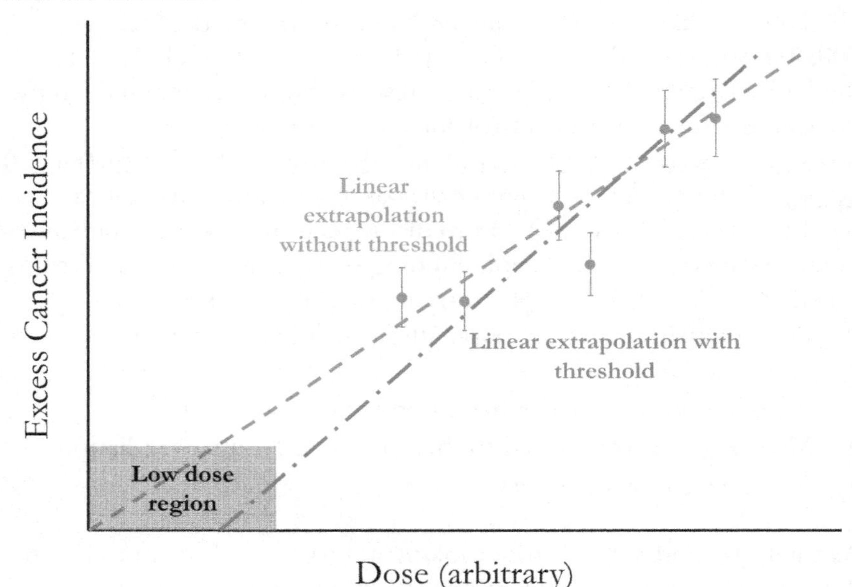

Figure 2-6. Illustrating the difficulty of extrapolating data from higher dose observations to lower doses. Most people are exposed in the low dose region for which the data are not conclusive.

When setting safety standards, however, we do not necessarily have to resolve these questions. The job of health physics and radiation protection is to protect people from harmful effects of radiation, even if that means sometimes taking a conservative approach. The ICRP in its recommendations endorses the Linear-No Threshold (LNT) model for purposes of setting standards. The LNT model assumes that all radiation doses cause some inherent risk, and the risk is linear without a threshold. This does not mean that the model is necessarily correct, but this model is straightforward to implement in assessing risk, even if it may be overly conservative in some situations. One major drawback to the LNT model is that it assumes that every radiation exposure, regardless of how small it is, carries some calculated risk. In practice, this can lead to extraordinary measures being taken to reduce radiation dose in situations where it may be questionable if the dose reduction truly gives additional protection from radiation effects. Understandably, this has caused a lot of debate about the validity of LNT and whether another model should be used for risk assessment.

Finally, as discussed in Chapter 1, the ICRP considered several other factors besides just the risk of fatal cancer and combined this into the concept of detriment when developing the tissue weighting factors. In ICRP 103, the ICRP bases its recommendations on risk factors expressed as a probability per Sv. These risk coefficients do not apply to individuals but are intended to apply to a population of individuals who all receive a comparable effective dose. In the next chapter, it will be easier to put these risks in perspective when we examine the magnitude of doses that people receive. Table 2-1 below summarizes the overall risk coefficients based on cancer and heritable effects.

Table 2-1. ICRP detriment-adjusted risk coefficients as presented in ICRP 103.

Group	Risk coefficient for cancer (Sv^{-1})	Risk coefficient for heritable effects (Sv^{-1})	Total risk coefficient (Sv^{-1})
Adult Workers	4.1×10^{-2}	0.1×10^{-2}	4.2×10^{-2}
General Population[a]	5.5×10^{-2}	0.2×10^{-2}	5.7×10^{-2}

a. The general population includes all ages and thus has a higher risk due to the inclusion of children.

In reviewing Table 2-1, we can try to put the risk of radiation-induced cancer in perspective with a very simplistic comparison. As mentioned earlier, roughly 20% of all annual deaths in the United States are due to cancer (a risk of 0.2). A total equivalent dose of 1 rem (0.01 Sv) would present an additional risk from cancer of 0.00055. Distinguishing this additional risk from the relatively large baseline probability of fatal cancer (which undoubtedly has error bars) is not a trivial task and underscores the challenge of trying to relate excess risk from common radiation doses to workers and the public, whose annual doses are often less than 0.0005 Sv per person. This also emphasizes the near impossibility of saying that any person's cancer is assuredly due to radiation exposure. We may calculate the probability, but we cannot say with certainty that radiation is the culprit.

With the information in Table 2-1, the concept of w_T also becomes easier to understand. For example, the value of w_T of 0.04 (from Chapter 1) for the bladder indicates that for the adult worker population, 4% of the total risk in Table 2-1 is attributable to the bladder (the risk is cancer-related in this case since the bladder is not responsible for heritable effects). The values of w_T for other tissues thus give an indication of the risk attributable to those tissues.

2.3.2 Deterministic Effects of Radiation

Unlike stochastic effects, deterministic effects are generally directly attributable to the radiation exposure. As defined in Section 2.3, deterministic effects are those for which the severity of the effect is influenced by the radiation exposure. While other toxins could cause a similar effect, the magnitude of the radiation exposure and its known mechanisms are sufficient to identify the radiation dose as the cause of the effect.

To understand this concept, an example that does not involve ionizing radiation may help. Exposure to sunlight outdoors in small amounts does not cause sunburn. However, if a person is exposed to too much ultraviolet (UV) light, a sunburn can develop. Of course, other things can cause burns on the skin as well, such as touching a hot surface. If the person is known to have been exposed to large amounts of sunlight, though, the burn on the skin can be tied directly to that UV exposure, both in terms of cause as well as severity.

In the following sections, we will look at some deterministic effects that can occur following irradiation of individual tissues/organs and the deterministic effects following whole-body irradiation.

2.3.2.1 Deterministic effects for individual tissues and organs

Table 2-2 below summarizes data published by the ICRP in Publication 118 to give a sense of some deterministic effects in specific tissues along with the magnitude of the tissue dose required to cause the effect. In this table, the quoted doses are referred to as the 1% dose, or the dose that would cause the effect to be observed in 1% of the exposed population, since different thresholds exist for different people. The effects in this table are not the only possible effects but give a general sense of some deterministic effects following irradiation of specific organs or tissues.

Table 2-2. Data on selected deterministic effects as presented by the ICRP in Publication 118. Absorbed doses are localized to the organ/tissue of interest and not to the total body to cause a 1% incidence.

Effect	Irradiated Organ/tissue	Approximate time for effect to develop	Approximate acute Absorbed Dose (Gy) to cause 1% incidence
Temporary sterility	Testes	3-9 weeks	0.1
Permanent sterility	Testes	3 weeks	6
Permanent sterility	Ovaries	< 1 week	3
Depression of blood-forming process	Bone marrow	3-7 days	0.5
Skin reddening	Skin (large areas)	1-4 weeks	<3-6
Skin burns	Skin (large areas)	2-3 weeks	5-10
Temporary hair loss	Skin	2-3 weeks	4
Cataracts	Eye	>20 years	0.5

2.3.2.2 Deterministic effects for whole-body irradiation

Large doses of radiation in a short timeframe (an acute exposure) to the whole body can cause life-threatening effects. In some cases, proper medical intervention can reduce the risk of death; however, in other situations, medical intervention is solely to provide palliative care. While it is not pleasant to discuss these effects, it is still incumbent on radiation protection personnel to be familiar with the possible outcomes following large radiation doses to the whole body.

When discussing doses that cause these effects, it is typical to express them in units of absorbed dose rather than equivalent dose or effective dose. As discussed earlier, at high doses, the radiation weighting factor w_R approaches unity for all radiation types so that the effects are dependent primarily on absorbed dose. While it is largely a judgment call, the author's preference is that acute absorbed doses above 100 rad (1 Gy) not be converted to equivalent dose or effective dose.

A common term that is used to characterize the mortality risk following acute irradiation is the lethal dose to 50% of the population (LD_{50}). In addition, a cutoff time for mortality is specified to facilitate scoring. Commonly, a 30-day or 60-day timeframe is used; for example, the $LD_{50,60}$ would be the absorbed dose necessary to be lethal to 50% of the affected population within 60 days. While there is a range of this value in the literature, the $LD_{50,60}$ is generally in the range of 4 Gy.

The source of information on these acute effects, unfortunately, is due largely to unintentional radiation incidents that have occurred. Some of them have occurred in the United States as a result of accidents. Others have occurred in countries where radioactive material may not have been controlled properly and an unsuspecting member of the public unknowingly acquired a large radioactive source and took it to their residence.

The following sections discuss three of the notable syndromes that occur. The absorbed doses leading to these syndromes are not absolute; rather, the doses listed here are intended to give a rough idea of the magnitude of dose that might be expected to cause the syndrome and are taken from ICRP Publication 103.

2.3.2.2.1 Bone Marrow/Hematopoietic Syndrome

As noted earlier, rapidly dividing cells are more radiosensitive than other cells; the bone marrow is the source of the blood cells in the body, including the red blood cells that carry oxygen and nutrition as well as white blood cells and others that ward off infection.

Acute doses of less than 1 Gy can cause measurable changes in the blood but are not generally life-threatening. However, acute doses of 3 to 5 Gy cause the marrow (the stem cells for the blood) to be greatly diminished and may kill the marrow entirely. As circulating blood cells in the body die naturally, the depopulation of the marrow means that replacement blood cells are not manufactured quickly enough to keep the person's oxygen level and immune system at peak values. Thus, the person is at risk of suffocation from lack of oxygen but is also at risk of dying from infection with little or no immune system.

The treatment of persons experiencing this syndrome is aimed at giving the person's marrow sufficient time to repair itself if possible. Blood transfusions can help raise the hemoglobin until the red bone marrow can recover and boost red blood cell production. Also, keeping the person in isolation to minimize infection is paramount because the immune system would be compromised, and the person could die from an infection that might be considered minor if the immune system were intact.

If the bone marrow is believed to be destroyed entirely by the acute radiation dose, bone marrow transplants may be an option. However, this introduces its own level of challenges due to the need to have a compatibility match between the donor and the recipient. Additional health challenges (infection, among others) following the transplant could occur as well.

2.3.2.2.2 Gastrointestinal Syndrome

At doses of 5-15 Gy, an additional syndrome becomes of concern due to the sensitivity of the epithelial lining of the gastrointestinal (GI) tract. Nutrition is absorbed by the GI tract through this lining that sloughs off naturally as food passes through. The lining is replaced by stem cells located in the folds of the intestine called crypts. Because the stem cells are rapidly dividing, they are more susceptible to radiation damage, which can cause denuding of the epithelial lining. Figure 2-7 below shows a rough illustration of the lining of the small intestine.

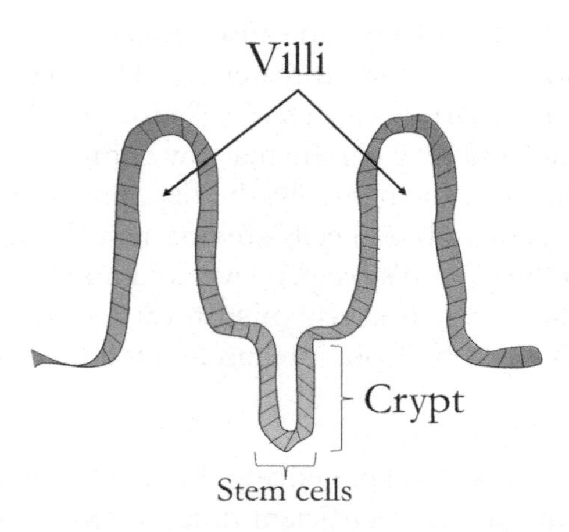

Figure 2-7. The epithelial cells of the small intestine line the villi. Stem cells in the crypt of the intestine produce additional cells to replace the normal wearing away of the epithelial lining.

The symptoms of GI syndrome are not pleasant. Nausea, explosive diarrhea, dehydration, and vomiting are common symptoms. If the stem cells have been totally destroyed, there would be no chance of recovery because there is no transplant option available for these cells at the present time. If the stem cells have not been destroyed, there is a chance of recovery with extreme medical care, but the person would still have to deal with the complications of the hematopoietic syndrome that could still cause death.

A high-profile case of GI syndrome occurred in 2006 with the death of Alexander Litvinenko, a former officer of the Russian Federal Security Service who had defected to Great Britain. Unfortunately, the physicians in his case did not know that he was experiencing GI syndrome until just before his death because they were unaware that he had been exposed to radiation. Litvinenko had been experiencing GI symptoms for several weeks; however, tests for biological agents and chemical toxins did not show the cause for the symptoms. Shortly before his death, it was determined that he had been poisoned with polonium-210, which emits primarily alpha particles. This was the first known case of a person dying from GI syndrome caused by the intake of radioactive material and highlights the subtlety of diagnosing GI syndrome when a large dose of radiation was not suspected.

2.3.2.2.3 Central Nervous System Syndrome

At doses of 15 Gy and higher, the radiation damages the central nervous system by destroying the ion balance in the body's electrical system. The affected person would typically lose consciousness shortly thereafter. At this level of absorbed dose, any medical care is palliative to keep the person comfortable while dealing with the symptoms of the GI syndrome and the hematopoietic syndrome. Death is inevitable; therefore, the emphasis is on treating symptoms and minimizing discomfort while awaiting the eventual outcome.

2.3.2.2.4 Summary of the Acute Radiation Syndromes

Table 2-3 below summarizes the acute radiation syndromes based on the discussion in the previous sections.

Table 2-3. Summary of acute radiation syndromes as presented in ICRP 103 due to low-LET radiation exposure uniformly across the body.

Whole-body absorbed dose (Gy)	Principal contributor to death	Approximate time of death following exposure (days)
3-5	Hematopoietic syndrome (bone marrow depopulation)	30-60
5-15	GI syndrome	7-20
>15	Central nervous system syndrome	<5, but dependent on the dose

2.4 Continuing Work on Bioeffects

In no way should this chapter be considered a comprehensive look at the biological effects of radiation. Many journal articles, books, and scientific presentations are focused on this topic as new information is analyzed. The reference list at the end of this chapter provides a limited list of resources that the reader can use to understand more of the complexity of this important topic. Also, in the United States, a comprehensive projected called the Million Person Study is ongoing and continues to publish analyses based on newer data. More information is available at https://www.millionpersonstudy.org.

Free Resources for Chapter 2

- Questions and problems for this chapter can be found at www.hpfundamentals.com
- CDC Information on Acute Radiation Syndromes available at https://www.cdc.gov/nceh/radiation/emergencies/arsphysicianfactsheet.htm
- International Atomic Energy Agency, *Medical Management of Radiation Injuries,* Safety Reports Series No. 101, IAEA, Vienna (2020)
- ICRP Publication 103, *The 2007 Recommendations of the International Commission on Radiological Protection*, Annals of the ICRP 37(2-4), 2007. Available at www.icrp.org
- ICRP Publication 118, *ICRP Statement on Tissue Reactions / Early and Late Effects of Radiation in Normal Tissues and Organs – Threshold Doses for Tissue Reactions in a Radiation Protection Context.* Annals of the ICRP 41(1/2), 2012. Available at www.icrp.org.
- Radiation Emergency Assistance Center/Training Site https://orise.orau.gov/reacts/index.html
- United Nations Scientific Committee on the Effects of Atomic Radiation (UNSCEAR) www.unscear.org has many reports and documents that are free for download.
- The National Research Council prepared the Biological Effects of Ionizing Radiation (BEIR) VII Report that is free for download at https://nap.nationalacademies.org/catalog/11340/health-risks-from-exposure-to-low-levels-of-ionizing-radiation
- The Million Person Study looks at effects on Americans from a variety of radiation exposure scenarios and is available at https://www.millionpersonstudy.org/

Additional References

Many health physics/radiation protection books discuss the biological effects of radiation. However, more detailed information can generally be found in books devoted to the topic of radiobiology, such as the following:

- Hall, EJ, and Giaccia AJ. Radiobiology for the Radiologist (8th edition), Walters Kluwer Health/Lippincott Williams & Wilkins, Philadelphia, PA, 2018.

Chapter 3 Radiation Exposures to People

As radiation protection professionals, it is important to be aware of the radiation exposure that people receive in everyday life. There are several benefits to being familiar with radiation exposure information:

- It helps people to understand that radiation is a normal part of life and that we already receive some radiation dose aside from what may be due to human activity.
- It helps people to put occupational radiation doses in perspective.
- It helps people to better gauge the units of radiation dose and what constitutes a large dose compared to a lesser dose.

The information in this chapter is summarized largely from Report 160 of the National Council on Radiation Protection and Measurements (NCRP), a US organization that makes radiation protection recommendations similar to what the ICRP does at the international level. Report 160 addresses data that the NCRP reviewed for 2006. At first glance, this may seem out of date. However, the radiation dose that anyone receives is always subject to some variation due to lifestyle and other factors. Therefore, most of the data from 2006 are relevant because the information for that year still gives a rough idea of the radiation exposures for most categories. However, the NCRP updated its information in 2019 regarding medical exposure to radiation; therefore, the more recent data is contained in the section dealing with those exposures.

This chapter presents information for the general public as well as summaries of occupational exposure (the radiation dose to people who work with radiation and radioactive materials). The various radiation sources include the following:

- General Population Exposure
    - Background Radiation Exposure

    - Consumer Products and Activities
    - Medical Exposure of Patients
    - Industrial, Security, Medical, Educational and Research Activities
- Occupational Exposure to Radiation Workers

The NCRP reports several different dose quantities in Publication 160; Section 3.6 in this chapter summarizes the following dose quantities:

- The total collective dose, which is the total dose for all people expressed in units of person-Sv.
- The average individual dose for everyone in the United States regardless of whether they were actually exposed, which is the total collective dose divided by the population of the United States.

3.1 Background Radiation Exposure

Like so many terms, the word "background" has a different meaning depending on context. For example, in radiation detection, the word background generally refers to any undesirable detected radiation that obscures the ability to detect the radiation of interest.

In discussing radiation exposure of the population, though, background radiation refers to radiation that comes from sources that are not produced or enhanced by human activities. The NCRP refers to this as ubiquitous background radiation. This can include radiation dose resulting from sources outside the body or as a result of inhaling/ingesting radioactive material that is present in the environment. The following sections summarize the major sources of background radiation exposure.

3.1.1 Space radiation

Sometimes referred to as cosmic radiation, space radiation has its origin outside our planet. Nuclear reactions in stars or other celestial bodies give rise to a radiation field that impinges on our atmosphere. This radiation field is comprised of mainly high energy charged particles along with photons. There is a baseline amount of space radiation from the sun (and increased radiation during solar events), but much of the space radiation has its origin outside our solar system.

The earth's magnetic field serves as a partial shield to deflect the charged particles so that they move around the earth in keeping with the force exerted on moving charges by a magnetic field. A large amount of the radiation is trapped in two belts that are shaped much like doughnuts; these are called the Van Allen Belts. Figure 3-1 below shows a rendition of the Van Allen Belts along with distances to several satellites and the International Space Station. As shown, the inner belt begins at roughly 1000 miles of altitude. However, there is a weak spot in the earth's magnetic field that leads to a high-intensity region of charged particles called the South Atlantic Anomaly (SAA) that extends as close as 120 miles from the earth's surface.

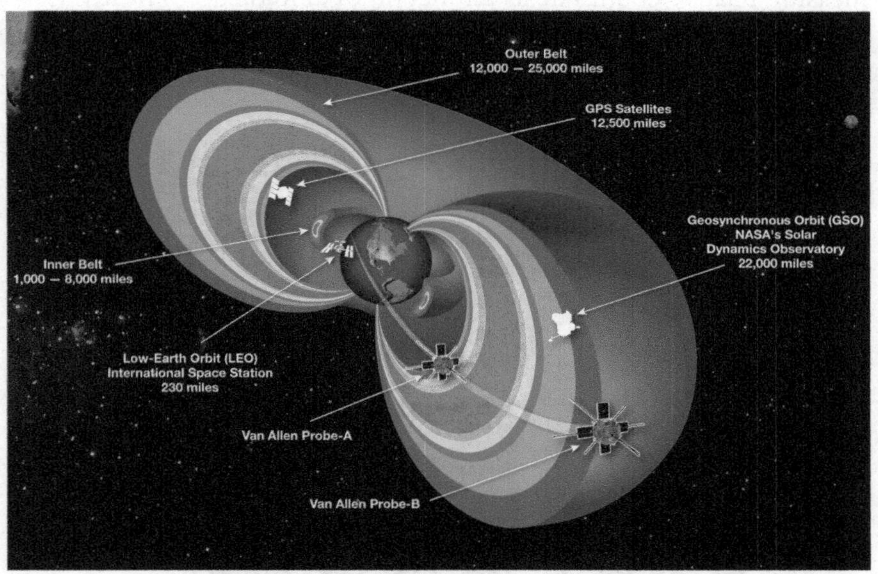

Figure 3-1. Rendition of the Van Allen Belts (Image courtesy of National Aeronautics and Space Administration www.nasa.gov)

The charged particles that make it through the magnetic field of the earth then encounter the shielding effects of earth's gaseous atmosphere. As the radiation strikes the atmosphere, nuclear reactions can result, causing the production of secondary particles such as neutrons, muons, electrons, and photons, some of which penetrate to the earth's surface.

Understandably, the thickness of the shielding afforded by the atmosphere affects the amount of space radiation detectable at ground level. Therefore, we should expect that locations near sea level (with more atmosphere above it) would have a lower radiation dose from space radiation while higher altitudes would have a higher radiation dose. In addition, the radiation dose has some variation with latitude.

The variation in space radiation dose with altitude is easily measurable with devices as simple as a Geiger counter. The author of this text has led numerous student groups who have conducted high-altitude balloon launches with an onboard Geiger detector to measure relative count rate as a function of altitude up to nearly 30,000 meters (over 18 miles). Figure 3-2 below shows some photos of the various aspects of the balloon project.

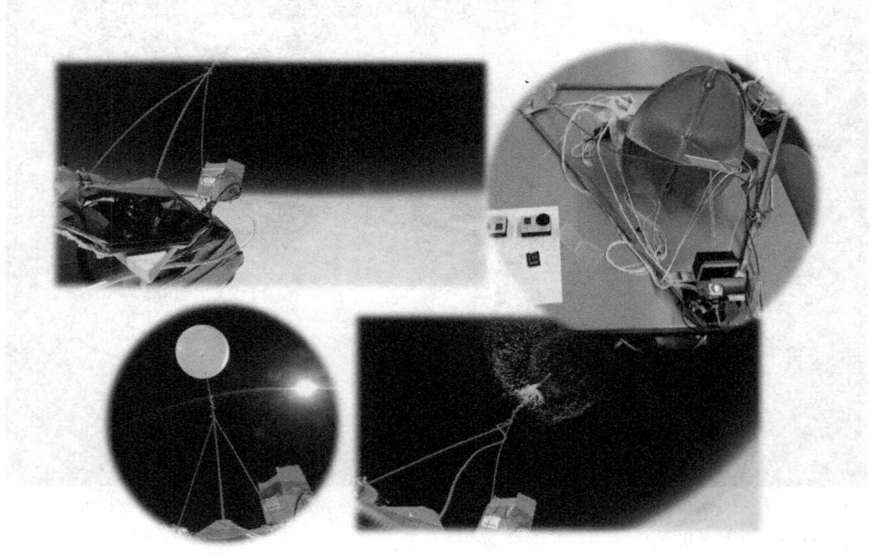

Figure 3-2. High altitude balloon launch using a weather balloon carrying aloft a windframe structure with video equipment and scientific measuring equipment, including a Geiger counter. The author has conducted numerous launches to measure space radiation.

Figure 3-3 shows the raw radiation data obtained with the equipment. An onboard global positioning system (GPS) sensor provided the altitude information, and the datalogging capability of the Geiger counter provided the information to correlate the count rate with the altitude. The feature identified in the figure as the Regener-Pfotzer maximum is the altitude (~ 20 km) where the primary and secondary radiation fluence reaches a maximum. At altitudes below this maximum, the shielding effect of the atmosphere reduces the primary radiation fluence sufficiently such that the total radiation fluence is reduced. Above this altitude, the primary radiation field has not interacted sufficiently in the atmosphere to produce the maximum amount of secondary radiation.

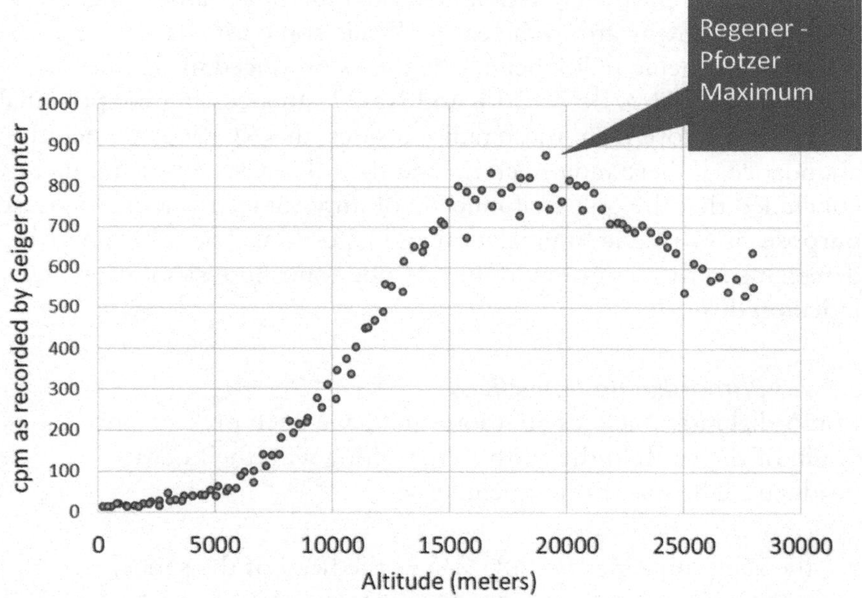

Figure 3-3. Plot of raw experimental data of counts per minute recorded by a Geiger counter in a high-altitude balloon launch.

More detailed information on the makeup of the radiation field and the dose rates at various altitudes has been reported by the National Aeronautics and Space Administration (NASA) as part of their Radiation Dosimetry Experiment (Rad-X) program, which can be found online through the NASA website.

Environmental measurements and calculations both show the variation with geographical location and altitude. NCRP Report 160 reported values of the effective dose varying from 0.28 mSv per year in Honolulu HI (essentially sea level) to 0.82 mSv per year in Colorado Springs CO (almost 6,000 feet altitude).

3.1.2 Cosmogenic Radionuclides and Primordial Radionuclides

3.1.2.1 Cosmogenic Radionuclides

Cosmogenic radionuclides are those that are formed in the atmosphere as the result of cosmic ray (space radiation) interactions in the atmosphere. As noted above for space radiation, the incoming particles are high energy and are capable of a variety of nuclear reactions, thus producing radionuclides in the upper atmosphere that can migrate downward to the earth. While some lists have over a dozen cosmogenic radionuclides, the ones produced in significant abundance are H-3, Be-7, C-14, and Na-22. In assessing the potential for radiation dose from these radionuclides, the NCRP examined the abundance of these radionuclides and their decay schemes and has concluded that the only radionuclide of importance for radiation dose purposes is C-14 due to inhalation and ingestion. The others may be detectable in the environment but do not contribute significant radiation dose.

3.1.2.2 Primordial Radionuclides

Primordial radionuclides are those that have been present since the origin of the earth in the earth's crust along with their decay products. The major ones include:

- the actinium series (with U-235 as the head of the series)
- the thorium series (with Th-232 as the head of the series)
- the uranium series (with U-238 as the head of the series)
- K-40 (decays to stable Ca-40)
- Rb-87 (decays to stable (stable Sr-87)

The three natural decay series listed above are of note for three reasons. First, many of the radionuclides in the three natural series listed above decay by alpha emission. Following alpha decay, the alpha particles travel a very short distance in the earth before acquiring two electrons from their surroundings and forming He-4 gas molecules that are trapped in the earth. This helium is extracted from the earth via wells that are drilled into geologic deposits. Thus, the helium that we use daily in activities is the result of radioactive decay.

Secondly, there is an isotope of radon in the uranium series (Rn-222), the actinium series (Rn-219), and the thorium series (Rn-220). Radon is a noble gas; as such, it is able to emanate from the earth and can migrate into buildings from the soil under the buildings. Of the three isotopes, Rn-222 is the one of most concern due to its longer half-life (3.8 days), which allows it time to migrate before decaying to non-gaseous progeny. Rn-220 (also called thoron) and Rn-219 have shorter half-lives (55.6 seconds and 3.96 seconds, respectively). While Rn-220 can still migrate appreciably, the Rn-219 is low in abundance in the earth's crust and has such a short half-life that its contribution to radiation dose is negligible.

Thirdly, since these decay series are present in the earth's crust in measurable quantities, it is possible for them to be present in drinking water that comes from wells.

The natural decay series nuclides and the other primordial radionuclides can cause radiation dose through three methods:

- Direct external radiation exposure from gamma rays and beta radiation
- Inhalation of the radionuclides (primarily Rn-222 and Rn-220) with dose caused mostly by the non-gaseous progeny that remain in the respiratory system
- Ingestion of the radionuclides in drinking water, milk or food; the radionuclides enter the food chain either through deposition of the radionuclides in the environment or by using groundwater with radionuclides for drinking water or irrigation.

One additional consideration with the natural decay series is the ability for these radionuclides to be concentrated as the result of normal work activities. For example, oil pipelines will accumulate a layer of material (called pipe scale) on the interior of the pipes. This scale can contain measurable amounts of radioactivity with elevated dose levels. While this is naturally occurring radioactive material (NORM) and is not intentionally concentrated, it is referred to as technologically enhanced naturally occurring radioactive material

(TENORM) and is the subject of much study and research in learning how to address the potential doses that arise from human exposure to these materials.

3.2 Consumer Products and Activities

Radioactive materials are present in a variety of products in use by the general public. In some cases, the radionuclides are present naturally (usually the primordial natural decay series) and are not essential to the operation of the product. In other cases (such as smoke detectors), the radioactive material is present intentionally and is necessary to the product function.

Because of the numerous consumer products with radioactive material, this section does not attempt to list every product. Table 3-1 below lists some common products and the source of potential radiation dose in the product.

Table 3-1. Listing of some common consumer products and the source of potential radiation dose (continued on next page).

Consumer Product or Activity	**Source of Potential Radiation Dose**
Tobacco products	Rn-222 is present in the soil naturally but may also be enhanced in concentration due to fertilizers that are used in farming, which contain small amounts of Ra-226. The radon progeny stick to the underside of the tobacco leaves and serve as a significant source of radiation dose. Pb-210 decays by beta emission, but Po-210 decays by alpha emission and thus can cause a relatively large dose per unit activity either from chewing tobacco products or from smoking.
Building materials	Primordial natural decay series and progeny in brick, stone, cement, concrete, and similar products.

Consumer Product or Activity	Source of Potential Radiation Dose
Commercial air travel	Cosmic/space radiation. Dose rate varies based on altitude and latitude. NCRP used a computer model to calculate effective dose rates of ~0.34-6.52 μSv per hour in the air.
Mining and Agriculture	Fertilizers contain trace amounts of K-40 and the natural decay series
Luminous watches and clocks	A small amount of radioactive material (Ra-226, H-3, or Pm-147) is mixed with a luminous agent such as ZnS to provide viewing in the dark
Smoke detectors	Alpha particles from Am-241 impinge on a radiation detector. The presence of smoke interferes with the detection of the alpha particles and triggers an alarm
Thorium welding rods	Natural thorium is used in some tungsten welding rods due to its electrical and metallic characteristics.
Combustion of Fossil Fuels	Primordial natural decay series present in coal; radon released when burning natural gas
Non-electrical Exit signs containing tritium	H-3 used in conjunction with a phosphor to make the signs visible in the dark with no electricity required
Highway and road construction materials	Aggregate may contain primordial natural decay series
Antique Glass and ceramics	Uranium used in some cases historically as a coloring agent in glass or in the glaze of ceramics.

3.3 Medical Exposure of Patients

Medical use of radiation that is electronically produced or emitted by radionuclides has risen dramatically since the 1980s when the NCRP previously issued its report on population exposure to radiation. While there are more procedures performed each year compared to the past, there have also been innovations to reduce unnecessary radiation dose in certain procedures.

Since the NCRP published Report 160 in 2006, there has been a significant reduction in the doses from many medical procedures due to the advances in technology. The NCRP issued Report 184 in 2019 to address the updated values, and the cited values in this chapter for patient doses are the newer values provided by the NCRP.

Radiation can be used for both diagnosis as well as therapy; however, this section will focus on diagnostic uses of radiation. The categories we will focus on are the following:

- Computed tomography (also called CT scan or "cat" scan) – a three-dimensional image that is produced by assembling slices of images obtained using an x-ray beam on one side of a patient and a sensor on the other side.
- Conventional radiography and fluoroscopy – Both these techniques use x-ray beams. Radiography is the traditional x-ray still image while fluoroscopy involves longer beam times to produce real-time video to observe movement or function. While traditional x-rays historically used film that had to be chemically developed, modern x-ray equipment uses digital sensors to obtain an optimum image.
- Interventional fluoroscopy – fluoroscopy as defined above with the purpose of using the video image to help place or use a medical device in the body (such as placing stents during heart catheterization).
- Diagnostic nuclear medicine – the use of gamma-emitting radionuclides in imaging to provide information on the function of organs or tissues. One particular type of scan, called a positron emission tomography (PET) scan, uses short-lived radionuclides that decay by positron emission and rely on detecting the 0.511 MeV photons resulting from annihilation radiation of the positrons. PET scans often are used in cancer diagnosis because the radionuclides can be incorporated into sugar molecules which are taken up in greater amounts by cancer cells compared to surrounding healthy cells. Other radionuclides can be used to obtain structural or functional information about organs in the body as well.

3.4 Industrial, Security, Medical, Educational and Research Activities

Many facilities use radiation and radionuclides as part of their operation. Facilities and sources of radiation include such things as:

- Hospitals (non-patient radiation dose)
- Universities
- Industrial facilities
- Nuclear power generation
- Transportation of radioactive material
- Waste disposal
- Department of Energy facilities, including production and research
- Particle accelerators
- Facilities with x-ray equipment or other radioactive material

Workers in those facilities understandably receive radiation dose as part of their work duties. In addition, as discussed in the last section, medical patients receive radiation dose in hospitals. However, the NCRP uses this category of exposure to address the radiation dose to the public who are not directly involved in the activities. The dose might arise from direct radiation from the facility or from emissions that might be distributed to the air via ventilation stacks or to water due to permitted discharges of small amounts of radioactivity that are then diluted. In the case of hospitals, it might arise because members of the public are present in the hospital and are exposed to small amounts of radiation even though they are not part of a procedure involving the use of radiation.

3.5 Nuclear Weapons Fallout

We also need to consider the issue of fallout from nuclear weapons testing. Prior to 1963, it was a common practice for countries to detonate nuclear weapons above ground as part of developing their nuclear arsenal; this understandably led to large amounts of radioactivity being spewed into the atmosphere. In the case of atmospheric weapons testing, the radioactive fallout could travel large distances and be deposited far from the country of origin. In August

1963, the United Kingdom, Soviet Union, and United States signed a treaty to ban testing of nuclear weapons in the atmosphere, in space, or underwater. Shortly thereafter, other countries joined in the treaty as well.

The treaty and ban in 1963 did not undo the fallout that had already been released, however. Following a nuclear detonation, fission products are deposited in the air and water. Many of them are short-lived and did not persist in the environment for long. The two isotopes that are longer-lived were Cs-137 and Sr-90, both with half-lives of approximately 30 years. Due to radioactive decay and weathering (cesium is particularly soluble and would wash downward into the soil and be diluted), the amount of fallout in the environment has been reduced greatly and is no longer factored into the NCRP dose estimates, even though fallout may still be detected in the environment.

Underground weapons testing was still permitted after 1963. In this case, a shaft thousands of feet deep in the earth (in the United States, it was at the Nevada Test Site) would house a nuclear weapon that would then be detonated underground such that measurements of yield and other parameters would be made but allow the earth above to be a filter to contain the fallout. The detonation would vaporize the soil immediately around the weapon resulting in a collapse of the soil above it to form subsidence craters that could be several hundred meters in diameter. Underground testing for the most part has not been performed since the mid-1990s. In 1996, the United Nations General Assembly adopted a treaty to ban underground nuclear detonations as well. With rare exception, the fallout from underground testing was contained in the immediate vicinity of the detonation and did not expose the public; thus, the NCRP has not included any public dose data in its report.

Figure 3-4 below shows the physical results of above-ground weapons testing and underground weapons testing.

Figure 3-4. The stereotypical mushroom cloud from an above ground weapons test prior to 1963(left) and the subsidence craters (right) at the Nevada Test Site from many underground nuclear tests (Photos courtesy of the United States Department of Energy-NNSA).

3.6 Summary of Dose Estimates for the General Population

Determining an average individual dose for the general population is not a trivial task. For each category of radiation exposure, it is understandable that there is variation in the amount of radiation that could be received. In addition, not every person is exposed to every single source of radiation. As discussed earlier, the NCRP has tabulated results for the collective effective dose to the US population. The NCRP also calculated an average effective dose by dividing the collective dose by every person in the United States. The average effective dose is not necessarily applicable to any single individual. However, it at least gives a rough order of magnitude.

Table 3-2 below summarizes the results for the categories discussed in Sections 3.1 through 3.4.

Table 3-2. Summary of NCRP estimates for radiation exposure of the public (NCRP Report 160 and NCRP 184). Bolded entries indicate totals for each category that are then summed to give the grand total.

Category	Collective Dose (person-Sv)	Average effective dose per person in the United States (mSv)
Background		
Space radiation	99,000	0.33
Inhalation (radon and thoron)	684,000	2.28
Ingestion	87,000	0.29
External terrestrial	63,000	0.21
Total from Background	**933,000**	**3.11**
Total from Consumer Products and Activities	**39,000**	**0.13**
Medical[a]		
Computed Tomography	469,000	1.45
Diagnostic Nuclear medicine	133,000	0.41
Radiography/fluoroscopy	71,000	0.22
Interventional fluoroscopy	82,000	0.25
Total from Medical	**755,000**	**2.3**
Total from Industrial, security, medical, and research	**1,000**	**0.003**
GRAND TOTAL (rounded)	**1,730,000**	**5.5**

a. Values in the medical category were updated substantially in NCRP Report 184; the values in this table reflect the newer data. To allow comparison to the other values in the table, the doses in the medical category use the 1990 values of w_T from ICRP 60 rather than the newer ones from ICRP Publication 103.

Clearly, the vast majority of the average dose is due to background (mostly radon) and medical procedures with very little from the other categories. This should help put radiation doses in perspective for members of the public, many of whom may assume that the radiation from human activities in industry and the government are a major contributor to our radiation dose.

3.7 Occupational Exposure of Radiation Workers

People who work with radioactive materials or whose jobs require them to be exposed to radiation represent a small percentage of the

US population. However, the dose to these workers is usually much smaller than commonly believed.

One interesting category of radiation worker is the air carrier aircrew. The Federal Aviation Administration (FAA) recognized aircrews as radiation workers in 1994; more information can be found in the FAA-published document *What Aircrews Should Know About Their Occupational Exposure to Ionizing Radiation* at https://www.faa.gov/sites/faa.gov/files/data_research/research/med_humanfacs/aeromedical/0316.pdf. Owing to the space radiation to which they are exposed at high altitudes, the radiation dose to these workers can exceed that of some workers in nuclear power plants or other nuclear facilities.

Both the Department of Energy (DOE) and the Nuclear Regulatory Commission (NRC) publish annual data on occupational doses at facilities that they regulate. Most workers at DOE and NRC facilities receive no measurable dose (i.e., any dose that they receive is below the ability of monitoring devices). More information is available at the following links:

- NRC Annual Dose Information: https://www.reirs.com/0713.html
- DOE Annual Dose Information: https://www.energy.gov/ehss/listings/annual-doe-occupational-radiation-exposure-reports

Table 3-3 below summarizes the NCRP estimates for the collective dose for all radiation workers in the US for 2006 along with the average effective dose per worker. The data in the table demonstrate that on average, occupational radiation workers receive less dose as the result of their occupation than they receive from background radiation or from medical exposures.

Table 3-3. Summary of annual occupational doses from NCRP Report 160.

Occupational Category	Collective Dose (person-Sv)	Average effective dose per exposed worker (mSv)
Medical workers	550	0.8
Aviation aircrews	530	3.1
Commercial Nuclear power	110	1.9
Industry and Commerce	110	0.8
Education and Research	60	0.7
Government, DOE, military	40	0.6
GRAND TOTAL (rounded)	**1,400**	**1.1**[a]

a. The total average effective dose is not the sum of the averages because the exposed population for each individual category is different.

3.8 Summary and Conclusions

As stated before, there is great variation in the amount of radiation dose that is received by individuals depending on their lifestyle, place of residence, and medical needs. Thus, the numbers in this chapter should be interpreted strictly as general values to give some sense of magnitude.

Several websites offer an online radiation dose calculator for individuals to assess their own annual dose and are listed below. While these sites may not all produce the same exact numbers, they are nonetheless useful and helpful.

- US Environmental Protection Agency https://www.epa.gov/radiation/calculate-your-radiation-dose
- US Nuclear Regulatory Commission https://www.nrc.gov/about-nrc/radiation/around-us/calculator.html
- American Nuclear Society https://www.ans.org/nuclear/dosechart/#dose
- Los Alamos National Laboratory https://envweb.lanl.gov/newnet/info/dosecalc.aspx

As shown by these dose calculators, a person's choice of residence can greatly influence their radiation dose. For example, it is conceivable for a person who chooses to live at a higher altitude in a mountainous region to receive more radiation dose from space radiation and radon than a radiation worker who deals with radioactive material daily. While some might perceive that the "natural" radiation is less harmful than the human-made radiation, all data contradict this perception. Thus, the radiation protection professional should be familiar with these ideas and use them to help communicate effectively when questions of radiation dose arise.

Free Resources for Chapter 3

- Questions and problems for this chapter can be found at www.hpfundamentals.com
- Online individual dose calculators
    - US Environmental Protection Agency https://www.epa.gov/radiation/calculate-your-radiation-dose
    - US Nuclear Regulatory Commission https://www.nrc.gov/about-nrc/radiation/around-us/calculator.html
    - American Nuclear Society https://www.ans.org/nuclear/dosechart/#dose
    - Los Alamos National Laboratory https://envweb.lanl.gov/newnet/info/dosecalc.aspx

Additional References

- NCRP Report 160, *Ionizing Radiation Exposure of the Population of the United States (2009)*, available for purchase through ncrponline.com.
- NCRP Report 184, *Medical Radiation Exposure of Patients in the United States (2019)*, available for purchase through ncrponline.com

Chapter 4 Dosimetry Calculations for External Sources of Radiation

This chapter is the first of several that give the fundamentals of calculating and using radiation dose quantities that were discussed in Chapter 1. Controlling radiation dose is the major driver behind radiation protection regulations and practices in the workplace. Consequently, familiarity with the basic principles of dosimetry calculations and measurements helps in understanding the implementation of radiation protection protocols.

This chapter covers the calculations that involve external radiation sources: that is, those sources of radiation that are outside the body. Chapter 7 will cover the calculational methodology for radiation sources that are inside the body. Despite psychological reactions, it is important to remember that whether the source of radiation is outside or inside the body, the same definitions of dose apply, and the consequences of the dose remain the same. The difference in external dosimetry and internal dosimetry is in the calculational methodology that is applied.

As discussed in Chapter 1, we must be specific about how we characterize a dose calculation. In this chapter we will discuss differing calculations including the following:

- The dose to human tissues or organs
- The dose to the human body as a whole
- The dose from charged particles
- The dose from photons
- The dose from neutrons
- The dose at a point in a material
- The average dose over a volume of material

It can be a challenge to organize all these types of calculations; however, our general approach will be to use the radiation types as major headings and then to tackle the various types of calculations that would be appropriate for that radiation type.

The starting point for the calculations will typically be manual (hand) calculations. This will help illustrate the principles that must be considered. For many calculations in the workplace today, hand calculations are still entirely acceptable and preferred. However, there are complexities to some problems that will make hand calculations laborious or untenable. In that case, it is typical to resort to computational methods (spreadsheets or other standalone computer programs) to help arrive at a solution. The computational methods help solve the problem generally by one of three methods:

- Repetitious calculations that could be performed by hand with a simple calculator can be entered into a spreadsheet or programmed to save time.
- Numerical methods can be programmed or entered into a spreadsheet to approximate a solution when an analytical solution may not be obtainable.
- Monte Carlo methods can be used that involve using random numbers and statistical methods to obtain a solution. In health physics/radiation protection, Monte Carlo usually takes on the form of a 3-dimensional simulation of radiation moving through matter. Each radiation is tracked as to its trajectory, interactions, energy deposition, etc., to perform a virtual experiment when an actual physical experiment would be impractical or impossible.

4.1 Absorbed Dose Calculations for Common Charged Particles

When considering charged particles, the fundamental calculation that is typically performed is for absorbed dose. As discussed in Chapter 1, exposure is not defined for any radiation type other than photons and only for air, and any of the other dosimetric quantities relies first on knowing the absorbed dose.

The common charged particles that are of concern in most workplaces would be alpha particles, electrons, and protons. The alpha particles and electrons would typically be the result of alpha decay and beta decay, respectively. The protons would usually be found in accelerator facilities where protons are accelerated to induce nuclear reactions that produce positron emitters or produce neutrons via spallation. The dose from other particle types (mostly subnuclear particles) is discussed in Section 4.4.

As it pertains to dose calculations for external sources of radiation, alpha particles typically pose little to no risk to humans. The reason for this is that alpha particles typically have energies of 4-6 MeV and are unable to penetrate the dead layer of skin (the epidermis) of the body to cause radiation dose to any sensitive layers. While the epidermis varies in thickness over the surface of the body, the usual assumption is a thickness of 70 μm (7 mg/cm^2), which is greater than the penetrating range of alpha particles from common isotopic sources. Of course, if the alpha particles are from an accelerator at high energies, then this generality would not hold true and would have to be reconsidered. In addition, it is still possible to calculate the alpha dose to thin absorbers such as air or other gases. Therefore, the following sections present information that includes alpha particles for completeness.

4.1.1 Point absorbed dose calculations

For a fluence at an infinitesimally small point in a material, the absorbed dose at that point is given by Equation 4.1 below:

$$D = \Phi(\frac{dE}{\rho dx})_{col} \quad (4.1)$$

Where

- D is the absorbed dose (MeV/gram)
- Φ is the fluence (cm^{-2})
- $(dE/\rho dx)_{col}$ is the collisional mass stopping power for the charged particle at that location (MeV/g/cm^2) or (MeV·cm^2/g)

Equation 4.1 can be modified to calculate the absorbed dose rate by using the fluence rate ϕ (cm^{-2}s^{-1}) instead of the total fluence Φ(cm^{-2}).

As a reminder, the mass stopping power of a material is the energy loss per unit density thickness of the material and includes collisional stopping power and radiative stopping power. The collisional stopping power is comprised of energy losses to the electrons and nucleus in the medium where it is all deposited relatively locally through ionizations and excitations. The radiative stopping power is comprised of energy losses due to hard nuclear collisions where bremsstrahlung photons are produced; this is only a practical issue for electrons. In these hard collisions, the incoming electrons experience an energy loss, but the energy given to the bremsstrahlung photons is not deposited locally because photons can travel some distance from the point of origin. Hence, we use the collisional stopping power for the absorbed dose calculations.

Examination of Equation 4.1 shows that the units for absorbed dose in this equation are given as MeV/g. This is a consequence of how fluence is commonly expressed and how reference values are tabulated. Conversion of the absorbed dose to SI units is readily achieved by converting MeV to J (1 MeV=1.6×10^{-13}J) and converting grams to kilograms. Thus, 1 MeV/g =1.6×10^{-10} J/kg.

Values for the mass stopping power for alpha particles, electrons, and protons can be found in variety of hard-copy and online references. However, one of the most convenient tools for determining the mass stopping power is provided by the National Institute of Standards and Technology (NIST). NIST is a United States federal agency that is responsible for measurement science, the development of measurement standards, and traceability of measurements to the known standards. NIST maintains a rigorous database of physical measurements, and scientists make use of such data.

At the time of this writing, the online databases for mass stopping power are located at the following web addresses:

- Alpha particles: https://physics.nist.gov/PhysRefData/Star/Text/ASTAR.html
- Electrons: https://physics.nist.gov/PhysRefData/Star/Text/ESTAR.html
- Protons: https://physics.nist.gov/PhysRefData/Star/Text/PSTAR.html

If these addresses are inactive, a simple web search involving the search terms NIST "physical reference data" will bring up a page that lists categories of data. On that page, there is a category entitled "Radiation Dosimetry Data." From that page, the various databases can be accessed.

The search form can be used to find the mass stopping power at standard reference energies. However, the form also allows input of a custom energy to accommodate values that do not lie at the reference energies. Typical output from the search form is formatted as shown in Figure 4-1, which shows the results for alpha particles in aluminum.

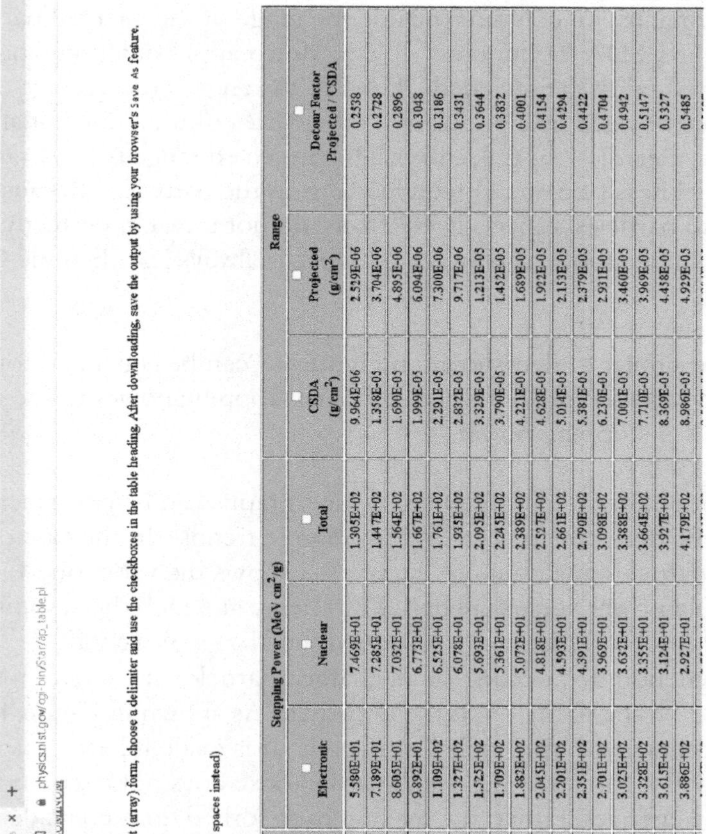

Stopping Power and Range Tables

physics.nist.gov/cgi-bin/Star/ap_table.pl

ALUMINUM

To download data in spreadsheet (array) form, choose a delimiter and use the checkboxes in the table heading. After downloading, save the output by using your browser's Save As feature.

Delimiter:
- space
- | (vertical bar)
- tab (some browsers may use spaces instead)
- newline

Download data | Reset

(required) Kinetic Energy (MeV)	Stopping Power (MeV cm²/g)			Range		
	Electronic	Nuclear	Total	CSDA (g/cm²)	Projected (g/cm²)	Detour Factor Projected / CSDA
1.000E-03	5.580E+01	7.469E+01	1.305E+02	9.964E-06	2.529E-06	0.2538
1.500E-03	7.189E+01	7.285E+01	1.447E+02	1.358E-05	3.704E-06	0.2728
2.000E-03	8.605E+01	7.032E+01	1.564E+02	1.690E-05	4.895E-06	0.2896
2.500E-03	9.892E+01	6.773E+01	1.667E+02	1.999E-05	6.094E-06	0.3048
3.000E-03	1.109E+02	6.525E+01	1.761E+02	2.291E-05	7.300E-06	0.3186
4.000E-03	1.327E+02	6.078E+01	1.935E+02	2.832E-05	9.717E-06	0.3431
5.000E-03	1.525E+02	5.693E+01	2.095E+02	3.329E-05	1.213E-05	0.3644
6.000E-03	1.709E+02	5.361E+01	2.245E+02	3.790E-05	1.452E-05	0.3832
7.000E-03	1.882E+02	5.072E+01	2.389E+02	4.221E-05	1.689E-05	0.4001
8.000E-03	2.045E+02	4.818E+01	2.527E+02	4.628E-05	1.922E-05	0.4154
9.000E-03	2.201E+02	4.593E+01	2.661E+02	5.014E-05	2.153E-05	0.4294
1.000E-02	2.351E+02	4.391E+01	2.790E+02	5.381E-05	2.379E-05	0.4422
1.250E-02	2.701E+02	3.969E+01	3.098E+02	6.230E-05	2.931E-05	0.4704
1.500E-02	3.025E+02	3.632E+01	3.388E+02	7.001E-05	3.460E-05	0.4942
1.750E-02	3.328E+02	3.355E+01	3.664E+02	7.710E-05	3.969E-05	0.5147
2.000E-02	3.615E+02	3.124E+01	3.927E+02	8.369E-05	4.458E-05	0.5327
2.250E-02	3.886E+02	2.927E+01	4.179E+02	8.986E-05	4.929E-05	0.5485

Figure 4-1. Screenshot of the output of the astar database provided by NIST.

The first column of the output table shows the incident radiation energy. The next two columns show the contributions of collisions between the alpha particles and the absorber electrons and the collisions between the alpha particles and the absorber nuclei. Since bremsstrahlung is not an issue for alpha particles, both of these are considered collisional stopping power, and the total stopping power for alpha particles is thus the collisional stopping power.

Additional useful columns include the range of the charged particle. Both the CSDA (continuous slowing down approximation) and projected range are included. The CSDA range is the average distance traveled by a population of particles of the same initial energy, regardless of trajectory. The projected range is the straight-line thickness of material required to stop the particles. Because charged particles, especially electrons, do not travel in perfectly straight trajectories, the projected range is always less than the CSDA range.

A table similar to that shown in Figure 4-1 can be generated for proton radiation as well where the total stopping power is also the collisional stopping power.

In addition to the output table, various graphs can be constructed from the data, which is helpful to observe trends. In the case of alpha particles in aluminum, Figure 4-2 shows the variation of stopping power versus energy. The energy at which the stopping power reaches a maximum is termed the "Bragg Peak," a phenomenon that occurs for all charged particles at energies that depend on the particle and the absorber. As shown in Figure 4-2, this occurs at a surprisingly low energy. It is evident based on Equation 4.1 that the highest absorbed dose does not occur at high particle energies; rather, the maximum absorbed dose coincides with the energy where the stopping power is a maximum, which is often where the particle energy is quite low. This may seem counterintuitive, and it is important to remember that it is the stopping power, and not the initial energy itself, that determines the value of the absorbed dose at a point.

The Bragg Peak has special application in radiation therapy involving proton beams. When the tumor location is known to be located at a certain depth in the body, the proton energy can be selected so that the protons are near the end of their tracks within the tumor where the Bragg Peak will cause the delivery of maximum dose. This way, the dose to the tissue at shallower depths is much lower, thus providing a higher dose to the tumor than to the surrounding tissue.

ASTAR : Stopping Power and Range Tables for Alpha Particles

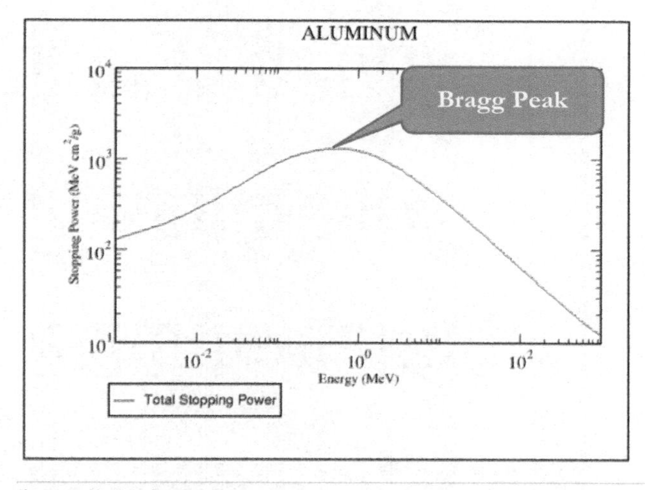

Figure 4-2. Graphical plot of total stopping power versus energy for alpha particles in aluminum with the Bragg peak at an energy less than 1 MeV. Source: NIST astar database at www.nist.gov.

For electrons, the output table is formatted differently owing to the radiative stopping power component of the total stopping power. As shown below in Figure 4-3 for incoming electron radiation in an aluminum absorber, the output table shows the collisional and radiative stopping power separately and then shows the total stopping power. For absorbed dose calculations, the collisional stopping power would be used and not the total stopping power.

The NIST databases have information for a variety of absorption media, including pure elements as well as compounds.

The calculation of absorbed dose due to charged particles as shown in Equation 4.1 may seem deceptively simple. However, in practice it can be a challenge to know the fluence at the point of interest due to charged particle interactions that may have occurred if the point is located some distance within the absorber. Secondly, knowing the energy of the incoming particles when they reach the point of interest can be a challenge, especially for electrons which can lose large amounts of energy in relatively few collisions. Thus, for a practical situation, hand calculations of absorbed dose at a point due to charged particles from external sources are mostly limited to heavy charged particles.

Stopping Power and Range Tables ×

physics.nist.gov/cgi-bin/Star/e_table.pl

ALUMINUM

To download data in spreadsheet (array) form, choose a delimiter and use the checkboxes in the table heading. After downloading, save the output by using your browser's Save As feature.

Delimiter:
- space
- | (vertical bar)
- tab (some browsers may use spaces instead)
- newline

Download data | Reset

(required) Kinetic Energy (MeV)	Stopping Power ($MeV\ cm^2/g$)			CSDA Range (g/cm^2)	Radiation Yield	Density Effect Parameter
	Collision	Radiative	Total			
1.000E-02	1.649E+01	6.559E-03	1.650E+01	3.539E-04	2.132E-04	3.534E-04
1.250E-02	1.398E+01	6.700E-03	1.398E+01	5.192E-04	2.583E-04	4.937E-04
1.500E-02	1.220E+01	6.798E-03	1.221E+01	7.111E-04	3.017E-04	6.538E-04
1.750E-02	1.088E+01	6.871E-03	1.088E+01	9.284E-04	3.435E-04	8.332E-04
2.000E-02	9.844E+00	6.926E-03	9.851E+00	1.170E-03	3.840E-04	1.031E-03
2.500E-02	8.338E+00	7.004E-03	8.345E+00	1.724E-03	4.616E-04	1.483E-03
3.000E-02	7.287E+00	7.059E-03	7.294E+00	2.367E-03	5.355E-04	2.005E-03
3.500E-02	6.509E+00	7.100E-03	6.516E+00	3.094E-03	6.058E-04	2.594E-03
4.000E-02	5.909E+00	7.133E-03	5.916E+00	3.900E-03	6.736E-04	3.246E-03
4.500E-02	5.430E+00	7.162E-03	5.437E+00	4.783E-03	7.390E-04	3.960E-03
5.000E-02	5.039E+00	7.191E-03	5.046E+00	5.738E-03	8.022E-04	4.732E-03
5.500E-02	4.714E+00	7.217E-03	4.721E+00	6.763E-03	8.636E-04	5.560E-03
6.000E-02	4.439E+00	7.243E-03	4.446E+00	7.855E-03	9.232E-04	6.440E-03
7.000E-02	3.998E+00	7.295E-03	4.005E+00	1.023E-02	1.038E-03	8.351E-03
8.000E-02	3.661E+00	7.350E-03	3.668E+00	1.284E-02	1.147E-03	1.045E-02
9.000E-02	3.394E+00	7.411E-03	3.401E+00	1.568E-02	1.252E-03	1.271E-02
1.000E-01	3.177E+00	7.476E-03	3.185E+00	1.872E-02	1.355E-03	1.513E-02
1.250E-01	2.781E+00	7.659E-03	2.789E+00	2.714E-02	1.593E-03	2.175E-02

Figure 4-3. Screenshot of the output of the estar database provided by NIST.

The following example shows the basics of how to do a point dose calculation for protons.

Example: What is the absorbed dose at a point due to protons with a fluence of 2.3×10^5 protons/cm^2 at the surface of a slab of polyethylene if the protons have a single energy of 1.62 MeV?

Solution: We will access the pstar database of stopping powers provided by NIST and use the total stopping power as the collisional stopping power. However, 1.62 MeV is not one of the reference energies; so, it is necessary to enter it as an additional energy as shown below in Figure 4-4.

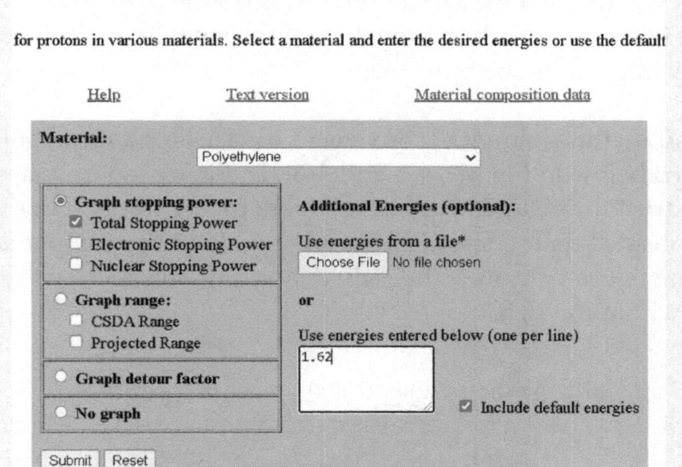

Figure 4-4. Screenshot of the pstar database input screen from NIST with an additional energy of 1.62 MeV entered to determine the stopping power of polyethylene for protons of that energy.

The NIST database provides an output table showing that the total stopping power is 204.2 MeV·cm^2/g. With the fluence rate given in the premise, the absorbed dose is thus:

$$D = \Phi(\frac{dE}{\rho dx})_{col} = (\frac{2.3\times10^5}{cm^2})(204.2\frac{MeV\cdot cm^2}{g}) = 4.7\times10^7\frac{MeV}{g}$$
$$= 0.00751\ Gy = 0.751\ rads$$

For completeness, the absorbed dose has been converted from its original units to other more commonly used units of gray and rads.

4.1.2 Average absorbed dose calculations

While the point dose calculation is used quite often, it is also common to need the average dose in a particular volume. Average dose calculations are important for calibrating and designing detector volumes for measuring absorbed dose as well as calculating average doses in body tissues or organs as discussed in Chapter 2 to understand biological impacts of radiation.

Hand calculations for average dose calculations are easiest generally if the volume is a regularly shaped solid. This requires a face that is perpendicular to the beam and is a fixed thickness. The exact shape of the face can vary so long as the area of the face is known. For purposes of discussion, we will refer to this as a "slab"; however, there is no requirement that the face of the solid be rectangular as is typically implied by the term "slab."

To illustrate this point, it is convenient to visualize a rectangular solid as shown below in Figures 4-5 and 4-6. In Figure 4-5, the face of the slab is struck by charged particles that are totally stopped in the slab and deposit all their energy. In Figure 4-6, the incoming particles penetrate entirely through the slab, depositing only some of their energy within the slab.

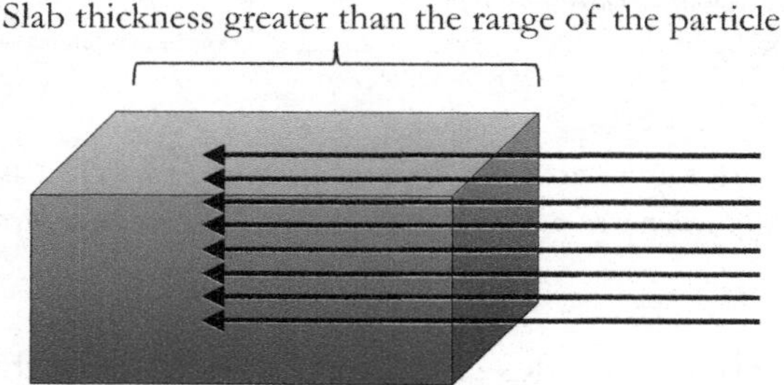

Figure 4-5. Incoming charged particles incident on a slab with a rectangular face. In this example, the charged particles stop within the volume.

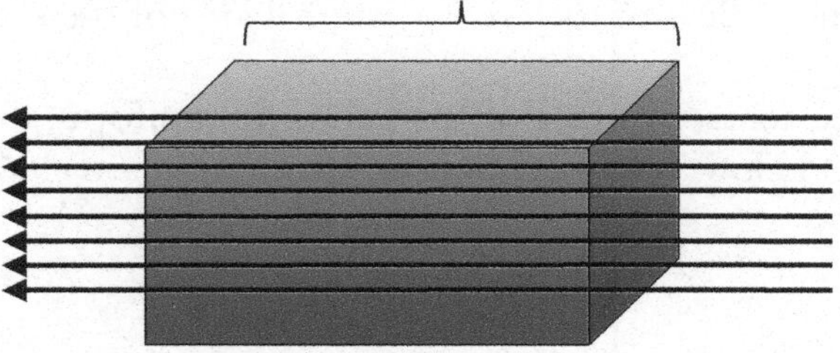

Figure 4-6. Incoming charged particles incident on the face of a slab where the charged particles pass through the slab while losing some of their energy within the slab.

The methods for determining the average dose are different for these two scenarios and are discussed in the following sections.

4.1.2.1 Average Dose Calculation When the Charged Particles Are Stopped Within the Volume

This scenario is the more straightforward of the two. So long as the energy of the particles when they enter the slab is known, then the energy deposited in the volume is the total energy of the particles, as seen above in Figure 4-5. It only remains to determine the mass of the material within the volume of interest. If the fluence is uniform across the face of the slab, then the exact cross sectional-area of the slab becomes superfluous, and Equation 4.2 below can be used to determine the average dose.

$$D = \Phi(\frac{E_i}{t\rho}) \quad (4.2)$$

Where

- D is the average absorbed dose within the volume (MeV/g)
- Φ is the fluence of the incoming particles (cm^{-2})
- E_i is the energy of the incoming charged particles (MeV)
- t is the linear thickness of the slab (cm)
- ρ is the density of the material (g/cm^3)

This equation is independent of the cross-sectional area of the face because if the fluence is uniform, then a larger cross-sectional area of the slab would mean more energy deposited within the total slab, but the larger area would also mean a proportionately larger mass would be irradiated with the end result being that the average absorbed dose is the same regardless of the cross-sectional area of the slab.

If the fluence across the face of the slab is not uniform, then Equation 4.3 below would be used instead.

$$D = \frac{N_i E_i}{At\rho} \tag{4.3}$$

Where

- D is the average absorbed dose within the volume (MeV/g)
- N_i is the number of charged particles incident on the slab
- E_i is the energy of the incoming charged particles (MeV)
- A is the cross-sectional area of the slab (cm^2)
- t is the thickness of the slab (cm)
- ρ is the density of the material (g/cm^3)

In this equation, N_i/A becomes the average fluence across the slab face so that Equation 4.3 is the same in concept as Equation 4.2.

An important reminder is that the average dose in this calculation is not uniform in the volume. If the particles stop far short of the back face of the slab, then there could be a significant portion of the slab that receives no dose. In addition, for heavy charged particles, the maximum dose in the slab will occur near the end of the range of the particles where the Bragg peak occurs; so the front face of the slab will receive less absorbed dose than those slices of the slab that are further within the slab.

Example: What is the average absorbed dose in a polyethylene slab that is 0.5 cm thick due to protons with a uniform fluence rate of 2.3×10^5 protons/cm^2 if the protons have a single energy of 1.62 MeV?

Solution: Looking back at the output table from the pstar database that was generated for our last example, the projected range (the thickness of material that stops the heavy charged particle), is 4.724×10^{-3} g/cm^2 (note that this is a density thickness). The NIST pstar database also includes a hyperlink to the material composition data that is used for the various materials, and the density of polyethylene is given as 0.94 g/cm^3. Thus, the projected range (in linear thickness) for 1.62 MeV protons in polyethylene is 0.00444 cm, which is far less than 0.5 cm. Thus, the protons will be totally stopped within the slab. Since a uniform fluence is given in the premise, Equation 4.2 can be used.

$D = \Phi\left(\frac{E_i}{t\rho}\right) = \left(\frac{2.3\times10^5}{cm^2}\right)\left(\frac{1.62\ MeV}{(0.5\ cm)\left(0.94\frac{g}{cm^3}\right)}\right)$ =792,765 MeV/g

=0.000127 Gy = 0.0127 rads

By comparing this example to the previous calculation for point dose, it is obvious that by averaging the dose over a larger volume where the dose is not uniform, the average dose is much less than the point dose in this case by a factor of roughly 60.

4.1.2.2 Average Dose Calculation When the Charged Particles Travel Through the Volume

As illustrated in Figure 4-6, charged particles could penetrate entirely through the volume, leaving only part of their energy within the volume. The calculation for absorbed dose in this case is a little more challenging but is not insurmountable. The main point is to recognize that the energy deposited in the volume is the difference in the energy of the particles when they entered the volume and when they left the volume.

As noted before, this calculation is of practical use primarily for heavy charged particles since their paths tend to be in straighter lines such that the projected range and the CSDA range are comparable.

Similar to Equation 4.2, the equation to determine the average dose in this case is expressed below in Equation 4.4

$$D = \Phi\left(\frac{E_{in} - E_{out}}{t\rho}\right) \tag{4.4}$$

Where

- D is the average absorbed dose within the volume (MeV/g)
- **Φ** is the fluence of the incoming particles (cm^{-2})
- E_{in} is the energy of the charged particles as they enter the slab (MeV)

- E_{out} is the energy of the charged particles as they leave the slab (MeV)
- t is the thickness of the slab (cm)
- ρ is the density of the material (g/cm^3)

Determining the value of E_{out}. is a multistep process that relies on understanding the concept of projected range and using some trial and error. Figure 4-7 below shows the conceptual idea behind the method. The underlying principle is to take the total range of the particle in the material and subdivide it into two parts:

1. the physical distance travelled by the particle in reality
2. a virtual slab comprised of the difference between the total range and the slab thickness.

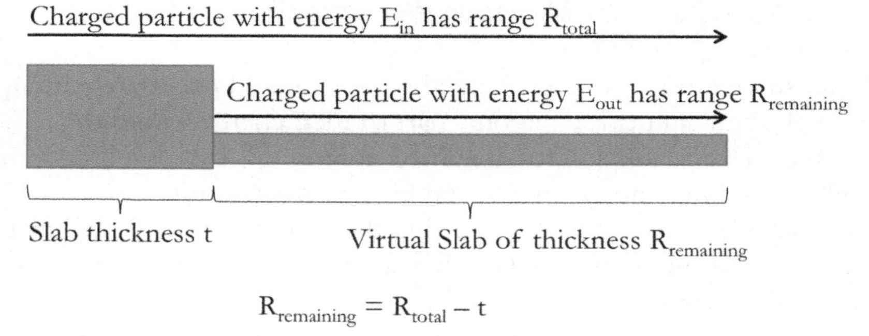

$$R_{remaining} = R_{total} - t$$

Figure 4-7. Illustration of determining E_{out} using the range of the particle.

The steps to determining E_{out} are as follows:

1. Determine the total range R_{total} for the particle in the slab material even though the material physically does not extend that far.
2. Subtract the thickness t of the slab from R_{total}. This gives the distance that the particle would still have to travel in the material to obtain the thickness of the virtual slab, $R_{remaining}$. Again, this is not to say that the particle physically will travel this far; it is a virtual technique.
3. Recognize that if the virtual slab were present, the particle <u>must</u> still have sufficient energy to travel the rest of R_{total}, which means that we need to find what value of E_{out} has a range equal to $R_{remaining}$.
4. Use trial and error with the NIST database tables to determine the value of E_{out}.

The following example illustrates how these steps can be used.

Example: A fluence of 2.3×10^5 protons/cm^2 with energy 100 MeV impinges on a slab of polyethylene of thickness 3.5 cm. What is the expected energy of the protons as they exit the slab? What is the average absorbed dose in the slab?

Solution: The slab thickness expressed in density thickness is 3.5 cm multiplied by the previously given density of polyethylene of 0.94 g/cm^3 for a value of 3.29 g/cm^2. Using the pstar database as before in the previous examples, we can see that the projected range of 100 MeV protons in polyethylene is 7.235 g/cm^2, which is greater than the slab thickness. Thus, the protons will travel all the way through the slab while depositing only part of their energy.

Therefore,

- R_{total}=7.235 g/cm^2.
- $R_{remaining} = R_{total} - t$ = 7.235 g/cm^2 – 3.29 g/cm^2 = 3.945 g/cm^2
- After exiting the slab, the protons must still have sufficient energy to have traveled another 3.945 g/cm^2 if the virtual slab were present. By inspection of the pstar output database table for default energies, 70 MeV protons have a range of 3.818 g/cm^2, and 75 MeV protons have a range of 4.322 g/cm^2. Thus, E_{out} must lie between 70 MeV and 75 MeV.
- By using the custom additional energies box in the pstar database entry form, we can enter 72.5 MeV first (halfway between 70 and 75 MeV) to see that the range of that energy is 4.067 g/cm^2, which is still too large. Then we could try values between 70 and 72.5 MeV to the nearest 0.1 MeV to obtain the closest value, which turns out to be 71.3 MeV, which is our best estimate of E_{out}.

Logically, then the energy deposited in the slab per particle must be $E_{in} - E_{out}$, which is 100 MeV-71.3 MeV=28.7 MeV.

Using Equation 4.4 to obtain the absorbed dose,

$$D = \Phi\left(\frac{E_{in} - E_{out}}{t\rho}\right) = (\frac{2.3 \times 10^5}{cm^2})(\frac{(100\ MeV - 71.3\ MeV)}{(3.5\ cm)\left(0.94\ \frac{g}{cm^3}\right)}$$

$$= 2.00 \times 10^6\ \frac{MeV}{g} = 3.2 \times 10^{-4} \frac{J}{kg}$$

$$= 0.032\ rads$$

Note that even though the proton energy is much higher in this case than in the previous example, the average absorbed dose is not orders of magnitude greater. Again, this indicates that when the particles are not totally stopped, the stopping power is the primary determinant of dose, not the initial energy of the particles themselves.

4.2 Absorbed Dose and Exposure Calculations for Photons

For photons, absorbed dose is the starting point for calculating all other dosimetric quantities, including exposure. However, despite the careful definition of absorbed dose by the ICRU, it is not possible to calculate the absorbed dose from photons directly without some assumptions. Instead, we start with a quantity known as the KERMA (Kinetic Energy Released per unit Mass). As the name implies, the KERMA for photons is the energy transferred to the electrons in the absorbing medium.

Once energy is transferred to the electrons via photon interactions, the electrons may dissipate their energy by either (a) collisional interactions that result in ionization and excitation or (b) radiative interactions that produce bremsstrahlung. Thus, the KERMA can be divided into two pieces:

- Collision KERMA K_c, where the energy transferred to the electrons is eventually given up as ionizations and excitations.
- Radiative KERMA K_r, where the energy transferred to the electrons shows up as photon radiation that may travel a great distance from the initial interaction location.

Since ionizations and excitations will happen near the volume where the initial interaction took place, the focus of our discussion of KERMA will be on the collision KERMA K_c.

To understand the distinction between collision KERMA and absorbed dose, it may be helpful to refer to Figure 4-8 below. This is a highly stylized figure for illustrative purposes. If photons impinge on a slab absorber from the left side, secondary electrons are liberated in the slab at various depths due to photon interactions and move to the right. The tracks of the individual liberated electrons are represented by the orange bars that move from left to right. Some are liberated near the surface and penetrate to greater depths, while others are generated further in the slab. As the electrons move deeper in the absorber, they deposit their energy along the tracks represented by the straight lines, even though in reality we know that electrons seldom move in straight lines.

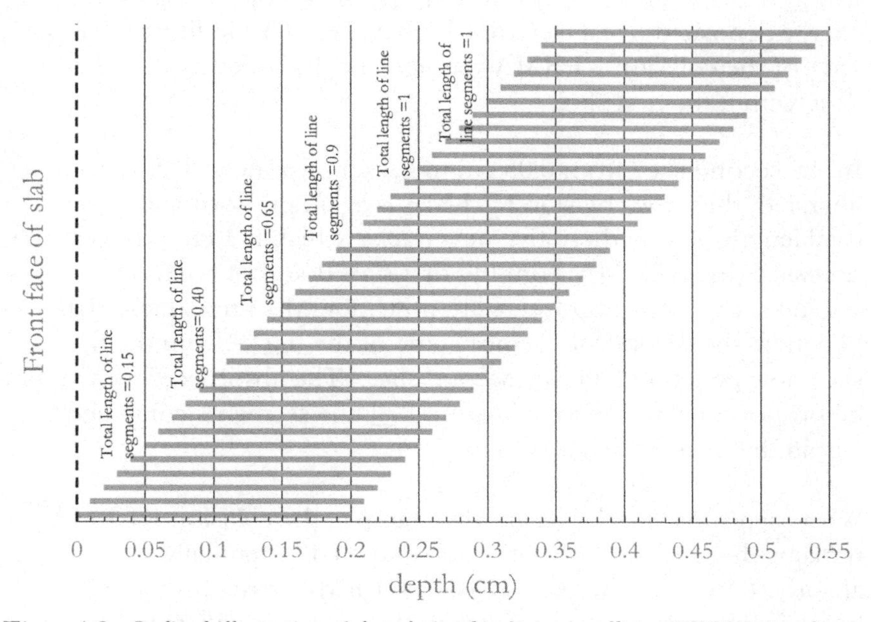

Figure 4-8. Stylized illustration of the relationship between collision KERMA and absorbed dose. In each slice, the KERMA is constant while the absorbed dose gradually increases to be equal to the KERMA when charged particle equilibrium is established.

If we look at the first slice from 0 cm to 0.05 cm in the slab, we see that five electrons are liberated and proceed further into the slab. The collision KERMA within that slice of the slab is proportional to the initial energy given to the electrons within that slice; however, the absorbed dose would be proportional only to the energy that is actually deposited within that slice. In other words, the collision

KERMA is the energy given to the electrons by the photons in a volume regardless of where that energy eventually is deposited. The absorbed dose is the energy deposited by the electrons within that volume, regardless of where it originated.

For our example, we can make a grossly simplifying assumption that the energy either given to or deposited by these electrons is proportional to the length of the electron track. In the first slice of the slab, five electrons are liberated, each with a track length of 0.2 cm. Thus, the energy given to these electrons is proportional to (5×0.2 cm) =1 cm. The collision KERMA is thus also proportional to 1 cm. However, the total length of the tracks that is contained within the first slice is only 0.15 cm, which is proportional to the energy deposited by all electrons in that slice. This indicates that only 15% of the collision KERMA was actually deposited as absorbed dose within the first slice.

In the second slice from 0.05 cm to 0.1 cm, five new electrons are liberated; thus, the collision KERMA is again proportional to the total length of only these five new tracks, which is 1 cm just as in the first slice (the KERMA from the first slice does not count in the second slice). However, there are now more than five tracks that penetrate the second slice, since some of the tracks from the first slice now penetrate into the second slice. The absorbed dose would be proportional to the total length of all line segments contained within this slice, which is 0.4 cm.

With each subsequent slice in this example, the collision KERMA remains the same and is proportional to 1 cm; eventually, the absorbed dose matches the collision KERMA at the fifth slice because the energy leaving a slice to the right is evenly matched by energy entering the slice from the left. The collision KERMA in this slice is still proportional to 1 cm, and the absorbed dose is proportional to the lengths of all line segments contained within the slice, which is also 1 cm. When this condition is met (energy leaving a volume is replaced exactly by energy entering the volume), we say that charged particle equilibrium (CPE) has been met.

Table 4-1 shows the numerical trend of KERMA and absorbed dose as the depth in the slab increases. For each slab depth, the total length of line segments corresponding to either the KERMA or absorbed dose is shown so that the reader can see that at the surface of the slab, the absorbed dose is quite small but eventually converges to be equal to the KERMA.

Table 4-1. Comparison of relative collision KERMA and absorbed dose with depth in the slab illustrated in Figure 4-8.

Depth (cm)	Relative Absorbed Dose (total length of line segments)	Relative KERMA (total length of line segments)
0-0.05	0.15	1
0.05-0.1	0.4	1
0.1-0.15	0.65	1
0.15-0.2	0.9	1
0.2-0.25	1	1
0.25-0.3	1	1
0.3-0.35	1	1
0.35-0.4	1	1
0.4-0.45	1	1
0.45-0.5	1	1
0.5-0.55	1	1

Figure 4-9 below shows a rough comparison of the KERMA and absorbed dose at depths within the slab. With smaller slices of thickness, it is possible to construct a curve showing an almost continuous trend to compare the KERMA and the absorbed dose. An example of this is shown in Figure 4-10, where it is more obvious how the absorbed dose converges to the same value as KERMA. At greater depths in the slab beyond what is shown, both the absorbed dose and KERMA would decrease due to the shielding effects of the slab.

One notable benefit of the CPE phenomenon is in the use of external photon beam therapy for cancer treatment. By selecting the proper photon energy to be sufficiently high, the entrance dose (the dose near the skin of the body) is minimized while the dose deeper in the body where the tumor is located is greater, thus minimizing the dose to the healthy tissue surrounding the tumor while providing the maximum dose to the tumor itself.

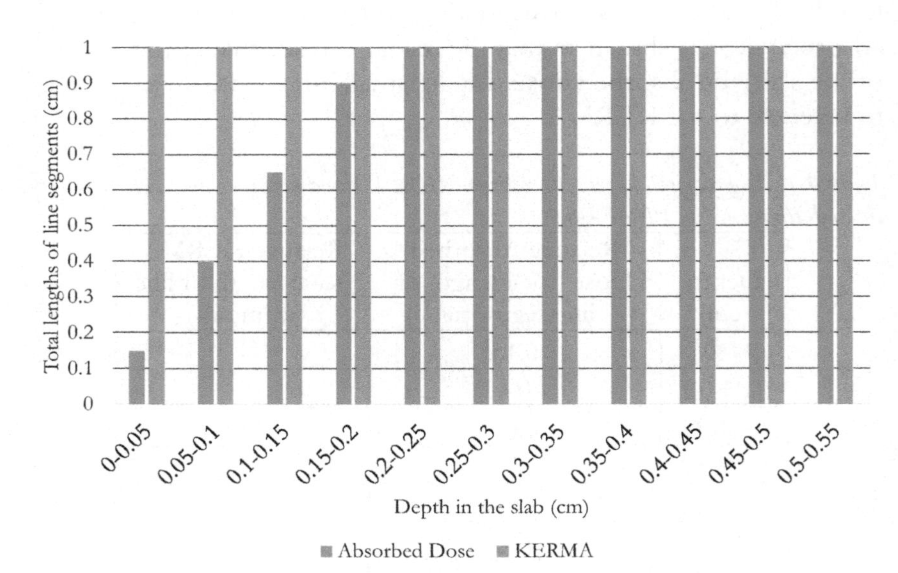

Figure 4-9. Comparison of KERMA and absorbed dose at various discrete depths in the slab illustrated in Figure 4-8. At the surface, the absorbed dose is lowest but increases with depth where CPE is achieved.

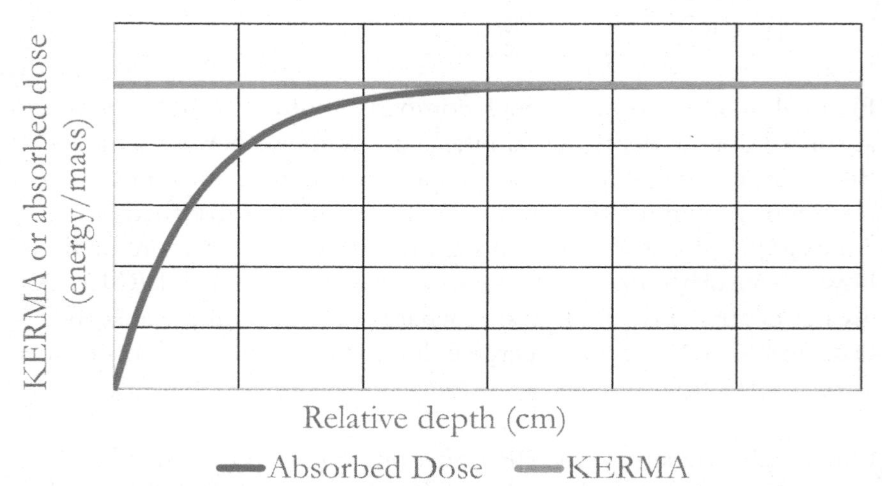

Figure 4-10. Comparison of KERMA and absorbed dose for infinitely thin slab slices showing how the absorbed dose converges to the same value as KERMA.

Eventually, at greater depths in the absorber, the KERMA would be reduced due to attenuation of the photon beam, since the absorber would provide a shielding effect. However, the absorbed dose would also be reduced as well and would still depend on the achievement of CPE.

The constraint of CPE is crucial for calculations and measurements involving both photons and neutrons. In practice, it is met at depths that are roughly equal to the range of the secondary charged particle that is produced. Referring again to Figure 4-9, it is easy to see that the collision KERMA and absorbed dose become equal at the depth of 0.2 cm in the slab. This is no accident, since the electrons all traveled a distance of 0.2 cm in this example; thus, 0.2 cm was the effective range of these electrons.

To achieve CPE in practice for photons, it is common to assume that a thickness is needed that is equal to the range of electrons with energy equal to the initial photon energy. For example, if 0.5 MeV photons are impinging on an absorber, then CPE is assumed to be reached at a depth equal to the range of 0.5 MeV electrons in that material, even though it is doubtful that a significant number of electrons would actually receive 0.5 MeV of energy. Thus, the amount of material needed for CPE will be dependent upon photon energy. At high photon energies in excess of several MeV, achieving CPE has some practical limitations and may not be achieved because the thickness required to achieve CPE also significantly reduces the fluence; this is a more advanced topic than is intended for this text. However, the practical side of dose measurements and the need for CPE in measurements is discussed in the next chapter.

4.2.1 Calculation of Collision KERMA and Absorbed Dose

Having introduced the concept of collision KERMA, we can look at how to calculate this quantity. Equation 4.5 below shows how collision KERMA for photons is calculated.

$$K_c = \Phi_i E_i (\frac{\mu_{en}}{\rho})_i \quad (4.5)$$

Where

- K_c is the collision KERMA expressed in units of MeV/g
- Φ_i is the photon fluence (cm^{-2}) of the photon "i" of interest
- E_i is the energy of the photon "i" of interest (MeV)

- $(\mu_{en}/\rho)_i$ is the mass energy absorption coefficient in units of cm^2/g for photon energy "i".

As the subscripts in the equation indicate, the collision KERMA must be calculated for each individual photon energy; therefore, if multiple photons are emitted by the same source, this calculation would be repeated for each energy. In addition, the collision KERMA rate can be obtained by using the fluence rate in the equation as opposed to the fluence.

The basis for this equation is in understanding the physical concept of the mass energy absorption coefficient. While it is not usually written this way, the mass energy absorption coefficient is conceptually as shown below:

$$\frac{\mu_{en}}{\rho} = (\frac{\mu}{\rho})_i \times \frac{\bar{E}_e}{E_i}$$

Where

- $(\mu/\rho)_i$ is the mass attenuation coefficient for photon energy "i" (the probability per density thickness of a photon interaction occurring)
- $\bar{E}_e$ is the average energy transferred to charged particles (electrons/positrons) that will eventually be deposited with no further photon production. In essence, it is the net energy that remains for collisions.
- E_i is the energy of the photon "i" of interest (MeV)

The fraction $(\bar{E}_e/E_i)$ is thus the fraction of the incoming photon energy that will be deposited by electrons/positrons, excluding bremsstrahlung. With this understanding, we can then rewrite Equation 4.5 and rearrange the terms as follows:

$$K_c = \Phi_i E_i \left(\frac{\mu_{en}}{\rho}\right)_i = (\Phi_i \times \frac{\mu}{\rho})(E_i \times \frac{\bar{E}_e}{E_i})$$

$$= \frac{number\ of\ interactions}{gram} \times \frac{net\ energy\ transferred}{interaction}$$

Logically, Equation 4.5 provides an estimate of the net energy per gram that is transferred to electrons that will then be dissipated through collisions with no radiative losses.

As discussed earlier, when CPE has been achieved, the collision KERMA and the absorbed dose are essentially the same, leading to the result shown in Equation 4.6.

$$D = \Phi_i E_i (\frac{\mu_{en}}{\rho})_i, \text{ assuming CPE} \quad (4.6)$$

One consequence of the calculation of absorbed dose that may seem counterintuitive is that for different materials placed in an identical radiation field, the absorbed dose will be different. This is due to the difference in (μ_{en}/ρ) at a given energy for different materials. To illustrate this, Table 4-2 below shows the values for (μ_{en}/ρ) for a few different materials at different energies.

Table 4-2. Values of (μ_{en}/ρ) for several different materials at various photon energies.

Photon Energy (MeV)	Mass energy absorption coefficient (cm^2/g)				
	Carbon	Aluminum	Iron	Air	Soft Tissue
2.00E-02	2.24E-01	3.09E+00	2.26E+01	5.39E-01	5.66E-01
3.00E-02	6.61E-02	8.78E-01	7.25E+00	1.54E-01	1.62E-01
4.00E-02	3.34E-02	3.60E-01	3.16E+00	6.83E-02	7.22E-02
5.00E-02	2.40E-02	1.84E-01	1.64E+00	4.10E-02	4.36E-02
6.00E-02	2.10E-02	1.10E-01	9.56E-01	3.04E-02	3.26E-02
8.00E-02	2.04E-02	5.51E-02	4.10E-01	2.41E-02	2.62E-02
1.00E-01	2.15E-02	3.79E-02	2.18E-01	2.33E-02	2.55E-02
1.50E-01	2.45E-02	2.83E-02	7.96E-02	2.50E-02	2.75E-02
2.00E-01	2.66E-02	2.75E-02	4.83E-02	2.67E-02	2.94E-02
3.00E-01	2.87E-02	2.82E-02	3.36E-02	2.87E-02	3.16E-02
4.00E-01	2.95E-02	2.86E-02	3.04E-02	2.95E-02	3.25E-02
5.00E-01	2.97E-02	2.87E-02	2.91E-02	2.97E-02	3.27E-02
6.00E-01	2.96E-02	2.85E-02	2.84E-02	2.95E-02	3.25E-02
8.00E-01	2.89E-02	2.78E-02	2.71E-02	2.88E-02	3.18E-02
1.00E+00	2.79E-02	2.69E-02	2.60E-02	2.79E-02	3.07E-02
1.25E+00	2.67E-02	2.57E-02	2.47E-02	2.67E-02	2.94E-02
1.50E+00	2.55E-02	2.45E-02	2.36E-02	2.55E-02	2.81E-02
2.00E+00	2.35E-02	2.27E-02	2.20E-02	2.35E-02	2.58E-02

Inspection of this table shows some trends. At energies between roughly 0.5 MeV and 2 MeV, the values of the coefficient are very similar to each other. However, at lower energies (most noticeably below 0.1 MeV), the coefficients are markedly different from each other. As we might expect, at the lower photon energies, the photoelectric effect is more likely to occur, and this interaction has a very strong dependence on atomic number (roughly Z^3) as well as photon energy (roughly $1/E_\gamma^3$).

The values for μ_{en}/ρ in Table 4-2 are taken from a database maintained by NIST. At the time of this text, the web address is https://www.nist.gov/pml/x-ray-mass-attenuation-coefficients, where values of both μ/ρ and μ_{en}/ρ can be found for individual elements and mixtures/compounds. Within this webpage, there are four tables of interest as follows:

- Table 1: Material constants (including density) for elements
- Table 2: Material constants for compounds and mixtures
- Table 3. μ/ρ and μ_{en}/ρ for elements
- Table 4. μ/ρ and μ_{en}/ρ for compound and mixtures

These tables will be used in the following chapters; so it is wise to become familiar with their location and how to quickly access them.

The dependency of absorbed dose on the mass energy absorption coefficient has some application in relating the absorbed dose in one medium to that in another. In general, if the absorbed dose in one material is known, the absorbed dose in a different material can be calculated using Equation 4.7 below, assuming that the same photon fluence would impinge on both materials and that CPE exists.

$$\frac{D_1}{D_2} = \frac{(\frac{\mu_{en}}{\rho})_1}{(\frac{\mu_{en}}{\rho})_2} \qquad (4.7)$$

Where

- D_1 is the absorbed dose in material 1
- D_2 is the absorbed dose in material 2

- $(\mu_{en}/\rho)_1$ is the mass energy absorption coefficient (cm^2/g) for material 1
- $(\mu_{en}/\rho)_2$ is the mass energy absorption coefficient (cm^2/g) for material 2

In Chapter 5, we will make use of the relationship in Equation 4.7 to relate the measured absorbed dose in a material (a dosimeter) to the absorbed dose that would be received by tissue.

Finally, it must be mentioned again that one of the main challenges with calculating absorbed dose lies with determining the fluence of the incoming radiation. In Chapter 1, we looked at solutions for unshielded point sources of radiation and unshielded line sources. In Chapter 6, we will apply shielding principles to see how the workplace can be designed to reduce dose rates to a desired level.

4.2.2 Calculation of Exposure

As discussed in Chapter 1, the radiation quantity exposure is used to characterize a photon radiation field and relies on determining the number of ion pairs produced per unit mass of air. To calculate exposure, we first calculate the absorbed dose to the air and then use a quantity called the $\overline{W}$-value to determine the exposure. $\overline{W}$ is the average energy needed to produce an ion pair in a particular gas. It is greater than the ionization potential because some interactions produce excitations instead of ionizations; so it may take multiple interactions with expenditure of energy before an ionization is realized. For air, the value of $\overline{W}$ is 33.97 eV/ion pair; this is also the same as 33.97 J/C.

Equation 4.8 below shows how to calculate the exposure in SI units for a given photon radiation field.

$$X = (1.6 \times 10^{-10} \frac{J \cdot g}{MeV \cdot kg}) \Phi_i E_i \left(\frac{\mu_{en}}{\rho}\right)_i \left(\frac{1}{\overline{W}}\right) = \frac{D_{air}}{\overline{W}} \qquad (4.8)$$

Where

- X is the exposure (C/kg)
- Φ_i is the fluence (cm^{-2}) of the ith radiation

- E_i is the photon energy (MeV) of the ith radiation
- $(\mu_{en}/\rho)_i$ is the mass energy absorption coefficient <u>in air</u> (cm^2/g) for the appropriate energy E_i
- $\overline{W}$ is the average energy expended per ion pair in air (33.97 J/C)
- D_{air} is the absorbed dose to the air (J/kg)

Example: What is the exposure rate from a 1 Ci point source of Cs-137 at a distance of 1 meter from the source due only to the 0.662 MeV photons?

Solution: Since this is a point source of radiation, we invoke the calculation of fluence rate from Equation 1.15 knowing that the yield of the prominent 0.662 MeV photon from Cs-137 is 0.851 γ/decay (we can look that up in the IAEA app again if needed). The activity of 1 Ci by definition is 3.7×10^{10}decays/second; we also express the distance in cm to obtain the proper units for fluence rate.

$$\phi_{point} = \frac{A \cdot Y_i}{4\pi r^2} = \frac{(3.7 \times 10^{10}\frac{decays}{second})(0.851\frac{\gamma}{decay})}{4\pi(100\ cm)^2}$$
$$= 250{,}566\frac{\gamma}{cm^2 sec}$$

The value of (μ_{en}/ρ) for air is not tabulated explicitly for 0.662 MeV. Using the values in Table 4-2 and doing either linear interpolation or logarithmic interpolation for values between 0.6 MeV and 0.8 MeV, we obtain a value of 0.0293 cm^2/g.

Finally, using Equation 4.8, we can determine the exposure rate:

$$X = \left(1.6 \times 10^{-10}\frac{J \cdot g}{MeV \cdot kg}\right)\Phi_i E_i \left(\frac{\mu_{en}}{\rho}\right)_i \left(\frac{1}{\overline{W}}\right)$$
$$= \left(1.6 \times 10^{-10}\frac{J \cdot g}{MeV \cdot kg}\right)\left(250{,}566\frac{\gamma}{cm^2 sec}\right)\left(0.662\ \frac{MeV}{\gamma}\right)\left(0.0293\frac{cm^2}{g}\right)\left(\frac{1}{33.97\frac{J}{C}}\right)$$

This gives a result of 2.29×10^{-8} C/kg/sec. In traditional units in the United States, this would be 0.32 R/hr.

Equation 4.8 is quite versatile in that it allows us to determine either the exposure or the absorbed dose to the air simply by knowing the other quantity. Further, we can then use Equation 4.8 in conjunction with Equation 4.7 to estimate the absorbed dose to soft tissue based on a measured exposure as shown below in Equations 4.9 and 4.10.

$$D_{air} = X\overline{W} \tag{4.9}$$

$$D_{tissue} = D_{air}\frac{(\frac{\mu_{en}}{\rho})_{tissue}}{(\frac{\mu_{en}}{\rho})_{air}} \tag{4.10}$$

In Equation 4.10, the ratio of the mass energy absorption coefficients for tissue and air must be known for the photon energy of interest. However, by inspecting Table 4-2, for energies between 0.1 and 2 MeV, this ratio is essentially a constant value of 1.1; for energy values from 0.01 to 0.1 MeV, it is no less than 1.05. Most photon energies in the workplace will be above 0.1 MeV, and it is thus a reasonable assumption to use 1.1 as a pseudo-constant value for the ratio of the absorbed dose in tissue to that of air: that is, $D_{tissue}/D_{air} \sim 1.1$. Hence, we can write Equation 4.11 as a shorthand version of Equation 4.10.

$$D_{tissue} \approx 1.1D_{air} \tag{4.11}$$

Example: Given a measured exposure of 1 R:

a) what is the absorbed dose to the air?

b) what would be the absorbed dose to soft tissue in that same location?

Solution: As noted earlier, we tend to do our calculations in SI units and then convert them if expedient to other units. In this case, we need to convert 1 R to C/kg; the conversion was given in Chapter 1 as $1R=2.58\times10^{-4}C/kg$.

a) We apply Equation 4.9 above to obtain:

$$D_{air} = X\overline{W} = \left(2.58 \times 10^{-4}\frac{C}{kg}\right)\left(33.97\frac{J}{C}\right) = 0.00876\frac{J}{kg}$$
$$= 0.00876\ Gy\ to\ the\ air = 0.87\ rad\ to\ the\ air$$

c) Without knowing the exact photon energy involved, we make an assumption that the energy lies within the range where Equation 4.11 is valid. Thus,

$$D_{tissue} \approx 1.1D_{air} = 1.1(0.00876\ Gy) = 0.0096\ Gy\ to\ tissue$$
$$= 0.96\ rad\ to\ tissue$$

The results of this example lead to the oft-repeated phrase in the workplace "1 R leads to 1 rad, which leads to 1 rem," since the radiation weighting factor for photons is 1. Strictly speaking, this is not true, since 1 R leads to 0.96 rad as a point dose to tissue, but as a rule of thumb, the rounding error is not significant. Again, it is crucial when reporting a dose to specify the medium as well as the dose quantity. In this case, the absorbed dose to the air is 0.87 rad while the absorbed dose to soft tissue would be 0.96 rad owing to the difference in material composition between air and soft tissue. Finally, we must remember that these are point doses and may not reflect the average dose for an organ or for the entire human body.

Often, we are interested in the exposure rate from a point source with known radionuclides. Rather than calculating this value from first principles every single time, it is convenient to tabulate values for a single defined scenario and then use our understanding of how fluence is affected by both distance and activity to scale the exposure rate for other scenarios.

It is common to define Γ as the Exposure Rate Constant (sometimes called the Specific Gamma Ray Constant); it is unique for each radionuclide because it requires applying Equation 4.8 to each photon emitted from the radionuclide of interest and summing the results for all photons from the given radionuclide. The units for Γ depend on the tabulated reference; conceptually, Γ has units of exposure rate per unit activity at a certain distance. When written, however, the units appear unwieldly with common units expressed below:

$$\Gamma\left(\frac{R \cdot cm^2}{mCi \cdot hr}\right)$$

which is the exposure rate in R/hr per mCi of activity at a distance of 1 cm. Note that this is simply a mathematical representation of Γ; the exposure rate at 1 cm from a source would probably not be the published value due to physical limitations of knowing the exact fluence at that location and the potential lack of CPE at that location.

Since exposure rate is directly proportional to fluence rate, and fluence rate is directly proportional to activity and (for a point source) inversely proportional to the square of the distance, we can use tabulated values of Γ to calculate the exposure rate for a variety of scenarios as shown in Equation 4.12 below:

$$\dot{X} = \Gamma \frac{A}{d^2} \tag{4.12}$$

Where

- $\dot{X}$ is the exposure rate in R/hr
- Γ is the tabulated exposure rate constant (R•cm^2/mCi•hr)
- A is the activity in mCi
- d is the distance from the point source in cm

A sample of values for Γ are presented below in Table 4-3 for common radionuclides. The complete journal article with a comprehensive list of values of Γ is located online and is included in the "Free Resources" list at the end of this chapter.

Table 4-3 Values of Γ for selected radionuclides (Smith and Stabin 2012).

Radionuclide	$\Gamma \left(\frac{R \cdot cm^2}{mCi \cdot hr}\right)$
Am-241	0.749
Ba-133	3.04
Cd-109	1.89
Co-60	12.9
Cs-137	3.43
I-131	2.2
Ra-226	0.0394
Tc-99m	0.795

While Γ has historically been called the exposure rate constant (or the gamma ray constant), ICRU Report 85 defines Γ as the Air-KERMA Rate Constant. As the name implies, in this case the value would be for the air KERMA rate, not the exposure rate. Equation 4.12 would still be valid, but the result of the calculation would result in a KERMA rate of Gy/s to the air. Hence, when researching values of Γ, it is important to read the context of the tabulated values to identify whether it is the exposure rate constant or the air KERMA rate constant.

Example: What is the exposure rate from a 0.5 Ci point source of Co-60 at a distance of 1.5 meters from the source?

Solution: With this being a point source, we are able to use the exposure rate constant from Table 4-3 for Co-60, which is 12.9 ($R*cm^2$)/(mCi*hr). With an activity of 0.5 Ci (500 mCi) and a distance of 1.5 m (150 cm), we substitute the values into Equation 4.12 as shown:

$$X = \Gamma\frac{A}{d^2} = \left(12.9\ \frac{R \cdot cm^2}{mCi \cdot hr}\right)\frac{500\ mCi}{(150\ cm)^2} = 0.287\frac{R}{hr}$$

4.3 Absorbed Dose Calculations for Neutrons

Before delving into the dose calculation process for neutrons, it is helpful to understand some background information on this radiation type.

4.3.1 Characterizing Neutrons by Energy

Unlike photons and charged particles, neutrons are seldom if ever monoenergetic. They are typically emitted as a spectrum from nuclear reactions such as fission, fusion, spallation, or absorption reactions that eject a neutron. Curiously, while neutrons are stable inside the nucleus, free neutrons have a radioactive half-life of approximately 10 minutes.

We tend to divide neutrons into rough energy groups for descriptive purposes. The boundaries between the groups are flexible to some degree; the scientific literature is not consistent with how many groups or the exact boundaries between the groups. For purposes of our discussion, we will tend to group neutrons into three categories as follows:

- **Thermal neutrons** have energy associated only with thermal motion of their surroundings. They follow the Maxwellian distribution just as gases do, and they therefore have no single value of energy. This should not be surprising given the mass of the neutron (its mass is roughly ½ of the diatomic H_2 molecule). At roughly 20°C, we typically refer to the energy of these neutrons as 0.025 eV (the product of Boltzmann's constant and the temperature) and their velocity as 2200 m/sec. The upper energy bound of thermal neutrons by convention is typically 0.5 eV.
- **Intermediate neutrons** have energy from about 0.5 eV up to around 10 keV.
- **Fast neutrons** have kinetic energy above 10 keV.

4.3.2 Neutron Interactions in the Human Body

While many reaction types are possible, the absorbed dose in tissue from neutrons is due in large part to the following reactions:

1. Intermediate and fast neutrons undergo elastic scattering with atoms in the body. Neutrons can transfer more of their kinetic energy in collisions with hydrogen because of their similarities in mass. Since the human body has lots of hydrogen, these collisions are predominant. This results in a proton being loosed from its molecular bond (proton recoil) and transferring energy to its surroundings. As the neutron collides with additional hydrogen atoms, it eventually slows to thermal energies.
2. Thermal neutrons undergo two primary absorption reactions:
    a. A reaction with nitrogen represented as $^{14}N(n,p)^{14}C$. This reaction causes the ejection of a proton that then transfers its energy to the surrounding tissue.
    b. A reaction with hydrogen represented as $^{1}H(n,\gamma)^{2}H$. This absorption reaction releases a gamma ray of 2.2 MeV which can travel far from the neutron interaction site but can still cause dose to the body.

4.3.3 Neutron Absorbed Dose Calculations

Neutrons pose a special challenge in the determination of dose quantities. This is due to several factors, some of which were mentioned earlier.

- Neutrons are typically emitted as a spectrum of energies.
- The spectrum of neutrons can change drastically between the location where they are emitted and the receptor location due to interactions in the air, scatter off other materials in the vicinity, etc.
- The cross sections for neutron interactions are generally not smooth functions of energy.
- The absorbed dose is highly dependent on the shape and size of the receptor because the number of scattering events that occur depends on the number of possible target atoms in the path. In addition, for the 2.2 MeV gamma ray emitted following thermal neutron absorption in hydrogen, some of the energy may escape smaller volumes.
- At energies above a few MeV, more interaction types become possible with the generation of many different types of secondary particles.

For these reasons, we are limited in being able to calculate neutron dose from first principles, although there were reasonably successful efforts in the past using what was called first-collision dose. In the modern era, Monte Carlo simulation of the neutron interactions in the body can be an effective method to estimate the absorbed dose to various tissues. The Monte Carlo method is typically used to simulate radiation interactions using statistical processes like what is observed in nature. Thus, it provides us with the ability to track individual particles.

For instance, a single neutron might be tracked in a three-dimensional virtual space (called a history) as it enters the body. Random numbers are used as inputs to well-defined mathematical equations describing the interactions of the neutrons to simulate how that particular neutron travels and deposits energy throughout its history until it terminates. Along its path, all the details of energy transfer, energy absorption, and trajectory are tracked in arrays to determine the absorbed dose at various locations. Once that neutron is terminated (by escaping the body or being absorbed by a nucleus), a new neutron history is started that will have different interactions and trajectory. This mathematical process is repeated over and over (typically millions of times) to obtain an average value of absorbed dose.

While Monte Carlo methods have a lot of utility, there are limitations on them. For example, a few million histories may sound like a lot; however, we must consider that a typical neutron fluence rate in the real world could easily be a modest number such as 10^8 neutrons/cm^2/sec. If a person were exposed to that fluence rate for just 1 minute, the total fluence would be 6 billion neutrons/cm^2. It is relatively easy to see that simulating a few million neutrons may not be a sufficient sample of the physical scenario. Thankfully, there are mathematical methods that can be used to reduce these errors so that we have greater confidence in the results obtained by the simulations.

Annex C of ICRP Publication 116 ("Conversion Coefficients for Radiological Protection Quantities for External Radiation Exposures") contains numerous tables with the results of Monte Carlo simulations for absorbed dose due to monoenergetic neutrons. These values are presented per unit fluence for various tissues for both females and males. Therefore, the absorbed dose can be calculated using Equation 4.13 below:

$$D = \Phi \, \mathrm{DCF} \tag{4.13}$$

Where

- D is the absorbed dose (pGy)
- Φ is the fluence in neutrons/cm^2
- DCF is the dose conversion factor from ICRP 116 (pGy•cm^2)

As mentioned earlier, neutrons are seldom monoenergetic; therefore, it is necessary to account for the variety of neutron energies that may be encountered, such as what is shown in Figure 4-11 below as an idealized neutron fission spectrum from U-235.

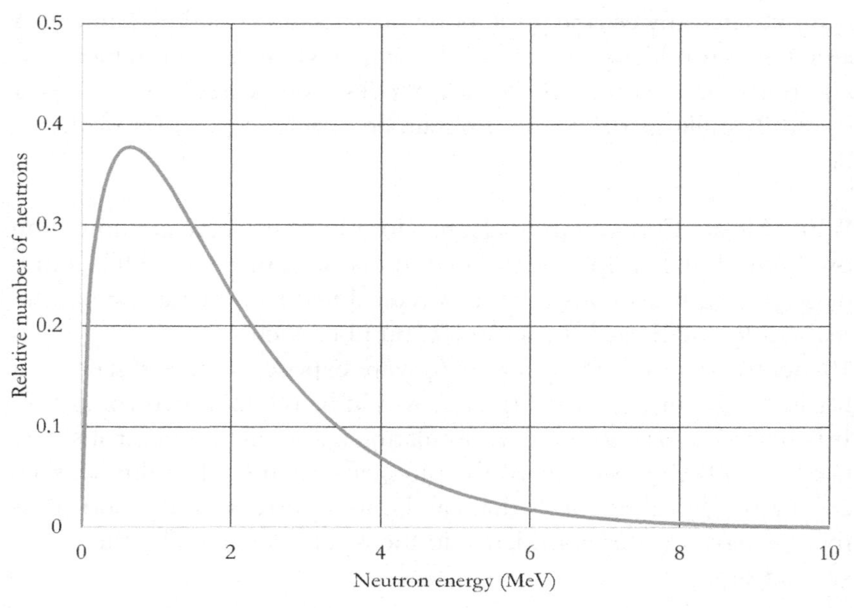

Figure 4-11. Theoretical neutron spectrum from U-235 fission illustrating a wide range of neutron energies.

There are two general approaches to dealing with a spectrum of energies. One approach would be to determine an average energy for the fluence and use that in conjunction with the appropriate DCF from ICRP 116. However, given the wide breadth of energies and the large variation in the DCF with energy, this technique would be prone to a large error. The other approach is to divide the spectrum into "energy bins": that is, we subdivide the entire spectrum into smaller groups while preserving the total number of neutrons. For example, in Figure 4-11, we might have an energy bin from 1-2 MeV that contains all the neutrons in that energy range, another bin containing all the neutrons from 2-3 MeV, and so on for all the energies down to thermal energies and for energies above 3 MeV. For each energy bin, we would then apply an average DCF that is appropriate for that energy group. The more bins that we use, the more accurate our final answer should be.

More information on using the information in ICRP 116 is presented in the next section.

4.4 Calculations of other dose quantities

When performing calculations for other dose quantities (equivalent dose, effective dose) or for specific targets (for example, the lens of the eye), it is impractical to do them from first principles by hand. Instead, Monte Carlo methods are used for the various particle types and for certain reference energies.

Authoring or using the most rigorous Monte Carlo codes can be a monumental task. However, to gain a rudimentary understanding of how the dose conversion factors (DCFs) can be calculated, there is a free alternative program called Visual Monte Carlo (VMC) available at www.vmcsoftware.com. Two different Monte Carlo programs are available for free download at that website. However, for our purposes, the VMC "Dose Calculation" program is especially useful because it can be used to understand conceptually how the photon dose conversion factors in ICRP 116 were developed (Note: the ICRP 116 DCFs were not calculated using VMC). Once installed, VMC allows the user to select the radiation source to be a radionuclide, an ICRP planar beam, or an X-ray beam. Once the number of histories has been set, the program tabulates the equivalent dose to each organ in the body (the user can select the male or female) based on interactions in a mathematical model (called a phantom) divided into 3-d voxels. Finally, the program will calculate the effective dose based on the organ dose equivalents and the tissue weighting factors.

VMC Dose Calculation uses an anthropomorphic phantom and allows the user to select various slices of the phantom to view, much like looking at cross sections of a computed tomography (CT) scan or a magnetic resonance imaging (MRI) scan. Figure 4-12 below shows a slice midway through the male phantom from the front.

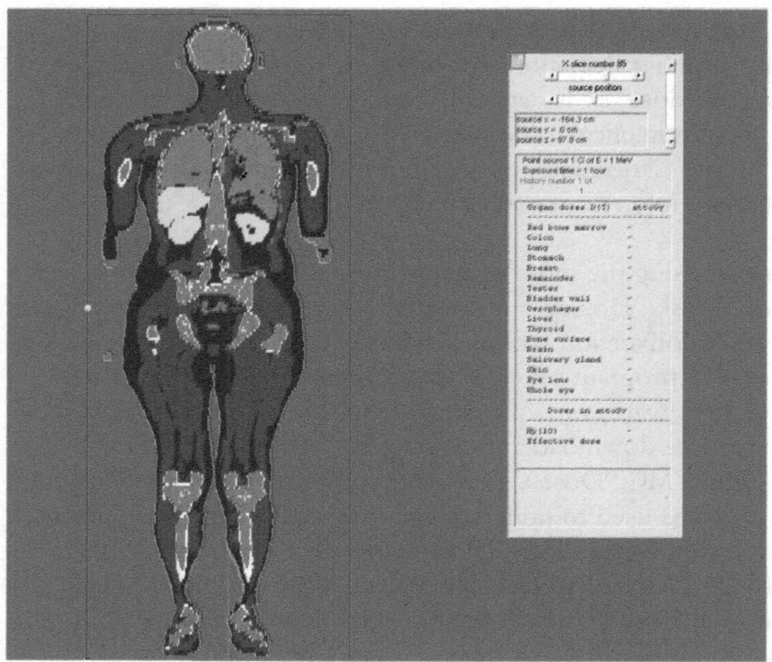

Figure 4-12. Voxel phantom used in VMC Dose Calculation (Image courtesy of John Hunt, author of VMC).

The software has a user manual for instructions, but much of the operation is intuitive. As the software runs its simulations, it updates the organ doses as well as the effective dose; observing these updates as the simulation is running can show how a few "hits" in certain tissues and organs can markedly change the effective dose. With more histories, there is greater opportunity for the effective dose to converge to a value that will not change drastically with additional histories.

An example of the output from a VMC run is shown below in Figure 4-13. For this scenario, a fluence of 1 MeV photons is incident on the phantom from the front to the back. The photons are from a planar source: that is, the photons are parallel to each other from a plane in front of the phantom and are not subject to the inverse square law or other geometry considerations. The simulation was performed with 1,000,000 histories with the results as shown below.

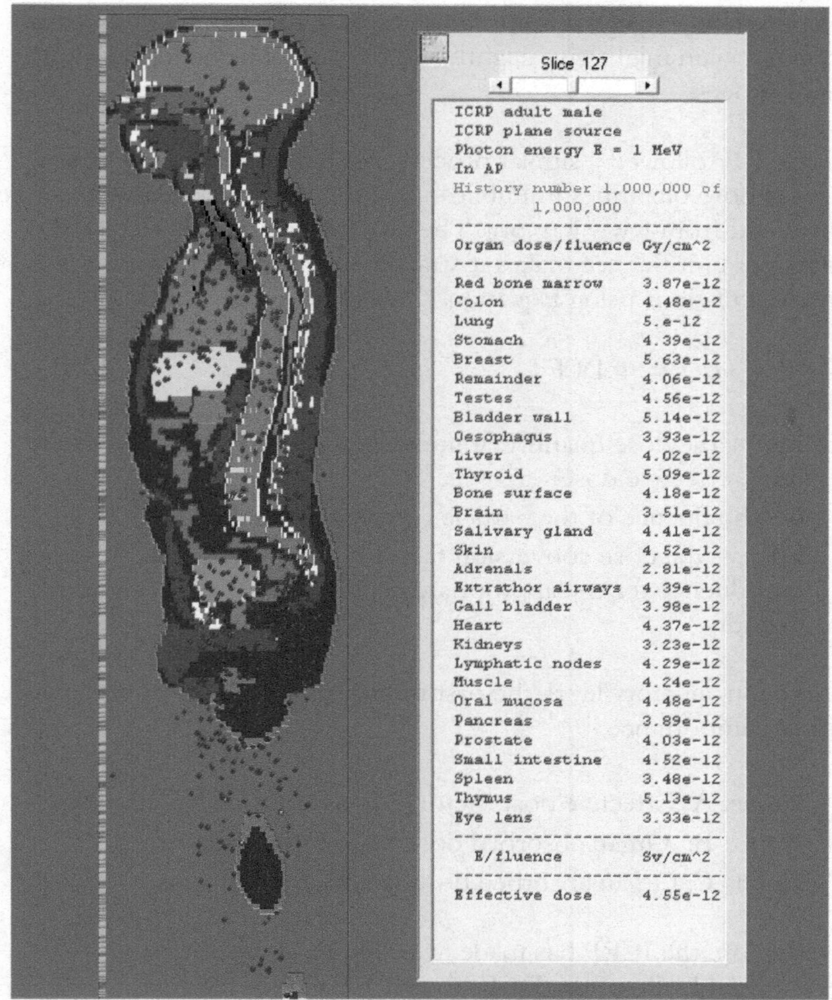

Figure 4-13. VMC output for 1 MeV photons impinging on the phantom from the front. The blue dots in the phantom show locations where interactions have taken place within the simulations (Image courtesy of John Hunt, author of VMC).

The output table in the figure shows several things. First, the reported doses are per unit fluence: that is, the dose is normalized to 1 photon per cm^2. This feature makes it easy to scale the Monte Carlo results for different fluences or fluence rates. Second, the tissue and organ absorbed doses vary from a low of 2.81 pGy to a maximum of 5.63 pGy, indicating the degree to which doses can vary across the human body. The weighting of the organ equivalent doses then presents an effective dose of 4.55 $pSv/\gamma/cm^2$ or

4.55 pSv•cm^2.Finally, it is important to note that running this same scenario again might give slightly different results due to the Monte Carlo process.

The ICRP followed a similar process in tabulating values for the various dose quantities in Publication 116, which is available as a free download from www.icrp.org. The normalizing of the calculated doses per unit fluence makes it straightforward to perform quick dose calculations using Equation 4.14 below.

$$DQ = \Phi \ \mathrm{DCF} \quad (4.14)$$

Where

- DQ = the dose quantity of interest (absorbed dose, equivalent dose, effective dose)
- Φ=the fluence of the particle type at a given energy
- DCF = the dose conversion factor that represents the dose per unit fluence for the appropriate particle type, energy, and dose quantity.

At the introductory level, the first three annexes of ICRP 116 are of primary importance.

- Annex A: Effective dose factors for a variety of radiation types
- Annex B: Organ absorbed dose factors for photons
- Annex C: Organ absorbed dose factors for neutrons

In addition, the ICRP has made available on its website a downloadable file of the DCFs in text file format for easy importation into spreadsheets and computer programs.

The DCFs are arranged in these annexes based on the orientation of the source with respect to the receptor using the following notation:

- AP: anterior-posterior (front to back)
- PA: posterior-anterior (back to front)
- LLAT: left lateral (left side to right side)
- RLAT: right lateral (right side to left side)
- ROT: rotational (the radiation field is uniform from top to bottom of the phantom in a 360-degree pattern when viewed

from above such as what might occur if a person were rotating within a directional source of radiation)

- ISO: isotropic (the radiation field is uniform on the phantom from all directions in three-dimensional space as though surrounded by a spherical source of radiation)

These orientations are illustrated below in Figure 4-14.

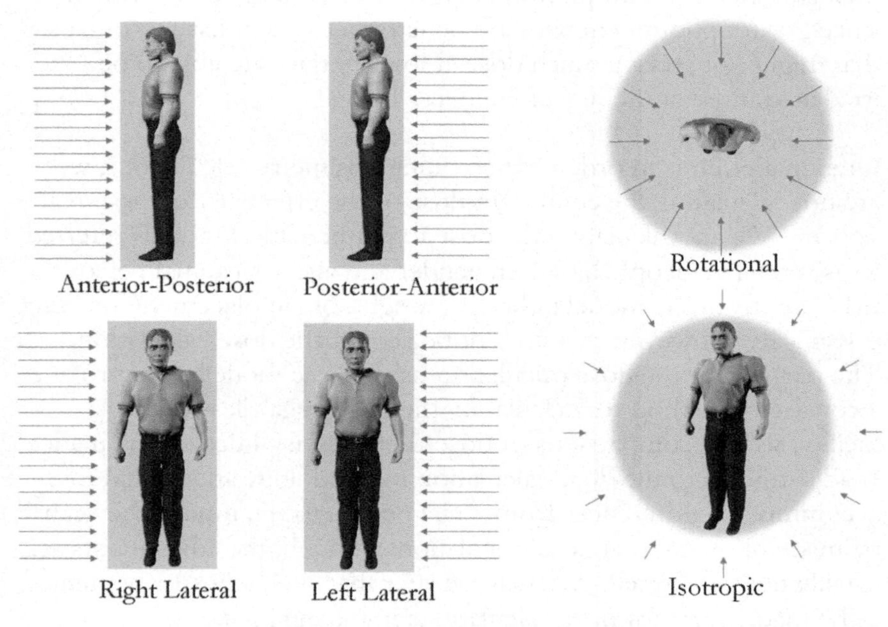

Figure 4-14. Irradiation geometries for which DCFs are tabulated in ICRP 116. The blue arrows represent the incident radiation field.

Table 4-4 shows a sample from ICRP 116 of the DCFs to calculate effective dose for photons. As can be seen, the irradiation geometry can markedly affect the effective dose, particularly for lower photon energies. This is because the various sensitive organs and tissues of the body receive differing amounts of shielding depending on the geometry. For example, organs near the front of the body receive more shielding in the PA geometry as opposed to the AP geometry. Since many of the sensitive organs of the body fall into this category, it is expected that the AP DCFs will be higher than the PA DCFs, as is borne out in Table 4-4. As a side note, the value for the effective dose per unit fluence shown in Figure 4-13 from a 1,000,000-history

VMC simulation (4.55 pSv•cm^2) can be compared to the DCF for the AP geometry at 1 MeV in Table 4-4 (4.49 pSv•cm^{-2}); this comparison shows the results to be very similar. Running more histories in VMC gives even closer results; the effective dose in VMC from using 10,000,000 histories gives a value of 4.51 pSv•cm^2.

In addition, the general trend in Table 4-4 is for the effective dose DCFs to increase with photon energy. This should be no surprise, since greater photon energies are penetrating so that tissues/organs that might not receive much dose at low energies are able to be irradiated more at the higher energies.

One final point is in order: when doing dosimetry calculations, we are not calculating the equivalent dose or the effective dose to a real person. We are calculating the dose to mathematical models referred to as reference people based on gender and age. An actual person may vary from the model in height, weight, organ placement, or exact organ size. However, this does not discount the dose calculation. The purpose of the dose calculation using these models is to provide a consistent method for calculating the dose, regardless of work facility, so that comparisons of doses between facilities can be made. If adjustments to the dose calculation methodology are needed to accommodate differences from a real person to the model, they can be made. For external dose calculations, though, the adjustments are usually not performed. We will see later that such adjustments tend to be made more for dose calculations involving intakes of radionuclides.

Table 4-4. DCFs for effective dose in units of pSv/ γ/ cm² (pSv•cm²) for photons in the ICRP-defined irradiation geometries from Table A.1 of ICRP 116

Photon Energy (MeV)	AP	PA	LLAT	RLAT	ROT	ISO
0.01	0.0685	0.0184	0.0189	0.0182	0.0337	0.0288
0.015	0.156	0.0155	0.0416	0.0390	0.0664	0.0560
0.02	0.225	0.0260	0.0655	0.0573	0.0986	0.0812
0.03	0.313	0.0940	0.110	0.0891	0.158	0.127
0.04	0.351	0.161	0.140	0.114	0.199	0.158
0.05	0.370	0.208	0.160	0.133	0.226	0.180
0.06	0.390	0.242	0.177	0.150	0.248	0.199
0.07	0.413	0.271	0.194	0.167	0.273	0.218
0.08	0.444	0.301	0.214	0.185	0.297	0.239
0.1	0.519	0.361	0.259	0.225	0.355	0.287
0.15	0.748	0.541	0.395	0.348	0.528	0.429
0.2	1.00	0.741	0.552	0.492	0.721	0.589
0.3	1.51	1.16	0.888	0.802	1.12	0.932
0.4	2.00	1.57	1.24	1.13	1.52	1.28
0.5	2.47	1.98	1.58	1.45	1.92	1.63
0.511	2.52	2.03	1.62	1.49	1.96	1.67
0.6	2.91	2.38	1.93	1.78	2.30	1.97
0.662	3.17	2.62	2.14	1.98	2.54	2.17
0.8	3.73	3.13	2.59	2.41	3.04	2.62
1.0	4.49	3.83	3.23	3.03	3.72	3.25
1.117	4.90	4.22	3.58	3.37	4.10	3.60
1.33	5.59	4.89	4.20	3.98	4.75	4.20
1.5	6.12	5.39	4.68	4.45	5.24	4.66
2.0	7.48	6.75	5.96	5.70	6.55	5.90
3.0	9.75	9.12	8.21	7.90	8.84	8.08
4.0	11.7	11.2	10.2	9.86	10.8	10.0

Free Resources for Chapter 4

- Questions and problems for this chapter can be found at www.hpfundamentals.com
- Online databases for alpha, electron, and proton stopping powers and ranges can be found at the following web addresses.
  - Alpha particles: https://physics.nist.gov/PhysRefData/Star/Text/ASTAR.html
  - Electrons: https://physics.nist.gov/PhysRefData/Star/Text/ESTAR.html
  - Protons: https://physics.nist.gov/PhysRefData/Star/Text/PSTAR.html
- Online database for mass attenuation coefficients and mass energy absorption coefficients for photons can be found at: https://www.nist.gov/pml/x-ray-mass-attenuation-coefficients
- Online database (XCOM) for individual interaction probabilities and total mass attenuation coefficients for photons can be found at: https://www.nist.gov/pml/x-ray-and-gamma-ray-data
- A website with many free dosimetry resources (external and internal) is the Radiation Dose Assessment Resource (RADAR) located at https://www.doseinfo-radar.com/
- A detailed table listing the Specific Gamma Ray Constants discussed in Section 4.2.2 is located at https://www.doseinfo-radar.com/Exposure_Rate_Constants_and_Lead_Shielding_Values%204.pdf
- ICRP Publication 116, *Conversion Coefficients for Radiological Protection Quantities for External Radiation Exposures.*, Ann. ICRP 40(2-5). Available at www.icrp.org
- VMC Monte Carlo software can be downloaded from www.vmcsoftware.com
- The RadPro Calculator is a free online resource for calculating a wide range of external dose quantities and is available at http://www.radprocalculator.com/

Additional References

- Smith, D.S. and Stabin, M.G. Exposure Rate Constants and Lead Shielding Values for Over 1,100 Radionuclides. Health Physics 102(3), 2012.
- Attix, F.H. Introduction to Radiological Physics and Radiation Dosimetry, John Wiley & Sons, Hoboken, NJ, 1986, 1991.

- Chilton, A.B., Shultis, J.K. and Faw, R.E. Principles of Radiation Shielding. Prentice Hall, Englewood Cliffs, NJ 1984.
- Shultis, J.K. and Faw, R.E Radiation Shielding, American Nuclear Society, La Grange Park IL, 2000.

Chapter 5 Dosimetry Measurements for External Sources of Radiation

In the workplace, we often make measurements that are relevant to reporting the dose to workers and regulators. We may prospectively calculate theoretical doses for certain situations to estimate what dose could be received using techniques described in the last chapter. However, those calculations have limitations in their accuracy and applicability and are not suitable for assigning official worker doses; therefore, we generally will make measurements in the workplace to evaluate our calculations and to estimate a more accurate dose for a worker for recording purposes.

Broadly speaking, we will refer to any device that is used to measure a dose quantity as a dosimeter (other parts of the world refer to it as a dosemeter). Later in this chapter, we will distinguish between portable survey meters that are used to measure dose quantities and dosimeters that a person wears.

Although there is ongoing research into using biological-based dosimeters, these are not practical at the present time. Therefore, we make dose measurements using dosimeters constructed of non-biological materials and then relate those measurements to what we expect the dose to be to a person.

As a reminder from Chapter 1, it is not possible to measure equivalent dose or effective dose; most measuring devices rely on a measurement related to absorbed dose and then use an algorithm to convert that measurement to the desired dose quantity. In addition, we must understand that any dose-measuring device responds to the absorbed dose that it receives, a concept discussed in Section 4.2.1. The dosimeter dose may be markedly different than what a person would receive; therefore, we must use our understanding of dose principles to relate the dosimeter measurement to what we expect the dose would be to human tissue.

This chapter focuses solely on the dose-measuring aspects of the devices and not on the programmatic or operational aspects of calibration, issuance, dose reporting, and other aspects of an external dosimetry program.

5.1 The Ideal Measuring Device

In a perfect world, there would be many traits of a measuring device for radiation dose. Among these traits might be some of the following:

- Same material composition as human tissue
- Same interaction mechanisms for all radiation types as tissue
- Same interaction probabilities over all energies as tissue
- Able to determine point dose and average dose
- Able to determine dose rate as well as total dose
- Able to determine equivalent dose and effective dose
- Rapid readout for the user
- Insensitive to environmental factors such as humidity and temperature

In reality, there are no devices that would meet all of these requirements (plus additional individual preferences) for all scenarios. Thus, we must try to minimize the shortcomings as we select which device to use for our dose measurements. In addition, there are some special considerations for each radiation type to be measured as discussed below.

Charged particles - As described in Chapter 4, the absorbed dose due to charged particles is proportional to the stopping power of the medium in which the particles deposit their energy. Thus, substantial differences in stopping power between the dose-measuring device and human tissue can result in a substantially different dose in the dosimeter compared to tissue. In addition, point doses from charged particles can be measured only with thin dosimeters where the stopping power does not change drastically over the thickness of the dosimeter. For alpha particles, this would be nearly impossible; thankfully, alpha radiation is not considered an external radiation hazard because of the shielding capabilities of the dead layer of skin surrounding the body (assumed to be an average of 70 μm). Even for electrons, measuring the point dose with a finite dosimeter can be a challenge; thus, corrections may be needed to mathematically compensate for a dosimeter that measures average dose over a finite thickness so that point doses can be estimated.

Photons – In Chapter 4, it was noted that the absorbed dose is proportional to (μ_{en}/ρ) of the absorber. Differences in the absorbed dose received by the dosimeter compared to tissue would thus be expected based on differences in the mass energy absorption coefficient. At low energies where the photoelectric effect is prominent, these differences in (μ_{en}/ρ) can be highly pronounced, giving rise to what is termed energy dependence of the dosimeter. Because many dosimeters have an effective atomic number that is greater than tissue, the dose to the dosimeter would be greater than that to tissue for low photon energies.

Neutrons – This is perhaps the most challenging radiation type when measuring dose because the interactions in the dosimeter typically are not the same as in tissue. In particular, the thermal neutron interactions are very different in neutron dosimeters compared to tissue. This means that the interaction probabilities and energetics of the interactions cause some difficulties in interpreting the measurements.

Consideration of these and other factors will be the focus of our discussion in the following sections.

5.2 Survey Meters

Survey meters are typically used in the workplace to characterize the radiological conditions. While this section is not intended to give a detailed look at how all survey meters operate, it is nonetheless helpful to be familiar with the operation of some of the common types used in the measurement of dose quantities.

5.2.1 Ion Chambers

In the workplace, an ion chamber is one of the most commonly used instruments. They are electronic devices primarily used to measure the dose quantity exposure from photons. As discussed in Chapter 1, exposure is defined as the charge produced per unit mass of air by photon radiation. This definition lends itself well to constructing a device with a detector that acts as a capacitor using air as the dielectric material between the electrodes. A high voltage is applied across the electrodes, but appreciable current does not flow if there are no ions present in the detector volume. When photons interact in the detector, producing free electrons and positive ions in the detector region, current can flow. The high voltage across the electrodes is sufficient to prevent the rapid recombination of the ions such that the electrons move quickly to the anode; thus, the magnitude of the current is proportional to the absorbed dose to the detector. A highly schematic illustration of an ion chamber is presented below in Figure 5-1.

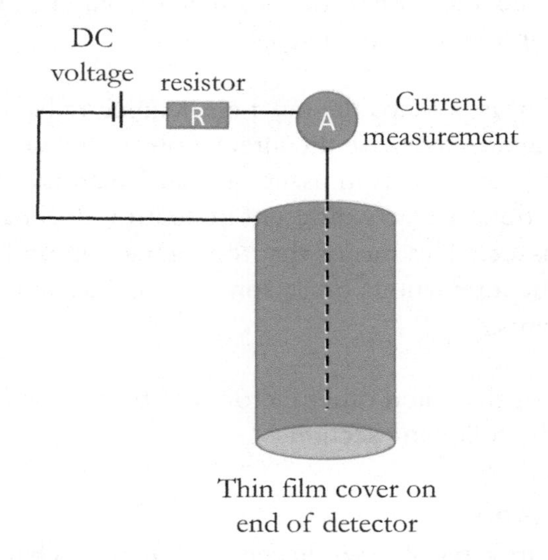

Figure 5-1. Schematic illustration of an ion chamber. The outer wall of the cylinder acts as the cathode for the circuit while a central wire protruding though the center of the cylinder (electrically insulated from the wall) acts as the anode.

Figure 5-2 shows an illustration of a common form factor for an ion chamber. The detector is inside the housing and is typically arranged in a vertical orientation; it is usually cylindrical with aluminized mylar covering the end of the cylinder. The display of the ion chamber can be either digital or analog and gives a readout of the exposure rate, typically in milliroentgens per hour (mR/hr) in the United States. On the underside of the instrument is a sliding cover that can be slid over the detector window to achieve CPE at the covered end of the detector. For low-energy photons, this window might be open while for higher energy photons it would be closed.

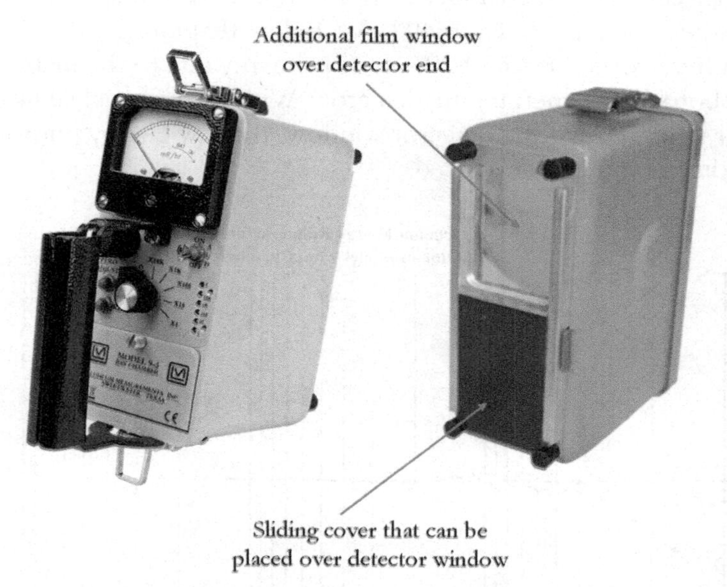

Figure 5-2. A common form factor for an ion chamber (Photos courtesy of Ludlum Instruments).

When considering photons, there are a few items to address. As mentioned earlier, the ion chamber fundamentally measures the electric current produced in the detector. This current is proportional to the absorbed dose to the detector <u>wall</u>, since the photons would interact mostly in the wall itself with only a small number of interactions in the air volume. However, we are interested in determining the absorbed dose to the air in the chamber based on our measurement of the absorbed dose to the wall of the detector so that we can determine the exposure. In general, the construction of the walls of the detector are such that $(\mu_{en}/\rho)_{wall}/(\mu_{en}/\rho)_{air}$ remains roughly constant over all photon energies. Thus, Equation 4.7 predicts that there would be a single constant value to relate the raw measurement of the ion chamber to the absorbed dose to air and consequently to the exposure in air.

Figure 5-3 shows a typical photon energy response curve for an ion chamber where the measurements at various energies are compared to the measurement at the 0.662 MeV photon energy of Cs-137. Ideally, the response at all energies would be the same as at Cs-137 (i.e., a value of 1), which would be called a "flat" energy response. As can be seen, the ion chamber response is in excellent agreement with the expected value down to 40 keV. Below that energy, the instrument would read 10-20% low, primarily due to the inability of the photons to penetrate the detector. With the external sliding cover in place over the detector window, the low energy photon shielding is more pronounced.

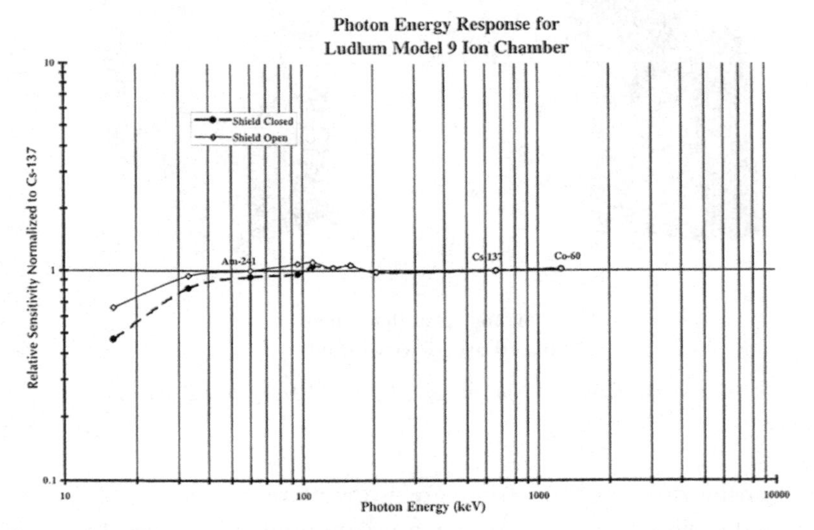

Figure 5-3. Photon energy response curve for a typical ion chamber. With the sliding cover over the window, the lower energy photons are shielded more heavily from entering the detector (Image courtesy of Ludlum Instruments).

Another use for the sliding cover is to help estimate a rough absorbed dose due to beta radiation. With the window open, the ion chamber would be sensitive to both photon and beta radiation entering the detector from the end. With the window closed, the detector would be sensitive primarily to photon radiation while shielding the beta radiation. The difference in these readings can be multiplied by what is termed a "beta correction factor" (BCF) to provide a rough estimate of the absorbed dose rate due to beta radiation. However, the beta correction factor is highly dependent on the geometry of the source (point, planar, etc.) and on the energy of the beta radiation. Low energy beta radiation does not penetrate the detector as easily and thus would require a higher BCF. Because of these dependencies, each facility would typically use its own experimentally determined BCF to give a rough estimate of the beta absorbed dose. In the literature, it is common to see BCF values that range from 2 to 6 depending on the application. For low-energy beta radiation, however, the BCF could be much higher.

Equation 5.1 shows how the beta absorbed dose can be estimated based on the readings of an ion chamber.

$$D_{\beta^-} = [OWR - CWR] * BCF \qquad (5.1)$$

Where

- D_{β^-} = estimated beta absorbed dose rate (mrad/hr)
- OWR = open window reading (mR/hr)
- CWR = closed window reading (mR/hr)
- BCF = beta correction factor (mrad/mR reading). This is a conversion between the reading of the meter and the estimated beta absorbed dose and not a conversion between exposure and beta absorbed dose.

A couple of additional notes about ion chambers are in order.

- Common ion chambers use a detector volume that is vented to the atmosphere. This means that the detector chamber is at the atmospheric pressure of its surroundings. Since exposure is defined based on the mass of the air (not the volume), this means that the readings of a vented ion chamber are altitude dependent because atmospheric pressure decreases as altitude increases, thus decreasing the mass of air in the detector. This altitude dependency is illustrated below in Figure 5-4, where it can be seen that an ion chamber at an altitude of 8000 feet would read approximately 80% of what it would read at altitude of 2000 feet. This altitude dependence does not apply to pressurized ion chambers, however, which are sealed detectors with an internal air pressure of several atmospheres.

- Since ion chambers operate in current mode, they are suitable for measuring high radiation fields. They are also suitable for measuring pulsed radiation fields such as might emanate from a particle accelerator.

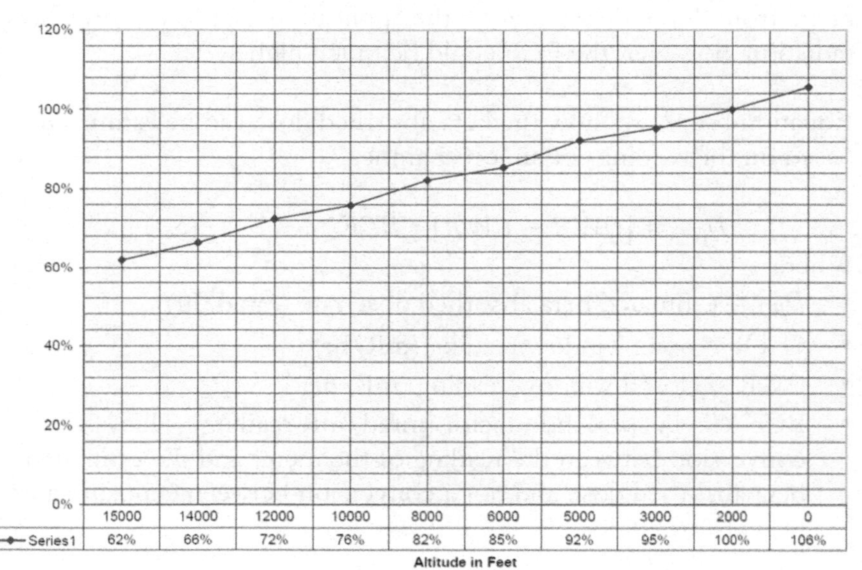

Figure 5-4. Change in readings of a vented ion chamber based on its altitude (Image courtesy of Ludlum Instruments).

5.2.2 Geiger Counters

Geiger counters (also referred to as GM counters) are gas-filled detectors that are useful for detecting the presence of radiation. In contrast to an ion chamber, the GM counter produces a voltage pulse instead of an electrical current. Also, the output pulse of a GM detector has no useful information about the energy of the radiation (i.e., the output pulse from a GM counter is the same regardless of the radiation event that caused it in the first place). This is achieved by using a high voltage to the GM tube that causes maximum electron multiplication inside the tube regardless of the energy of the initiating ionizing event. One might rightly question how such a detector could be useful for measuring dose quantities when in principle it can do nothing more than give an indication of the fluence rate of the radiation field. Of course, if the fluence is comprised of monoenergetic photons, this would be no problem. The count rate of the GM counter could just be multiplied using internal circuitry by a constant factor that would give an output in exposure rate. This is in keeping with Equation 4.8 that shows for a constant photon energy, the exposure rate would be directly proportional to the fluence rate.

However, if the photon field is comprised of multiple energies of photons, then even a fluence rate measurement would be suspect for two reasons:

- The mass energy absorption coefficient varies with energy.
- The detection efficiency of the GM counter varies as a function of photon energy.

With regard to the second point above, the GM counter detects photon radiation primarily by interactions in the side wall (cathode) material because the atom density is higher there than in the gaseous detector volume. The probability of interaction in the side wall is related to the mass attenuation coefficient (μ/ρ) for the material of interest. As a reminder from Chapter 4, the mass attenuation coefficient is the probability of interaction per unit density thickness of material; therefore, greater values of (μ/ρ) imply a greater probability of interaction and consequently a greater detection efficiency. Figure 5-5 shows the variation of the mass attenuation coefficient with energy for aluminum, a common material from which to make detector cathodes. As shown in the figure, the interaction probability is not constant with energy; therefore, lower energy photons will tend to be detected more efficiently than higher energy photons.

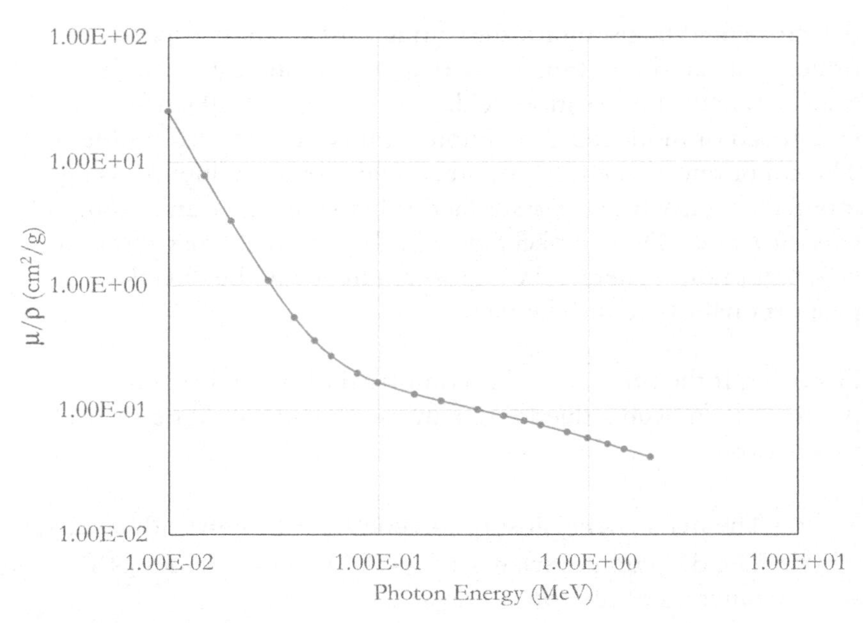

Figure 5-5. Variation of the mass attenuation coefficient of aluminum as a function of photon energy.

There is also another consideration in that Equation 4.8 shows the dependency of exposure on the value of (μ_{en}/ρ) for air, which is also not constant with energy as shown in Table 4-2.

With these considerations, it would seem impossible to measure exposure rate with a GM detector, but this is not the case. Energy-compensated GM detectors have been developed that use metallic filters around the GM tube to decrease the detection efficiency at lower energies. Careful construction can result in a detector where the number of pulses is directly proportional to the exposure rate regardless of the photon energy. Figure 5-6 below shows the response of an energy-compensated GM detector-based survey meter used for measuring exposure. The graph shows the measured exposure rate at various photon energies compared to the rate at 662 keV (the photon energy of Cs-137). Thus, a value of 1 is desirable as discussed for ion chambers in Figure 5-3.

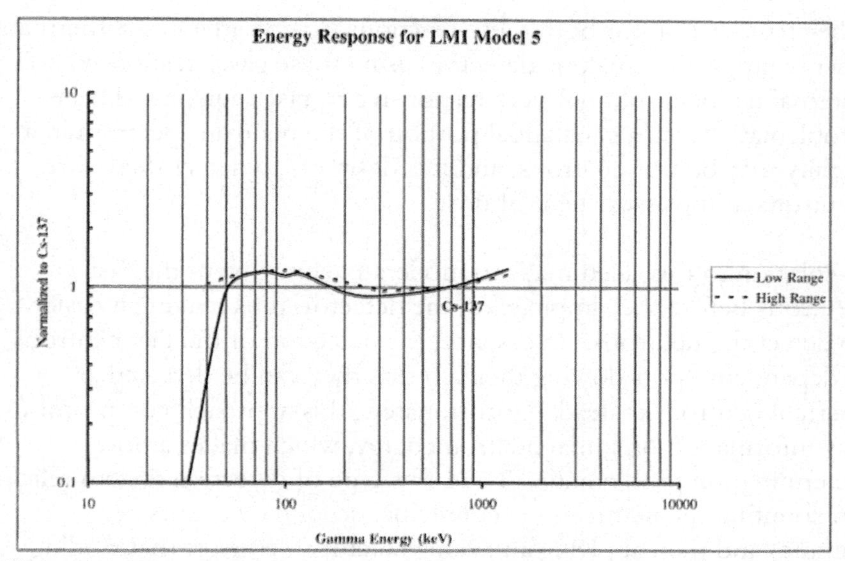

Figure 5-6. Response of an energy-compensated GM survey meter (Image courtesy of Ludlum Instruments).

5.2.3 Survey Meters for Neutron Measurements

As discussed earlier, neutrons pose a special problem for dose measurements, primarily because they do not liberate electrons in a single step like photons or charged particles might do. Since neutrons have no electrical charge, there are only two mechanisms for liberating charged particles that can then interact with electrons in a medium: generating recoil protons by mechanical collisions with hydrogen-rich media or having the neutrons undergo a nuclear reaction where the nucleus then ejects a charged particle. The former mechanism applies primarily to fast neutrons while the latter mechanism applies primarily to thermal neutrons.

If a sealed gas-filled detector is filled with either ^{3}He or BF_3 as the detector gas, it will be sensitive to neutrons due to either the $^{3}He(n,p)^{3}H$ reaction or the $^{10}B(n,\alpha)^{7}Li$ reaction as follows:

$$^{3}_{2}He + ^{1}_{0}n \rightarrow ^{3}_{1}H + ^{1}_{1}p \ (Q = 0.764 \text{ MeV})$$

$$^{10}_{5}B + ^{1}_{0}n \rightarrow ^{7}_{3}Li + ^{4}_{2}\alpha$$
$$(Q = 2.792\ MeV\ for\ ground\ state, 2.310\ MeV\ for\ excited\ state)$$

The cross section for both these reactions is very large in the thermal energy range, thus making detectors using these gases responsive to thermal neutrons but not very responsive to fast neutrons. This is problematic since a substantial portion of the neutron spectrum in a facility may be fast neutrons, and it is desirable to detect these fast neutrons in the assessment of dose.

A solution to this dilemma is to moderate (slow down) the fast neutrons before they impinge on the detector, thus making it easier to detect the neutrons. In essence, the detection of the fast neutrons is dependent upon slowing them so that they can be detected as thermal neutrons instead. Unfortunately, this approach compromises any information on initial neutron energy, which makes a dose determination problematic. Thus, this type of detection system relies on counting the neutrons (as counts per second or counts per minute) and then applying an average factor to convert that reading to a dose equivalent (typically millirem per hour). Therefore, these detectors do not measure an absorbed dose or an equivalent dose. The readout of equivalent dose is achieved simply by taking a thermal neutron count rate and applying an average calibration factor that is specific to the facility where the instrument is used because of the unique energy spectrum for that facility.

Moderating the neutrons involves surrounding the detector with a hydrogen-rich material (typically polyethylene) so that the neutrons transfer more energy per collision and thus slow to thermal energies quickly. A metallic cover for structural support or additional shielding may be present as well. Figure 5-7 below shows two common configurations for these detectors. In the system on the left (sometimes called a "rem ball"), the detector is inserted into a sphere of polyethylene with the electronics mounted on the top. In the system on the right, the detector is inserted into a cylinder of polyethylene with the electronics on top (an additional detector is on top of the electronics in the figure so that other measurements can be made for alpha, beta, and gamma radiation).

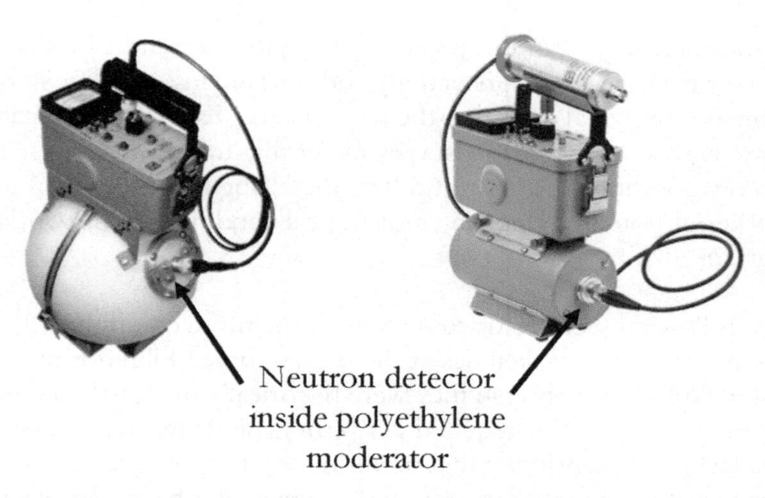

Figure 5-7. Neutron survey meters based on a thermal neutron detector placed inside a moderator. These specific instruments both use ^{3}He as the detector gas (Photos courtesy of Ludlum Instruments).

A more recent advance in neutron survey meters is the tissue equivalent proportional counter (TEPC), a detector with its walls and the detector gas comprised of tissue equivalent material. This type of detector can collect a spectrum of the recoil protons produced by the incoming fast neutrons, determine the absorbed dose from the recoil protons, and calculate the quality factor based on the linear energy transfer of the protons. Thus, this instrument can measure absorbed dose from fast neutrons and give an estimate of dose equivalent based on applying the appropriate quality factor. However, the use of this instrument is not as straightforward as the ^{3}He or BF_3 detector system. First, it requires the collection of a recoil proton spectrum, which takes much more time than using a dose rate instrument. Second, interpreting the recoil proton spectrum requires a basic understanding of microdosimetry principles, which are beyond the scope of this textbook. Thus, the TEPC is mentioned here for completeness but will not be discussed in depth.

5.3 Personal Dosimeters

While survey meters can be used to qualify and quantify the radiation field in the workplace, radiation safety programs will assign personal dosimeters to workers to wear throughout the course of their work activities. The dosimeters are usually worn on the trunk of the body in the chest region; in this position, they move around with the worker regardless of work activity and thus are in a good position to estimate the actual dose that the worker receives. This is typically more representative than a survey meter that may be used in several discrete locations that do not reflect the changing positions of the worker or potential changing radiological conditions while work is occurring.

The following sections discuss some of the more common dosimeters in use at facilities at the present time. Film dosimeters are not included even though they were historically the most common dosimeter type. They required very stringent chemical processing and lacked the sensitivity that newer dosimeters have achieved. Further, the sensitive part of film dosimeters is a halide crystal in the film emulsion which has an atomic number much higher than tissue, thus requiring additional interpretation of the results at low photon energies. As a result, these dosimeters have gradually been replaced with newer types of dosimeters.

Personal dosimeters generally are designed to measure the dose at three different depths in tissue as follows:

- 7 mg/cm^2, which is the assigned depth for the basal layer of the skin.
- 300 mg/cm^2, which is the assigned depth for the lens of the eye.
- 1000 mg/cm^2 (1 g/cm^2), which is the assigned depth for what is termed the deep dose.

The following sections will discuss some of the common dosimeters in use. Admittedly, a large amount of discussion will focus on thermoluminescent dosimeters, since they are still very common. However, some of the design features and dosimetric considerations for this dosimeter type will be applicable to other dosimeters as well.

Personal dosimeters used for an official dose report in the United States must be accredited, which is an involved process. For NRC Licensees, their programs must follow the requirements of the National Voluntary Laboratory Accreditation Program (NVLAP). For DOE facilities, their programs must follow requirements of the Department of Energy Laboratory Accreditation Program (DOELAP). More information is available at:

- NVLAP - https://www.nist.gov/nvlap/ionizing-radiation-dosimetry-lap
- DOELAP - https://www.id.energy.gov/resl/doelap/doelap.html

5.3.1 Thermoluminescent Dosimeters

Thermoluminescent dosimeters (TLDs) are solid-state dosimeters that rely on the general principle of band theory for their operation. Band theory is based on solid-state physics and uses the premise that electrons in a solid-state matrix do not behave as if they are attached to individual atoms; rather, the electrons occupy bands as shown in Figure 5-8. As this diagram illustrates, the energy levels that are usually associated with individual atoms merge into bands that represent energy levels in the entire solid matrix. However, the electrons are not capable of moving to a higher energy band unless energy is imparted to the electrons.

As shown in the figure, the highest band that contains electrons is referred to as the valence band. A forbidden energy gap separates the valence band from the conduction band. Electrons are not allowed to exist in the forbidden gap because ordinarily there are no energy levels within that region; however, electrons may translate to the conduction band through the forbidden gap if they are given sufficient energy.

The separation distance between the valence band and the conduction band is the usual criterion for classifying a material as an insulator, a semiconductor, or a conductor. For an insulator, the forbidden gap is on the order of a few eV, which is so great that only after a tremendous amount of energy is applied (e.g., high voltage or temperature) will the material promote electrons to the conduction band. The forbidden gap in a semiconductor is much smaller and is typically 1 eV or less. As might be expected, for a conductor, the forbidden gap is nonexistent, and electrons are easily promoted from the valence band to the conduction band.

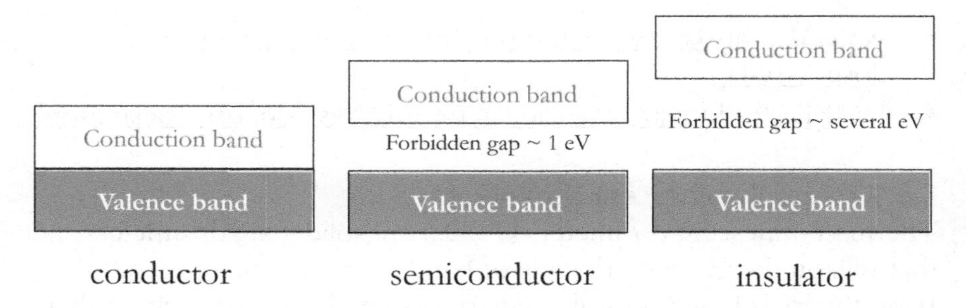

Figure 5-8. Generalized diagram of the band structure in a conductor, semiconductor, and insulator.

Although normally there are no energy levels in the forbidden gap, energy levels can be created within the gap by two means: physical defects in the material and chemical impurities that are present in the material. For both cases, the newly created energy levels may serve as "traps": that is, these levels may provide metastable states in which electrons may reside for an extended period of time. Therefore, electrons that are promoted from the valence band may be retained in the traps until they are released from the traps by the addition of energy. Thermoluminescent phosphors are intentionally constructed to have traps in their crystalline structure by incorporation of impurities (dopants), and this forms the basis for the thermoluminescent (TL) phenomenon, as shown in Figure 5-9 below.

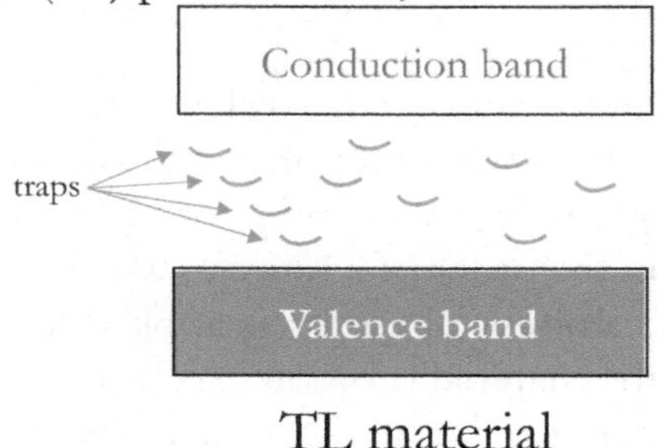

Figure 5-9. Chemical impurities or physical defects can provide metastable traps in the forbidden energy gap.

When a TL material is exposed to ionizing radiation, the following steps occur:

1. The interaction of the radiation with the crystal causes the promotion of electrons from the valence band to the conduction band.

2. Many of the promoted electrons return to the ground state almost immediately with the emission of light. This light is not detected and is not part of the dose-measuring process.
3. A fraction of the promoted electrons will be retained in the metastable traps. The total number of promoted electrons is proportional to the absorbed dose in the crystal; therefore, the number retained in the traps is also proportional to the absorbed dose.

Following irradiation, the phosphor is placed in a device called a TLD reader to extract the dose information as follows:

1. The phosphor is heated under controlled repeatable conditions to release the electrons from the traps.
2. Light is emitted as the result of the electrons' transition back to the ground state.
3. A light detection system measures the amount of light that is emitted.
4. The amount of emitted light is proportional to the number of populated (and subsequently emptied) traps, which is proportional to the absorbed dose to the crystal. Thus, the term "thermoluminescent dosimeter" appropriately refers to a dose-measuring medium that relies upon the detection of light from a heated crystal that has been previously exposed to radiation.
5. The heating of the TL phosphor empties all the traps and resets the crystal so that it can be reused many times, which is a practical advantage to a dosimetry program. This was one of its primary advantages when it became the dosimeter of choice in the 1970s and 1980s.

Several different heating methods have been employed by various manufacturers of TLDs. Some use direct contact heating with a heating element, some use a hot gas, and some use a heat lamp. There are some advantages and disadvantages of each, but the critical point is for the heating method to be controlled <u>and</u> reproducible for each individual dosimeter.

The amount of light emission from the TL phosphor upon heating follows a mathematical prediction that can be best understood by comparing it to radioactive decay. In radioactive decay, the decay probability for a given radionuclide is a single fixed value λ. In a TL phosphor, the decay probability for a given metastable state (which we will call "p") is temperature-dependent and is thus not a single fixed value. We can write this as shown below in Equation 5.2.

$$p = se^{-\frac{E}{kT}} \tag{5.2}$$

Where

- p is the probability of decay per unit time from the trap
- s is a proportionality constant
- E is the escape energy associated with the trap
- k is Boltzmann's constant (8.629×10^{-5} eV/K)
- T is the temperature in kelvin of the material

If a certain number n_0 traps are initially populated due to the interaction of ionizing radiation with the dosimeter, then the number of traps that remain populated is a function of time as predicted by Equation 5.3, which is similar in form to radioactive decay, except that the decay probability p is not constant as is λ.

$$n = n_0 e^{-pt} = n_0 e^{-\left(se^{-\left(\frac{E}{kT}\right)}\right)t} \tag{5.3}$$

Where

- n is the number of traps that remain populated at some time t
- n_0 is the initial number of traps populated at t=0
- t is the elapsed time
- the other variables are as described before.

To predict the number of traps emptied per unit time, we would multiply both sides of Equation 5.3 by the decay probability p to obtain Equation 5.4 as shown below.

$$pn = (se^{-\frac{E}{kT}})(n_0 e^{-\left(se^{-\left(\frac{E}{kT}\right)}\right)t}) \tag{5.4}$$

Where

- pn is the trap emission rate per unit time and is proportional to the light emitted.
- the other variables are as described before.

It should be noticed that Equation 5.4 has a similar structure to the equation for radioactive decay from Chapter 1 ($A=\lambda N=\lambda N_0 e^{-\lambda t}$), but the decay probability for thermoluminescence is a temperature-dependent quantity instead of a fixed constant value. If Equation 5.4 is plotted as a function of temperature in kelvin, the traps will be emptied in a fashion similar to what is illustrated below in Figure 5-10. This graph was prepared assuming a trap energy E=1.25 eV and a proportionality constant $s=1.0\times10^{11}\ s^{-1}$.

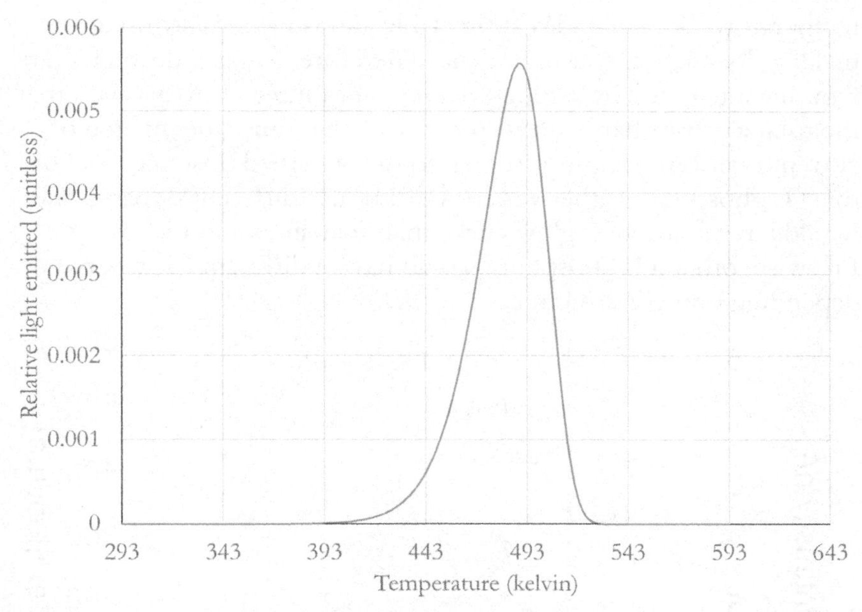

Figure 5-10. Predicted emptying of the traps in a TL phosphor assuming a linear heating rate of 3K/s, a trap energy of 1.25 eV, and proportionality constant $s=1.0\times10^{11}\ s^{-1}$

As can be inferred from Equation 5.4, traps with a lower trap energy (which are termed as shallow traps) will be emptied at lower temperatures whereas deeper traps with higher trap energies will require higher temperatures for the traps to empty. Again, this is because the trap decay probability is both temperature-dependent as well as trap energy-dependent.

Of course, Figure 5-10 is predicated on a single trap level. As described earlier, there may be multiple levels of traps in a specific

TL phosphor; so, there could be multiple distributions like what is shown in Figure 5-10, giving rise to a convolution of peaks in what is termed a glow curve. An example of a glow curve from lithium fluoride doped with magnesium and titanium (LiF:Mg,Ti) is shown below in Figure 5-11. The y-axis of the graph presents the electrical current measured by a photomultiplier tube as a function of time. There are theoretically 5 peaks in LiF that can be measured; however, Peak 1 is a shallow trap with a small trap energy that often causes it to deplete quickly (within minutes) at room temperature before it can be measured. Peak 2 has a somewhat longer half-time at room temperature, but still fades appreciably and is not usually used for making dosimetric measurements. Therefore, Peaks 3 through 5 are typically integrated by setting up a region of interest (ROI) such that the total electric charge measured during that time (the integral of current over time) is proportional to the absorbed dose received by the TL phosphor. Other varieties of LiF use different dopants and would have a different glow curve than that shown in Figure 5-11. Likewise, other TL phosphors would have a different glow curve depending on their structure.

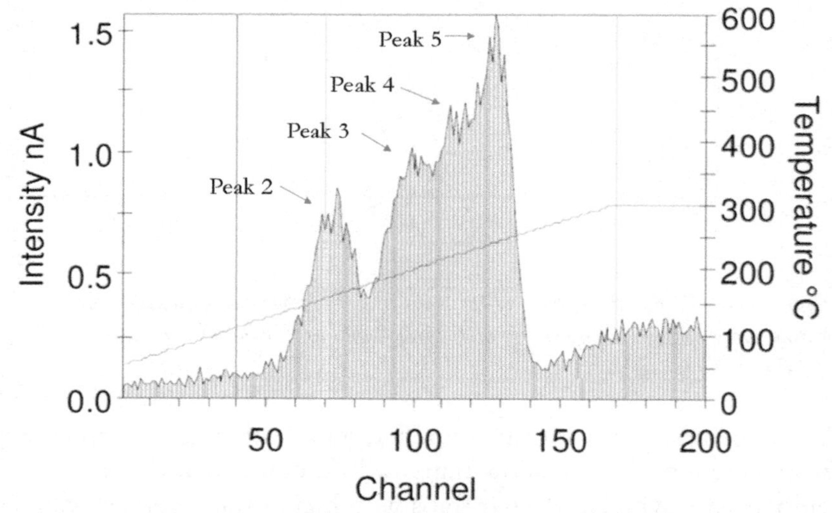

Figure 5-11. Typical glow curve for LiF:Mg,Ti showing 4 of the 5 peaks, indicating the various trap levels in the phosphor. The y-axis shows the electrical current of the photomultiplier tube used for measuring the light intensity. Integrating the current over time gives the total electric charge measured by the photomultiplier tube, which is proportional to the dose received by the dosimeter.

5.3.1.1 TL Response with Photon Energy

A common issue with TL phosphors is that they are not exactly tissue-equivalent. In the case of photons, this means that the value of the mass energy absorption coefficient for the TL phosphor is not the same as for tissue. It must be remembered that the light output from the TL phosphor is proportional to the dose that the dosimeter receives, whereas we are interested in reporting the dose that tissue would receive if it was in the same radiation field.

For photons, the comparison of TL phosphors to tissue can be most readily obtained by referring again to Equation 4.7 and modifying it to address the current situation for TL measurements as shown in Equation 5.5.

$$\frac{D_{TLD}}{D_{tissue}} = \frac{(\frac{\mu_{en}}{\rho})_{TLD}}{(\frac{\mu_{en}}{\rho})_{tissue}} \qquad (5.5)$$

As a first approximation, we can look at the ratio of the mass energy absorption coefficient of the TL material to that of tissue as described in Equation 5.5. If we plot that ratio as a function of photon energy, we obtain a graph like that shown in Figure 5-12.

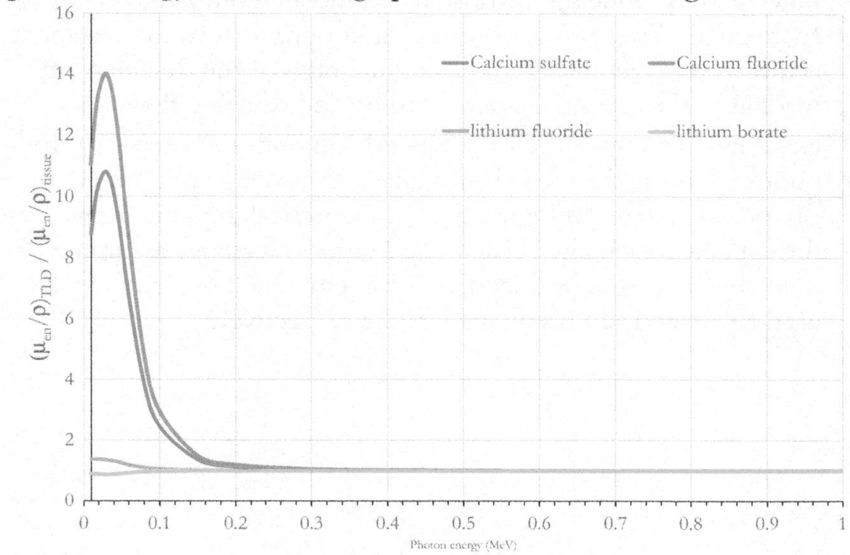

Figure 5-12. Ratio of the (μ_{en}/ρ) for TL phosphor to (μ_{en}/ρ) for tissue for four common TL phosphors normalized to the value at 1 MeV.

Above 0.2 MeV, the ratio of the mass energy absorption coefficients remains relatively constant for all the phosphors. However, at the lower energies where the photoelectric effect becomes prominent, the difference in atomic number between the TL phosphor and tissue becomes evident. Both lithium fluoride and lithium borate are similar to tissue in their dosimetric response; however, the calcium phosphors both have a tremendous overresponse at lower energies: that is, the dose received by the phosphor is much higher than the dose that would be received by tissue at that same photon energy.

The TLD reading can be corrected for its response at low energies if the energy of the photons is known so that the dose to tissue can be properly reported. There are generally two approaches to determining the photon energy as described below. Both involve using multiple TLD elements in a single badge that would be worn by the worker.

1. Two elements of the badge are placed behind two different thicknesses of filter material as shown below in Figure 5-13. For low photon energies, Filter 2 will provide more shielding than Filter 1, thus reducing the dose to Element 2 compared to Element 1. After being worn, the light output from each element is measured individually. By taking the ratio of the first element reading to the second element reading and realizing that exponential attenuation ($e^{-\mu x}$) depends on both the energy of the photon and the thickness of the filter, the average photon energy can be estimated. In Figure 5-13, a theoretical graph is shown for the expected Element 1/Element 2 ratio of readings assuming polyethylene filters of thickness 0.3 g/cm^2 and 0.6 g/cm^2 were used to cover TLD Elements 1 and 2 respectively.

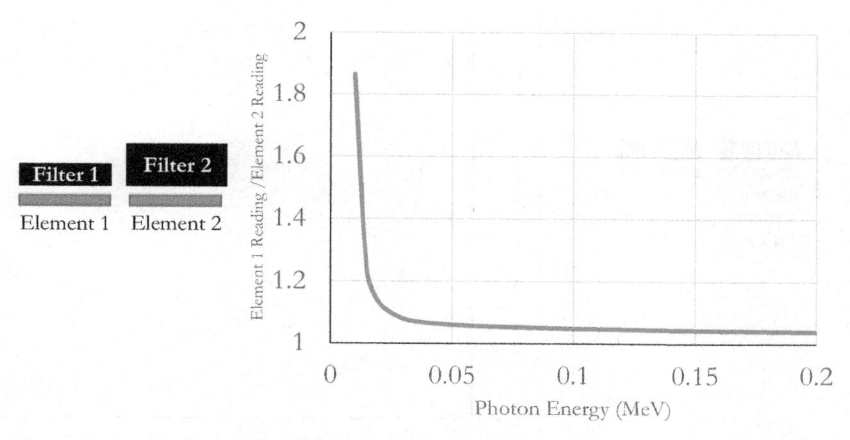

Figure 5-13. Illustration of using differing filter sizes to estimate the average photon energy impinging on a dosimeter.

2. Alternatively, two elements of the badge made of different TL phosphors could be placed under the same thickness of filter. As in the first case, the light output from each element would be measured individually. Since the two phosphors would have differing atomic numbers, each element would receive a markedly different absorbed dose at low photon energies; thus, the ratio of their readings would be indicative of the photon energy. Figure 5-14 below shows an example using calcium sulfate and lithium borate, where the ratio of the calcium sulfate reading to the lithium borate reading can be used to estimate the photon energy. The theoretical curve was obtained by using the mass energy absorption coefficient for each phosphor type and calculating the ratio at each photon energy.

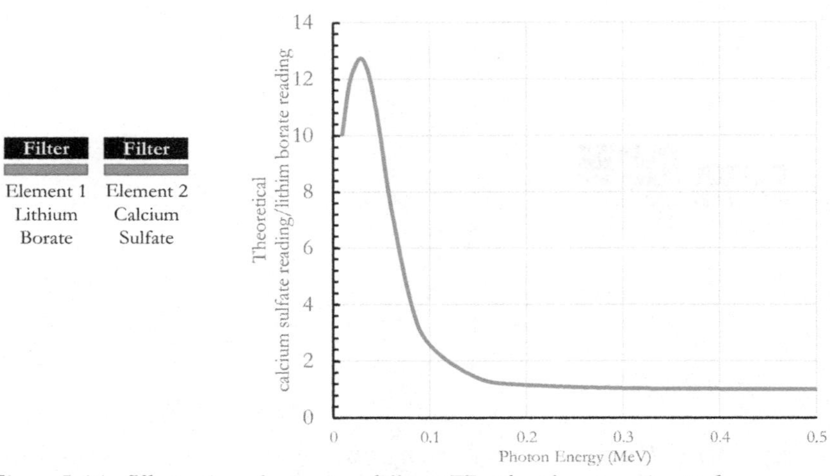

Figure 5-14. Illustration of using two different TL phosphors to estimate the average photon energy to which the dosimeter was exposed.

5.3.1.2 TL Response to Beta Radiation

For beta radiation, the main difficulty of using TLDs is to design the dosimeter such that it is capable of measuring the absorbed dose at a point as opposed to an average dose. As discussed in Section 4.1.1, the point absorbed dose is proportional to the mass stopping power of the material at that point. However, a dosimeter is not a true point; it has a finite thickness. The only way for the finite thickness of the dosimeter to act practically as a point is for it to be sufficiently thin that the beta radiation passes through the dosimeter without losing appreciable energy. Thus, the mass stopping power will be relatively constant across the dosimeter from front to back such that Equation 4.1 can be satisfied with a finite-sized dosimeter despite the fact that the dosimeter is not a true point.

When evaluating whether TL phosphors are tissue-equivalent with regard to beta radiation, we compare the mass stopping power of tissue to that of the TL phosphors. Similar to what was done with the mass energy absorption coefficients for photons, we can take the ratio of the stopping power for the TL phosphor to that of tissue as a function of electron energy. That will give a crude indication of whether an energy sensitivity exists for the TL phosphors. Figure 5-15 shows this ratio as a function of energy normalized to the ratio at 1 MeV.

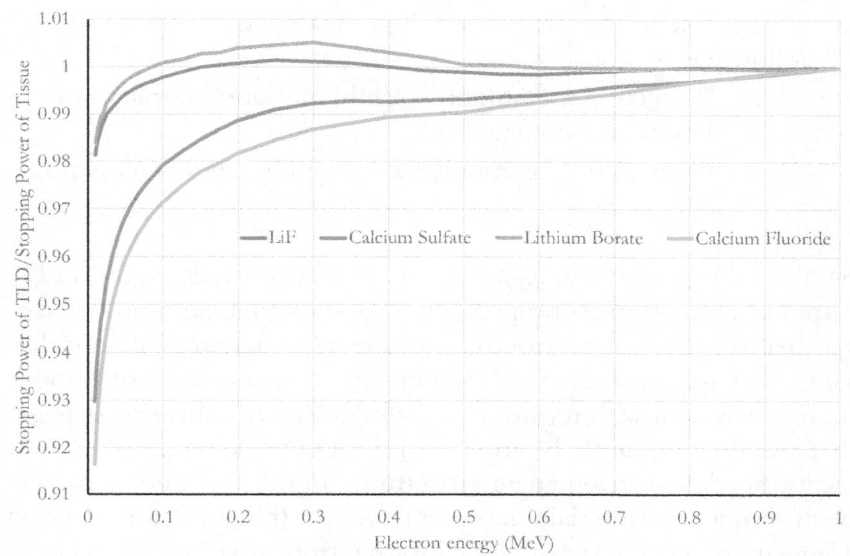

Figure 5-15. Ratio of TLD stopping power to tissue as a function of electron energy normalized to the value at 1 MeV.

One feature of this graph is that it shows that the ratio of the TLD stopping power to tissue does not vary widely (notice that even at energies below 0.1 MeV, the stopping power ratio varies by less than 10%). Even if the graph were extended up to 5 MeV, the ratios remain relatively constant. Therefore, the primary challenge with low energy electrons would be the physical thickness of the dosimeter rather than the difference in material between the dosimeter and tissue. Beta dosimeters are typically very thin to minimize errors.

5.3.1.3 TL Response to Neutron Radiation

The area where TLDs are least like tissue is when we try to do neutron dosimetry. In Chapter 4, we discussed that fast neutrons interact primarily via elastic scattering with the atoms in the body (primarily hydrogen), and thermal neutrons interact via the $^{14}N(n,p)^{14}C$ reaction and $^{1}H(n,\gamma)^{2}H$ reaction. TLD materials have none of these reactions. Instead, we must consider the following factors when trying to use TLDs for neutron dosimetry:

1. Fast neutrons have essentially no direct interaction in TLD materials.

2. Thermal neutrons interact with TLD materials via one of the following reactions:
    a. $^{6}Li(n,\alpha)^{3}H$ (commonly for lithium fluoride or lithium borate that contain ^{6}Li)
    b. $^{10}B(n,\alpha)^{7}Li$ (commonly for lithium borate that contains ^{10}B)

Similar to the discussion regarding neutron survey meters, the TLD response to neutrons is proportional to the thermal neutron fluence. Further, the thermal neutron reactions in TLD materials are not the same as those in tissue; thus, TLD materials are incapable of being tissue-equivalent with regard to neutron radiation. This necessitates that a calibration factor be applied to the TLD reading based on the facility-specific neutron energy spectrum. In this way, the thermal neutron fluence essentially acts as a tracer for the fast neutrons using a similar approach to what is used for neutron survey meters. The TLD output is proportional only to the thermal neutron fluence and not to a neutron dose that can be scaled to determine the dose to tissue.

As might be surmised, TLD materials with no ^{6}Li or ^{10}B have essentially no response to neutrons. For example, this would commonly include the following, although there could be others:
- lithium fluoride that is made with ^{7}Li
- lithium borate that is made with ^{7}Li and ^{11}B
- calcium fluoride
- calcium sulfate

Thus, it is possible to construct a dosimeter badge with multiple TLD elements in which some elements respond to photons only (by using TL phosphors that are insensitive to thermal neutrons) and some elements that are sensitive to both neutron and photon radiation. This makes it possible to separate the dose components because of the need to apply different dose conversion factors to the TL signal due to photons compared to the TL signal due to neutrons. As an example, a dosimeter might use lithium fluoride enriched in ^{7}Li to detect only photons while using natural enrichment lithium fluoride to detect photons and neutrons. Other combinations of phosphors are also possible.

The TLD response to neutrons can be enhanced by taking advantage of the albedo effect, whereby fast neutrons incident on the human body are moderated to thermal energies and then reflected back to the dosimeter because of the collisions and natural drift of thermal neutrons. This has the effect of increasing the number of thermal neutrons incident on the dosimeter. However, for the albedo neutrons to be properly used for dosimetry purposes, two additional considerations apply:

1. The dosimeter must be worn in a reproducible position on the body consistent with how the calibration of the dosimeters was performed. The usual requirement is that the dosimeter be placed in close contact with the trunk to maximize the albedo. Thus, the dosimeter may be clipped to clothing that is worn on the body but should not be hung from a lanyard. In some cases, the dosimeter may be required to be worn with a special belt to ensure its close proximity to the body.
2. Since the TLD will detect thermal neutrons from two sources (those from the original spectrum that were already thermal and fast neutrons that have been thermalized by the human body), it is sometimes helpful to separate the number of each. This can give additional information on the neutron spectrum. Some dosimeters will incorporate cadmium filters (since cadmium has a very high cross section for absorbing thermal neutrons) in such a way to achieve this separation as shown below in Figure 5-16. In this illustration, all three TLD elements are comprised of a neutron-sensitive phosphor.

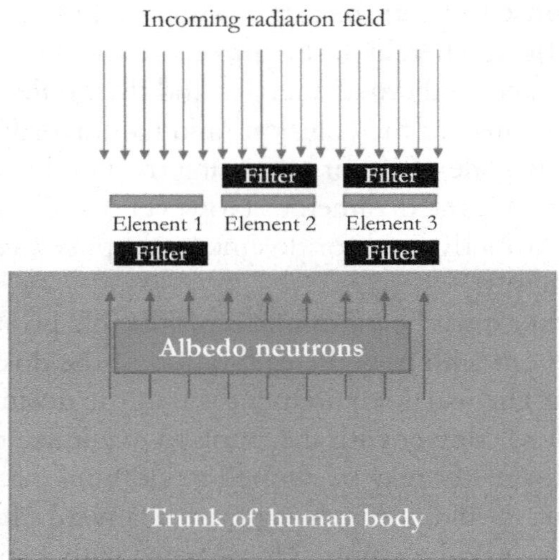

Figure 5-16. Illustration of how cadmium filters with TLDs can be used to address the albedo of neutrons.

As mentioned above, the filters in the illustration are comprised of cadmium which will act as a thermal neutron shield (fast neutrons will pass through largely unimpeded). The use of the filters on three elements allows for separation of the various radiations that are incident on the body. Assuming that the incoming radiation field is composed of both photons and neutrons (seldom would neutrons be present without photons as well):

- Element 1 would respond to the photons and thermal neutrons that are incident from the front while the filter would block albedo neutrons from the trunk of the body.
- Element 2 would respond to photons from the front and to albedo neutrons from the body, but the filter would block thermal neutrons from the front.
- Element 3 would be insensitive to thermal neutrons from any direction owing to the filters on both sides of the TLD element and would respond only to the photon radiation in the radiation field, allowing the gamma contribution to the response in Elements 1 and 2 to be subtracted.

5.3.1.4 Designing a Dosimeter with Badge Holder

The design of a dosimeter is an engineering task that is partly based on theoretical considerations and partly based on experiment. The purpose of the dosimeter and the radiation environment in which it will be used will put some constraints on any dosimeter design.

In addition to the hardware design of the dosimeter, there must be software applied as well. We have already seen how individual element readings in a dosimeter are used to determine photon energy and separate photon dose from neutron dose. The individual readings of the TLD elements in a badge must be processed quickly, and the readings obtained from them must be used in a decision tree process to identify the type and energy of radiation to which the dosimeter is exposed. Then the readings must be used to assess the dose from each of the radiation types at the depths specified for the skin, lens of the eye, and the deep dose. This lends itself well to a computer algorithm that is often modified with time as data are gathered and as testing is performed.

Obviously, we cannot cover every single type of badge holder for every radiation environment. However, we can consider a generic four-element TLD to illustrate some of the principles that would be considered in designing a dosimeter. Figure 5-17 below is an example of such a dosimeter.

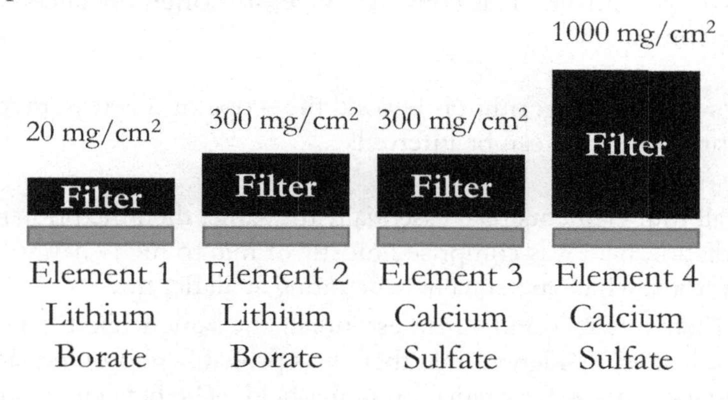

Figure 5-17. Generic TLD with four dosimeter elements and various filters.

The materials from which the filters are constructed varies by manufacturer. For thin filters (such as what covers Element 1), a thin plastic film could be used. However, using plastic as a filter for all four elements is impractical. As an example, for Element 4 above, a 1000 mg/cm^2 thick filter would require roughly 1 cm of plastic, which would make a very thick dosimeter badge. Therefore, metallic filters are commonly employed to achieve the same density thickness without having such a large linear thickness.

In examining Figure 5-17, there are some generalities that can already be determined:

- Element 4 will be sensitive primarily to photons. The lack of a thermal neutron reaction mechanism in calcium sulfate makes neutron detection impossible, and the thickness of the filter in front of this element makes beta detection improbable except for those electrons with more than 2 MeV of energy based on the CSDA range of electrons.
- Element 3 will be sensitive to photons and high energy beta radiation only.
- Element 1 and Element 2 are sensitive to beta, photon, and neutron radiation.
- A conversion factor will be needed to determine the skin dose, since it should be assessed at a depth of 7 mg/cm^2, but our badge holder has a filtration of 20 mg/cm^2 due to the necessary thickness of plastic film needed to protect the TLD element from the environment. This conversion factor is often obtained experimentally.

While a complete algorithm is beyond the scope of this text, there are some principles that can be inferred:

- If all four elements read essentially the same, then the primary radiation field was comprised mostly of mid to high-energy photons, which are equally penetrating at all depths.
- If Elements 2, 3, and 4 read essentially the same while Element 1 is substantially higher, then there was probably photon radiation and low-energy beta radiation in the field. The beta energy must be relatively low; otherwise, the readings from Element 2 and Element 3 would be higher than Element 4 but would be lower than Element 1. Neutron radiation was not present; otherwise, Element 2's reading would be comparable to Element 1.
- If Element 3's reading is substantially higher than Element 2's reading, then that implies a low-energy photon radiation field, since the absorbed dose to calcium sulfate would be higher than

to lithium borate for the same photon fluence, as shown in Figure 5-14.
- If Element 1 and Element 2 are both substantially higher than Element 3 and Element 4, then neutron radiation may have been present in the radiation field.

The thresholds for what constitutes "higher" or "lower" in comparing the element readings is part of the experimental process that is obtained over time and under a variety of tests. This requires exposing the dosimeter to known mixtures of radiation fields multiple times and identifying trends regarding element readings. Once the radiation types are determined, then the algorithm must be able to separate the components of the radiation field and determine the energy (for beta and photons) to apply the proper dose conversion factor.

There are also failures that can cause errant readings, such as moisture damage (from dropping the badge in a mud puddle), exposure to oils or other organic materials, and a host of confounding factors. Therefore, the readings of the dosimeter cannot be accepted blindly without evaluating the reasonableness of the numbers; further, the computer algorithm must be designed to try to catch as many errant readings as possible. This drives home the point that the professional-level health physicist must understand the measurements and the principles of dosimetry in order to report meaningful results that are as representative as possible.

5.3.2 Optically Stimulated Luminescence

Optically Stimulated Luminescence (OSL) is a newer dosimetry technology and is quite similar to TLD technology. OSL uses a phosphor (typically aluminum oxide or beryllium oxide) with the same physics phenomena of solid-state physics and traps as TLD. However, the readout of OSL dosimeters is accomplished with light as opposed to heat. Thus, light serves as the excitation source to promote electrons to the conduction band so that light from the phosphor is emitted when those electrons return to the valence band.

The other major difference with OSL is that the readout process does not empty all the traps; in fact, only a tiny percentage of available traps are emptied. This makes it possible to repeat the readout process of an OSL dosimeter if there are suspect results. This is in contrast to the TLD readout process which empties essentially all the traps in the phosphor, thus making readout a one-time process. However, an erasing protocol can be used with OSL dosimeters to totally reset the dosimeter when desired, although even the deepest traps may not be reset in that case.

Much of the discussion of TLD applies equally well to OSL in terms of tissue-equivalence, design of a badge holder, etc. Therefore, it will not be repeated in this section. Because aluminum oxide has a higher effective atomic number than tissue, OSL phosphors made of aluminum oxide will receive a higher dose than tissue at lower photon energies, as shown below in Figure 5-18. Thus, a correction is needed to adjust the readings of aluminum oxide phosphors at low energies to determine the absorbed dose to tissue. Beryllium oxide phosphors are closer to the effective atomic number of tissue and thus would not exhibit a large overresponse at low photon energies.

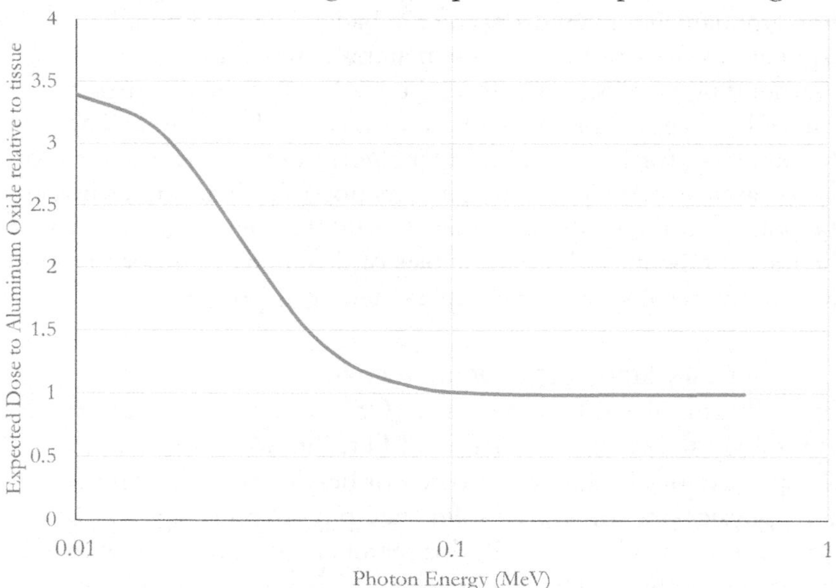

Figure 5-18. Theoretical expected response of aluminum oxide relative to tissue for low photon energies.

One challenge with OSL is the prospect of neutron dosimetry. Aluminum oxide does not have a ready-made thermal neutron detection mechanism via nuclear reaction. However, the aluminum oxide can be coated with lithium carbonate enriched in ^{6}Li, thus providing an (n,α) reaction whereby the alpha particle would be ejected into the aluminum oxide where it can be detected. As with other common neutron detection methods, this still requires a calibration factor to convert the reading from the thermal neutron response of the dosimeter to a dose equivalent from the neutron spectrum in the facility.

5.3.3 Nuclear Track Etch Dosimeters

As noted earlier, detection of fast neutrons by survey meters or by dosimeters typically relies on detecting the thermal neutron component of the neutron spectrum and then inferring the dose due to the fast neutrons. Track-etch dosimeters, on the other hand, provide a method to detect the fast neutron component of the spectrum without first thermalizing the neutrons.

The most common form of this type of dosimeter uses a plastic called CR-39 (short for Columbia Resin 39) in thin small pieces of plastic film as shown below in Figure 5-19.

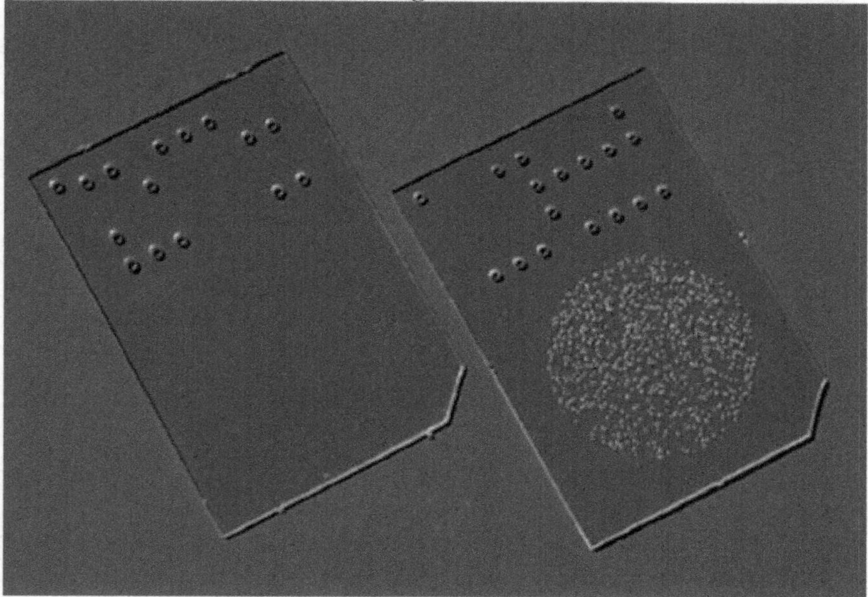

Figure 5-19. CR-39 track etch dosimeter film (Photo courtesy of Instrument Plastics - https://www.instrumentplastics.co.uk/products/cr39-dosimeter-grade-padc). The holes in the top of the films are used for badge identification purposes.

When issued to a person to wear, the foil has the appearance like that on the left in Figure 5-19. Because CR-39 is a plastic, it has lots of hydrogen atoms in the polymer structure. Fast neutrons colliding with these hydrogen atoms cause recoil protons to move through the film, causing damage to the polymer structure. This damage is not visible, however, without chemical treatment. Following exposure, the dosimeter is placed in a chemical bath of sodium hydroxide, which etches the areas of the film where the recoil protons caused damage, thus giving rise to the appearance of the tiny pits in the film on the right in Figure 5-19.

The number of damage sites per square millimeter is proportional to the dose. Of course, counting the damage tracks with the naked eye would be tedious and prone to error; therefore, the dosimetry lab incorporates a digital video system that uses a computer algorithm to search for the damage tracks. The average number of damage tracks per square millimeter is determined and related to the neutron dose by a conversion factor.

The chemical processing of the film means that the dosimeter cannot be reused; however, the dosimeter can be inspected and evaluated multiple times, which is an advantage if questions arise to an assigned dose.

Some facilities use track etch dosimeters in conjunction with TLDs or with OSL dosimeters to perform neutron dosimetry. The disadvantage of track etch dosimeters is the requirement of a carefully controlled chemistry laboratory for consistent etching; however, for facilities with appreciable neutron dose at a variety of energies, this type of dosimeter is very useful.

5.3.4 Criticality Dosimeters

In addition to the usual sources of neutrons that may be present at many nuclear facilities, some nuclear facilities use fissile material such that an accidental criticality event is a possibility. If an event happens, the usual neutron dosimeter is insufficient to assess the dose because the spectrum of neutrons would be markedly different from everyday operations and would vary spatially depending on the amount of moderating materials present in the vicinity.

Workers in facilities where a criticality event is possible may be issued criticality dosimeters. Despite the name, these are not true dosimeters. Rather, they are materials placed into a holder for the person to wear that rely on neutron activation analysis to determine the fluence of neutrons at certain energies. Thus, the criticality dosimeter is intended to estimate a crude spectrum of neutron energies. Specific materials are selected based on threshold neutron energies that are needed to produce a radioisotope that can be quantified following irradiation.

If a criticality event occurs, the components in the criticality badge can be assessed via a GM counter or laboratory detection system to determine the amount of radioactivity in each material. This information can then be used to determine the fluence of neutrons. Since some of the reactions have energy thresholds, the induced activity measurements can be used to estimate the shape of the neutron spectrum so that appropriate dose conversion factors can be folded into the calculation to estimate the neutron dose from the criticality event.

Table 5-1 below lists some materials that are commonly incorporated into a criticality dosimeter along with details on some reactions of interest.

Table 5-1. Examples of reactions used in criticality dosimetry (data from PNNL 2017)

Material	Reaction	Threshold Energy	Half-life of activated species
In-115	$^{115}In(n,\gamma)^{116m}In$	thermal	54.2 minutes
Au-197	$^{197}Au(n,\gamma)^{198}Au$	thermal	2.696 days
Cu-63	$^{63}Cu(n,\gamma)^{64}Cu$	thermal	12.701 hours
Cu-63	$^{63}Cu(n,\gamma)^{64}Cu$	580 eV[a]	12.701 hours
Au-197	$^{197}Au(n,\gamma)^{198}Au$	4.9 eV[a]	2.696 days
In-115	$^{115}In(n,n')^{115m}In$	1.2 MeV	4.486 hours
S-32	$^{32}S(n,p)^{32}P$	3.3 MeV	14.29 days

a. This reaction has a resonance at the stated energy in addition to a significant thermal neutron absorption cross section; the thermal component of the spectrum is removed by placing a cadmium filter around this element.

5.3.5 Electronic Dosimeters

One major disadvantage to all the personal dosimeters listed thus far in this section is that they are passive: that is, they require analysis in a laboratory to obtain the results. Because of this delay in processing the dosimeter, a worker would have no real-time indication of the dose. In situations where the radiation environment is relatively constant, this might not present a major issue. However, when workers are in a facility where radiological conditions may change unexpectedly or where the conditions may vary dramatically spatially, having a real-time indication of the dose is important.

As electronic circuitry has gotten smaller and microprocessors have gotten more powerful, it is no surprise that electronic dosimeters with numeric readout and alarming functions have become common. These are not worn in place of passive dosimeters but are in addition to the passive dosimeters; the official dose for recordkeeping is determined by the passive dosimeter and the electronic dosimeter is used to keep workers apprised of their dose while they are working in a radiation field. Total dose and dose rate alarms can be set before work begins, and the worker can check their dose throughout the work activity by looking at the electronic readout and make any adjustments in work pattern or can exit the area if unexpected dose has accrued. Data logging of the dose over time allows radiation protection personnel to reconstruct when particular doses were received. In addition, some electronic dosimeters have wireless capability so that radiological work can be monitored by radiation protection personnel who are located in a lower-dose area.

Several different companies manufacture electronic dosimeters with a variety of features. As with passive dosimeters, the components of the radiation field and the response of the dosimeter will play a role in which dosimeter is selected for a particular application.

The detector type for the electronic dosimeter is commonly the silicon pin diode. As might be anticipated, this material is not tissue-equivalent for photons; this can be confirmed by calculating the relative dose to silicon compared to tissue as shown below in Figure 5-20.

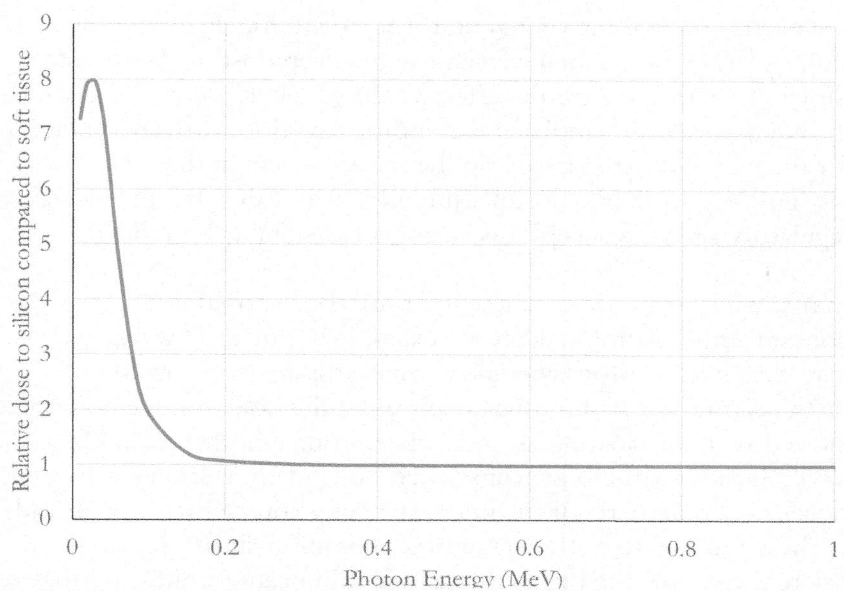

Figure 5-20. Relative dose to silicon compared to tissue showing a greater dose at low energy photons.

As is the case for the passive dosimeters, the lack of tissue equivalence usually requires using multiple pin diodes and also incorporating metallic filters to reduce the overresponse at low photon energies. The readings can then be used to estimate the average photon energy and apply the appropriate dose conversion factor to the reading. Electronic dosimeters are commonly available for beta, x-ray, and gamma radiation. Electronic dosimeters that are sensitive to neutrons are a relatively new entry to the field, but typically come at a greater financial cost.

5.3.6 Estimation of Effective Dose

The concept of effective dose was discussed in Chapter 1, and now we come to the difficult question of estimating the value of this quantity based on dosimetry measurements. As has been pointed out, it is not possible to directly measure the equivalent dose or the effective dose, yet we need to be able to ensure adequate protection of people based on the risk estimates that led to the creation of the effective dose concept.

If a person has a more or less uniform radiation field across the entirety of the body, then we can use the deep dose as discussed earlier as a surrogate for the effective dose. Thus, we would assess the equivalent dose (applying the appropriate dose conversion factor for the radiation type) based on the measurement at the 1000 mg/cm^2 depth. While this measurement is not the effective dose, regulatory agencies accept this surrogate as suitably bounding.

In many instances, though, a worker may be exposed to a nonuniform radiation field. One example is a medical worker who may wear a lead apron when performing fluoroscopy. In this situation, the exposed unshielded areas of the body receive noticeably more dose than those areas under the apron. Another example would be a scenario in a facility where the radiation source is much closer to certain parts of the body, such as a source that is overhead, as shown in Figure 5-21. Yet another common situation occurs where the source is a highly directional thin beam of radiation that does not irradiate the entire body such as shown in Figure 5-21.

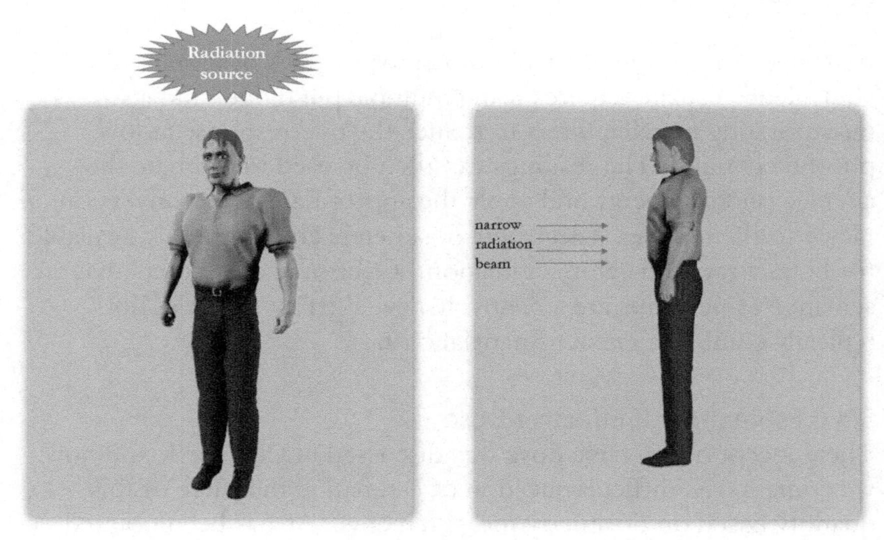

Figure 5-21. Examples of nonuniform external radiation fields on a person. On the left, the radiation source is overhead, thus exposing the upper body to a greater dose than the lower body. On the right, a narrow beam of radiation exposes the abdomen to a higher dose than the rest of the body.

If multiple dosimeters are worn at various locations on the body, then the effective dose may be estimated. The American National

Standards Institute (ANSI) published ANSI N13.41, which outlines a method to use compartment weighting factors as shown below in Table 5-2. These factors were derived for common locations where multiple dosimetry might be worn and are based on the tissue weighting factors from ICRP Publication 103 for the tissues/organs contained within these compartments.

Table 5-2. Compartment weighting factors from ANSI N13.41 based on the tissue weighting factors of ICRP Publication 103.

Compartment/area of the body	Compartment weighting factor W_c
Head and neck	0.12
Thorax, above the diaphragm	0.40
Abdomen, including the pelvis	0.46
Right arm, including the elbow	0.005
Left arm, including the elbow	0.005
Right leg, including the knee	0.005
Left leg, including the knee	0.005

The effective dose is estimated as the sum of the products of equivalent dose (determined at 1000 mg/cm^2) at each location and the appropriate compartment weighting factor as shown in Equation 5.6 below:

$$E = \sum W_c H_{p,c}(10) \quad (5.6)$$

Where

- E is the effective dose
- W_C is the appropriate compartment weighting factor
- $H_{p,c}(10)$ is the equivalent dose at the 1000 mg/cm^2 depth assigned to the particular compartment

The use of Equation 5.6 does not necessarily mean that a dosimeter must be placed at each of the seven locations listed in Table 5-2. However, all seven compartments must be addressed in the calculation of the effective dose. If a dosimeter is not at a particular location, then a nearby dosimeter that is deemed representative of that compartment may be used. Further, if multiple dosimeters are placed on a single compartment (for instance, it would not be uncommon to have a dosimeter on the chest and on the back, both of which are measuring the dose to the trunk of the body), those dosimeter measurements would be averaged.

Example: A worker is present in an area where a radiation source is overhead similar to the illustration in Figure 5-21. The worker wears two dosimeters. A dosimeter on the head indicates an equivalent dose of 500 mrem. A dosimeter on the chest indicates an equivalent dose of 200 mrem. What is a suitable estimate of the effective dose?

Solution: Despite having only two dosimeters, we still must assign doses to all seven compartments with those limited readings. The head dosimeter is most likely representative of the head and neck. The chest dosimeter reading can be assumed to be representative of the other compartments, even though it is suspected that this would overestimate the actual dose to the abdomen and the legs. The calculation would proceed as follows:

Compartment/area of the body	**Compartment weighting factor W_c**	**Assigned $H_{p,c}(10)$**	**$W_c\, H_{p,c}(10)$**
Head and neck	0.12	500	60 mrem
Thorax, above the diaphragm	0.40	200	80 mrem
Abdomen, including the pelvis	0.46	200	92 mrem
Right arm, including the elbow	0.005	200	1 mrem
Left arm, including the elbow	0.005	200	1 mrem
Right leg, including the knee	0.005	200	1 mrem
Left leg, including the knee	0.005	200	1 mrem
Effective Dose			236 mrem

Note that no part of the body was actually measured to receive 236 mrem. Consistent with the definition of effective dose, this scenario gives a risk to the worker as though a uniform whole-body dose of 236 mrem was received. In reality, this would be reported to only two significant figures, or 240 mrem.

5.4 Final Notes on External Dosimetry

Having now completed a chapter on external dosimetry calculations and a chapter on external dosimetry measurements, it should be apparent that dosimetry is not an exact science. Indeed, the inherent assumptions involved in the calculations and the limitations of measurement clearly show that dose estimates require professional judgment at times.

As mentioned before, dose estimates are for an idealized model and should not be interpreted as an actual dose to a real person. However, the consistent handling of the data and assumptions gives us confidence that we are providing adequate protection for people exposed to radiation. We will see this same approach as we move forward to internal dose assessment in Chapter 7.

Free Resources for Chapter 5

- Questions and problems for this chapter can be found at www.hpfundamentals.com
- PNNL (Pacific Northwest National Laboratory), *PNNL Measurement Results for the 2016 Criticality Dosimetry Exercise at the Nevada National Security Site (IER-148)*, PNNL-26497, Richland WA, 2017 available at https://www.pnnl.gov/main/publications/external/technical_reports/PNNL-26497.pdf

Additional References

- ANSI/HPS. Criteria for Performing Multiple Dosimetry. McLean, VA: Health Physics Society and New York: American National Standards Institute; ANSI/HPS N13.41-2018.
- Attix, F.H. Introduction to Radiological Physics and Radiation Dosimetry, John Wiley & Sons, Hoboken, NJ, 1986, 1991.
- Cameron, J.R, Suntharalingam, N., and G. N. Kenney. Thermoluminescence Dosimetry, University of Wisconsin Press, Madison, WI, 1968.
- Horowitz, Y.S. (editor), Thermoluminescence and Thermoluminescent Dosimetry, CRC Press, Boca Raton, FL, 1984.

Chapter 6 Controlling External Radiation Dose in the Workplace

The three most common dose reduction techniques for external exposure to radiation are time, distance, and shielding as shown below in Figure 6-1. The implementation of each of these techniques can lead to innovative ways to reduce radiation dose. While a significant amount of discussion in this chapter will revolve around shielding because of the calculational techniques that are used, it should be recognized that the principles of time and distance often are easier to implement for short-term work and can significantly reduce the external dose that people receive.

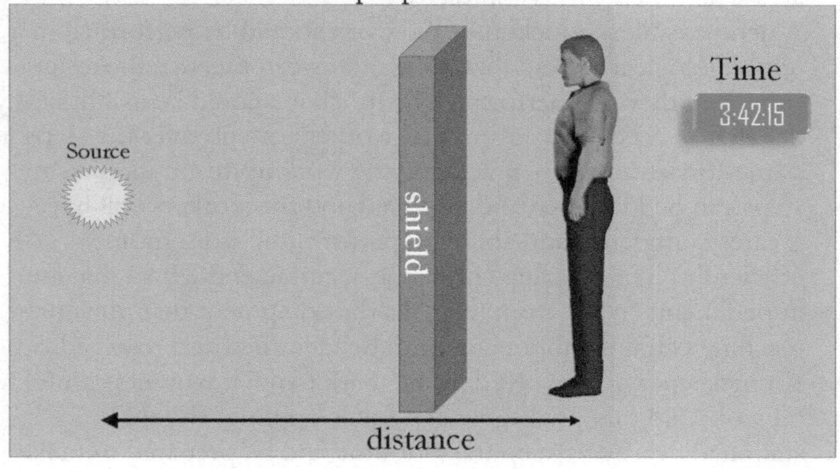

Figure 6-1. Illustration of the principles of time, distance, and shielding in the reduction of dose from an external source of radiation.

6.1 Time

The total dose received by a person is proportional to the instantaneous dose rate and the exposure time; therefore, reducing the exposure time can have a significant effect on the dose that a person receives. While it sounds simple enough to say that the time should be reduced, it often involves some human factors engineering to plan radiological work to be as efficient as possible. Some possible techniques include the following:

1. Workers performing radiological work should be familiar with the work they are about to perform without having to refresh their memory while in the radiation field. They should have studied any applicable procedures, reviewed the layout of the area, recognized where all work is to be performed, and have confidence that they can complete all steps with a minimal amount of delay.
2. Unnecessary materials and tools should not be brought into the work area. For example, a mechanic working on a piece of equipment should not bring a heavily loaded all-purpose toolbox with a tool for every occasion. Rather, the exact tools needed for the job should be brought and organized so that minimal time is spent searching for the proper tool. This will also reduce the time required to gather all tools and equipment before leaving the area when the work is complete.
3. When possible, a mockup of the work should be performed in a nonradiological area so that the workers can practice the exact steps that they will perform. The mockup should be as physically detailed as needed to ensure that workers can go over the steps in a true-to-size area. By practicing the work up front, inefficient steps can be identified and removed and the workers will have greater confidence and ability to perform the work more efficiently. The mockup should be accurate enough so that any impediments in the work (e.g., nearby equipment that prevents reaching certain components) can be identified and resolved.
4. A timekeeper not involved in the work (so that workers can focus on work and the timekeeper can focus on time) should be stationed nearby to keep track of time and be prepared to make decisions on when the workers should exit, even if work has not been completed. Sometimes when encountering an unexpected problem (e.g., a bolt shears off on a piece of equipment), it may be advantageous to exit the area, regroup, and develop a new solution rather than remain in the area while trying to solve a newly discovered problem impulsively.

Other techniques may be employed to reduce the time of exposure as circumstances and the facility layout permit. Again, the key is to evaluate each potential scenario with creativity while maintaining a culture of overall safety (not just radiological safety) to achieve the goals.

6.2 Distance

In Chapter 1, the various equations for the fluence rate were presented as a function of distance. For a point source, the inverse square law was shown to give substantial reduction in fluence rate with increased distance. For other source geometries, the reduction may not be as drastic, but increased distance from the source generally does reduce the fluence and hence the dose rate from the source.

As with the time aspect of radiation protection, the distance aspect can be enacted in several ways. Some ideas might include:

1. Using remote equipment to perform work where necessary, such as tools with extensions. In some situations, increasing the distance to the radiation source by even a few feet can significantly reduce the dose rate. However, if the time to perform the work increases as a result, then that may not be a viable solution. The total dose projected to be received by the worker should be estimated based on the anticipated dose rate and the exposure time to determine the best course of action.
2. Ensuring that workers not actively involved in the radiological work stay distant from the source. When considering radiological work, only the person performing the actual hands-on work should be close to the radiation source while others should be in a lower dose area.
3. Using more technological means such as robotics so that the worker does not have to be close to the source. While it may be necessary for the worker to enter the radiation field on occasion, the use of robotics extends the distance while reducing the time that the worker would be in the radiation field.

6.3 Shielding

When possible, the use of time and distance can serve as effective controls to reduce radiation dose. However, there are times where these two considerations by themselves are not sufficient due to the high dose rates that may come from the source. In that case, shielding may be used to supplement the time and distance controls that are in place.

Shielding can be accomplished with a variety of materials; the choice of material can depend on several factors. For instance, many people will automatically think of lead as the best shielding material, but that is not always the case. Because of its density, lead places a large weight load on any floor structure; moreover, lead can be toxic if ingested. In addition, lead is a soft metal and may need additional material around it for structural support. Last, but not least, moving lead bricks poses occupational safety hazards because the bricks are heavy and if one is dropped on an individual's foot or other body part, serious injury can result. Therefore, lead may be used, but it is not to be considered the universal shield for every occasion.

When considering the material from which to construct the shield, some factors to consider are:

- Radiation type to be shielded. Considerations for gamma, beta, and neutron radiation will necessitate different shield materials
- Ease of installation
- Weight on any floor structure
- Cost
- Long-term versus short-term use of the shield

Some common shield materials in addition to lead are water (less expensive, portable, and easy to install in containers that can form a wall for temporary use), iron/steel (strong, readily available, structurally sound, relatively easy to install), concrete (structurally sound, pourable, less expensive).

An additional use for shielding that is sometimes overlooked is in the establishment of a low-level background area for radiation detection instrumentation. When trying to detect the presence of radioactivity in a sample, radiation from other sources in the area can confound the results. Constructing a shielded counting chamber can help increase the sensitivity of the detector to low levels of activity.

In rare instances, the design of shielding can be achieved through experimental trial and error where a random amount of shielding is put into place, and the user performs measurements to see if it is sufficient. Additional shielding can then be added if necessary. However, the usual (and most defensible) technique for shielding is to calculate the necessary shielding in advance while applying a safety margin into the calculations. Once the design has been reviewed, it can be installed and then evaluated as to its effectiveness. This approach is particularly necessary when attempting to shield high dose rate sources that could cause an acute dose to individuals in the area where trial and error obviously is not possible. In addition, permanent shielding often involves room or facility design, and the shielding must be installed as part of construction or renovation where the trial-and-error approach is not practical.

The following sections discuss shielding considerations and techniques for the three main radiation types: photons, beta radiation, and neutrons.

6.3.1 Shielding for Photons

Before delving into the details of shielding calculations for photons, we need to quickly review the main photon interaction mechanisms because understanding those mechanisms will help us make sense of some decision-making that must take place in shielding design.

For photon energies below about 10 MeV, the three most common photon interactions for health physics purposes are the following:

Photoelectric effect: An incoming photon of energy E_γ is completely absorbed by an absorber atom. The atom ejects an inner shell electron (usually K or L) with kinetic energy equal to E_γ - BE_i, where BE_i is the binding energy of the ejected electron. Because of this vacancy that is created, a cascade of characteristic x-rays and Auger electrons is emitted as the atom fills the original vacancy and subsequent vacancies in the atom. This interaction is most likely for low photon energies (~100 keV and below) and for high atomic number materials.

Compton effect: An incoming photon of energy $E\gamma$ has a billiard-ball type collision with an essentially free electron (usually satisfied by a loosely held outer shell electron). The photon scatters with a new energy $E\gamma'$ and the electron scatters with an energy equal to $(E\gamma - E\gamma')$. This interaction is prominent for photon energies between 150 keV and 2 MeV and does not depend too strongly on atomic number.

Pair production: An incoming photon of energy $E\gamma$ interacts with the multipole fields of a nucleus and produces two particles: an electron and a positron. The photon disappears, and any excess energy it had above 1.022 MeV (the total rest mass energy of the two particles) goes into the kinetic energy of the two particles. When the positron slows to essentially zero kinetic energy, it collides with an electron, annihilating in the process and giving off two photons of essentially 0.511 MeV each. Pair production has a threshold energy of 1.022 MeV but becomes more likely at photon energies above 2 MeV and for high atomic number materials.

Figure 6-2 below shows how the various cross sections for these three interaction types vary for water and for iron.

There are several features to note. First, the Compton effect probability/interaction coefficient in both materials is quite similar, although not identical. Secondly, the photoelectric effect interaction coefficient is noticeably higher in iron compared to water; similarly, the pair production interaction coefficient is noticeably higher in iron compared to water. Finally, as noted earlier, the photoelectric effect interaction is greater at the lower photon energies, and the pair production coefficient is greater at higher photon energies. In iron, there is a noticeable discontinuity in the photoelectric interaction coefficient slightly below 0.01 MeV. This discontinuity is called the "K edge" and occurs when the incoming photon energy exactly matches the binding energy of the K shell in iron. In other higher atomic numbered absorbers, there would be discontinuities for the K shell, L shell, and M shell.

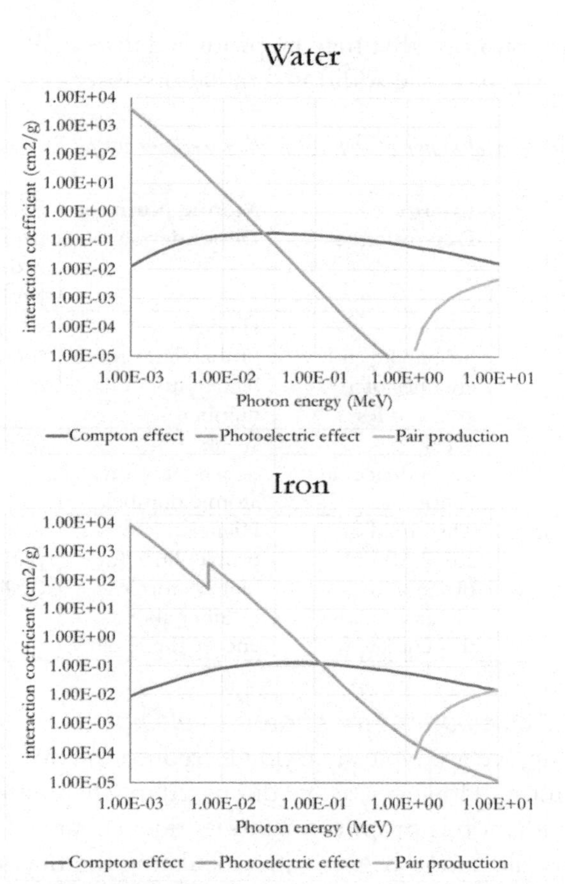

Figure 6-2. Interaction coefficients for photon interactions in water and iron.

Finally, to relate this discussion to our earlier discussions of interaction probabilities in Chapter 4, the sum of the individual interaction coefficients at any given energy is the total mass attenuation coefficient (μ/ρ) for that energy. Thus, for low photon energies, μ/ρ may be dominated by the photoelectric effect; for medium photon energies, μ/ρ may be dominated by the Compton effect; and for higher photon energies, μ/ρ is eventually dominated by pair production.

Table 6-1 below gives a quick summary of the information that is relevant for these interaction mechanisms when designing shielding for photons. As we will see, the choice of material affects the ability of the shield to reduce the primary photons from the source but also the secondary photons that may be produced inside the shield that can then emerge to cause additional radiation dose.

Table 6-1. Summary of major photon interaction mechanisms for energies of 10 MeV or less

Mechanism	**Energy Dependence**	**Atomic Number Dependence**	**Secondary photon radiations produced**
Photoelectric Effect	Higher probability at low energies of 100 keV and less	Higher probabilities for higher atomic number	Characteristic x-rays and Auger electrons
Compton Effect	Weak dependence on energy	Weak dependence on atomic number	Scattered photons
Pair production	Threshold of 1.022 MeV; probability increases above this energy	Higher probabilities for higher atomic number above energy threshold	Annihilation photons of 0.511 MeV

6.3.1.1 Initial calculations for photon shielding

With shielding, we are typically trying to reduce the dose rate in a specific location. However, as we discussed in Chapter 4, the dose rate due to radiation is proportional to its fluence rate. Therefore, the shielding calculation begins with calculating the fluence rate at the point of interest.

Referring again to the discussion in Chapter 4, the mass attenuation coefficient is the probability per density thickness that an interaction will occur. Using first order differential equations, the fluence of a beam of photons as a function of absorber thickness is shown in Equation 6.1 below, which can be written in terms of either fluence or fluence rate:

$$\phi_{shielded} = \phi_0 e^{-(\frac{\mu}{\rho})(\rho x)} \tag{6.1}$$

Where

- $\phi_{shielded}$ is the attenuated fluence rate after traversing through density thickness ρx

- ϕ_0 is the initial unattenuated fluence rate at the same point as ϕ but with no absorber in place
- μ/ρ is the mass attenuation coefficient in units of cm^2/g
- ρx is the density thickness of the absorber in units of g/cm^2

There are two important limitations to Equation 6.1:

- Although not explicitly called out in the symbolic notation, this equation applies to a fluence of a fixed single energy of photon because μ/ρ is energy-dependent. If a radiation source has multiple photon energies, then this calculation must be performed for each photon energy, which quickly becomes tedious in many situations.
- The attenuated fluence rate predicted by Equation 6.1 is only that from the primary radiation; it does not include secondary photon radiation produced by interactions in the shield. As shown in Table 6-1, all three interaction mechanisms produce secondary photon radiation that can still penetrate through the absorber, and Equation 6.1 by itself does not address this phenomenon. We will show how to handle that in the next section of this text, but it must be remembered that Equation 6.1 predicts the fluence rate of photons that had no interaction in the absorber enroute to the receptor.

Even with these limitations, Equation 6.1 is still very useful in that it can give us an initial estimate of shielding thickness that would be necessary. The final thickness that we employ in a given situation will almost always be thicker than that predicted by Equation 6.1, but a first approximation using Equation 6.1 can help us begin the process. The following examples are intended to get some familiarity with how Equation 6.1 can be used.

Example: An unshielded fluence rate of 3.4×10^4 $\gamma/cm^2/sec$ of 1 MeV photons impinges on a receptor. What is the attenuated fluence rate of 1 MeV photons after placing 1.5 cm of iron between the source and the receptor?

Solution: As discussed in Chapter 4, the NIST database that gives values for both μ_{en}/ρ and μ/ρ can be found at https://www.nist.gov/pml/x-ray-mass-attenuation-coefficients and will provide the value of μ/ρ that is needed to solve this problem. Using Table 3 from that webpage shows that for iron at 1 MeV, the value of μ/ρ is equal to 0.05995 cm^2/g. The density for iron also can be found using Table 1 of that same webpage, which is given as 7.874 g/cm^3. Thus, the density thickness of the iron shield is ρx, or (7.874 g/cm^3)(1.5 cm)=11.811 g/cm^2.

Applying Equation 6.1 results in the following:

$$\phi_{shielded} = \phi_0 e^{-\left(\frac{\mu}{\rho}\right)(\rho x)}$$

$$= \left(3.4 \times 10^4 \frac{\gamma}{cm^2 sec}\right) e^{-\left(0.05995 \frac{cm^2}{g}\right)\left(11.811 \frac{g}{cm^2}\right)}$$

$$= 1.67 \times 10^4 \ \gamma/cm^2/sec$$

Example: An unshielded fluence rate of 2.6×10^6 $\gamma/cm^2/sec$ of 0.5 MeV photons is calculated to impinge on a receptor. It is desired to reduce this to 4.8×10^5 $\gamma/cm^2/sec$. Using Equation 6.1, what is a first approximation of the thickness of ordinary concrete that could achieve this goal?

Solution: Referring again to the NIST tables referenced in the previous example, Table 4 provides values of μ/ρ for compounds and mixtures. For ordinary concrete at 0.5 MeV, the value is 0.08915 cm^2/g. Table 2 from that same page gives a density of ordinary concrete of 2.3 g/cm^2 (concrete can be custom-mixed to a variety of densities, and it is always important to identify which density is used).

Equation 6.1 can be rearranged to solve for ρx by a combination of algebraic rearrangement and taking the natural logarithm to obtain the following:

$$\rho x = -\frac{\ln\left(\frac{\phi_{shielded}}{\phi_0}\right)}{\left(\frac{\mu}{\rho}\right)} = -\frac{\ln\left(\frac{4.8 \times 10^5 \frac{\gamma}{cm^2 sec}}{2.6 \times 10^6 \frac{\gamma}{cm^2 sec}}\right)}{\left(0.08915 \frac{cm^2}{g}\right)}$$

$$= 18.95 \frac{g}{cm^2}$$

By carrying the units through, this reminds us that we have solved for the density thickness of the material. The linear thickness of concrete that is needed is then obtained by dividing the density thickness by the density to obtain (18.95 g/cm^2)/(2.3 g/cm^3) = 8.24 cm of concrete.

Again, this is a first estimate of the magnitude of thickness of concrete that would be needed in this case. Based on that initial result, it could be decided to use a different material that may not require as much linear thickness if that amount of concrete would not be practical due to space or other limitations.

We will now combine our knowledge from Chapters 1 and 4 along with our current chapter to do a more complex problem.

Example: A 0.8 Ci point source of Cs-137 is to be placed in a room where an individual receptor could be located as close as 1.25 meters from the source. Considering only the 0.662 MeV photons from the source, calculate the unshielded absorbed dose rate and then perform a first order estimate of what minimum thickness of steel would be necessary to be placed between the source and receptor to reduce the dose rate to 10 mrad/hour (0.0001 Gy/hour) from only the 0.662 MeV photons.

Solution: We will be using Equations 1.15, 4.5, and 6.1, and it is critical to keep track of the variables in these equations. We will calculate the absorbed dose rate to tissue, meaning we need μ_{en}/ρ for tissue. Our shield will be constructed of steel (we will use iron as a surrogate), meaning we need μ/ρ for iron (keeping track of materials and thinking our way through the problem is crucially important to avoid errors).

We will start by constructing a list of the information we have suitably converted to proper units.

- Activity (A) = 0.8 Ci = 2.96×10^{10} decays/second
- Distance (r) = 1.25 m = 125 cm
- Yield of the photon of interest Y = 0.851 γ/decay from the IAEA Isotope Browser App
- Tables 3 and 4 of the NIST Mass Coefficients Database provide the following information for soft tissue and iron:

Energy (MeV)	μ_{en}/ρ for soft tissue (cm^2/g)	μ/ρ for iron (cm^2/g)
0.6	3.254E-02	7.704E-02
0.8	3.176E-02	6.699E-02

Using linear interpolation, we can get the appropriate values at 0.662 MeV to be:

μ_{en}/ρ for soft tissue = 0.0323 cm^2/g

μ/ρ for iron = 0.0739 cm^2/g

- Density of iron ρ = 7.874 g/cm^3

Having obtained and organized the information, it is time to begin the calculation. The first step is to calculate the fluence rate from the point source using Equation 1.15 as shown below:

$$\phi_{point} = \frac{A \cdot Y_i}{4\pi r^2} = \frac{(2.96 \times 10^{10}\,\frac{decays}{sec})(\frac{0.851\,\gamma}{decay})}{4\pi(125\ cm)^2}$$

$$= 128{,}290\ \frac{\gamma}{cm^2 sec}$$

The unshielded dose rate can be obtained by using Equation 4.5; we will calculate the collision kerma to soft tissue and assume that charged particle equilibrium exists so that the kerma and the absorbed dose are the same. The unit conversions are particularly important for the energy, mass, and time units.

$$K_c = \Phi_i E_i (\frac{\mu_{en}}{\rho})_i$$

$$= \left(128{,}290 \frac{\gamma}{cm^2 sec}\right)\left(0.662 \frac{MeV}{\gamma}\right)\left(0.0323 \frac{cm^2}{g}\right)$$

$$= 2743 \frac{MeV}{g \cdot sec} = 1.58 \frac{mGy}{hour} = 158 \frac{mrad}{hour}$$

The premise of the question calls for a shield design to reduce the dose rate to 0.1 mGy/hour, which means a dose reduction factor of (0.1 mGy/1.58 mGy) =0.0633. Since this problem is based on a single photon energy, reducing the absorbed dose rate is achieved by reducing the photon fluence rate: that is, we need to reduce the fluence rate by the same factor of 0.0633 such that $\phi_{shielded} = 0.0633\phi_0$, or $\phi_{shielded}$=8121 $\gamma/cm^2/sec$.

As in the last example, we rearrange Equation 6.1 to solve for the density thickness of the shield as follows:

$$\rho x = -\frac{\ln\left(\frac{\phi_{shielded}}{\phi_0}\right)}{\left(\frac{\mu}{\rho}\right)} = -\frac{\ln\left(\frac{8121 \frac{\gamma}{cm^2 sec}}{128290 \frac{\gamma}{cm^2 sec}}\right)}{\left(0.0739 \frac{cm^2}{g}\right)} = 37.35 \frac{g}{cm^2}$$

This would then be a linear thickness of approximately 4.75 cm of steel/iron as a first estimate of the minimal shield thickness to reduce the primary photon fluence.

From the preceding examples, it should be clear that the main challenge in shielding calculations is attention to detail and documenting what assumptions are used to develop parameters. This is often the case with any radiation protection calculation, though. Hasty calculations tend to give more opportunity for avoidable errors in selecting parameters.

With additional photon energies (either because of multiple photons from the same radionuclide or individual photons from multiple radionuclides), the calculations become more numerous, because each photon energy will have its own unique fluence rate and its own unique mass attenuation coefficient. A given shield thickness will tend to attenuate the lower energy photons more severely than the high energy photons; therefore, solving for a shield thickness explicitly as we did in these examples is not as direct. Instead, a first approximation to a shield thickness may be obtained, and then the dose calculation would be performed again with that thickness to determine its efficacy.

For shielding calculations other than point sources, the mathematics gets more complex because of a simple geometry issue. This can be understood by referring to Figure 6-3 below, which shows a line source of radioactivity. If the line source is viewed/approximated as a series of individual point sources, the points along the line furthest from the receptor have a greater effective shield thickness between them and the receptor compared to the actual physical shield thickness. Thus, those points contribute less to the dose at the receptor for two reasons:

- the points are further away from the receptor
- any shield thickness will be effectively thicker as the distance along the source is increased along this line.

This is good news in a sense because we expect then that the portion of the line source closest to the receptor should contribute the greatest fraction of the total dose.

Of course, the shielding problem for this line source could in theory be solved using hand calculations and invoking calculus and integrating along the line. However, the resulting integral cannot be solved analytically, and numerical solutions must then be applied to it. In addition, consideration of the buildup factor (discussed in the next subsection) adds another layer of complexity. Thus, for other than point source calculations, computational methods are usually applied for shielding calculations, as discussed in Section 6.3.1.3.

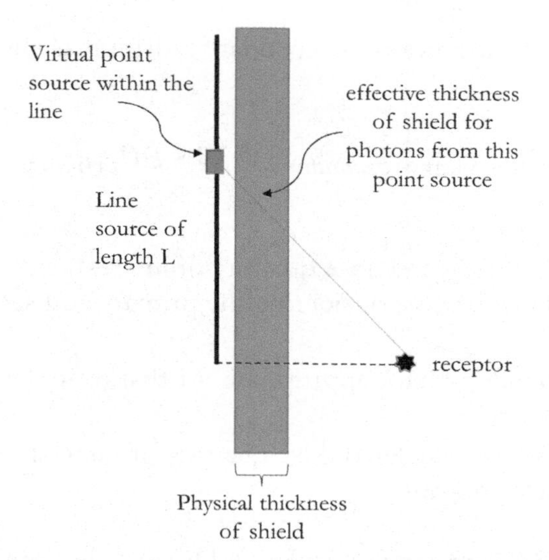

Figure 6-3. Illustration of the effective shield thickness provided to sources that are offset from the receptor in a line source owing to simple geometry considerations.

6.3.1.2 Buildup Factor

Now that we have addressed the basic idea of shielding, we can introduce some realistic complications to the calculations, one of which is a concept called buildup factor. As discussed in the previous section, Equation 6.1 predicts the fluence rate of photons that have not interacted in the shield; in essence, it assumes that when photons interact in the shield, they are totally removed from the fluence and do not contribute to the fluence on the receptor side of the shield. However, the interaction of photons in the shield can produce secondary photons that can penetrate through the shield, and these must be addressed because of their potential to cause additional radiation dose.

To account for these secondary photons, we introduce the buildup factor B, which is defined as shown in Equation 6.2

$$B = \frac{amount\ due\ to\ primary\ and\ secondary}{amount\ due\ to\ primary} \tag{6.2}$$

Conceptually, the buildup factor is a value that is multiplied by a calculated dose quantity considering only the primary radiation to adjust for the contribution of secondary radiation as shown in Equation 6.3.

$$DQ_{primary+secondary} = B * DQ_{primary} \tag{6.3}$$

Where

- $DQ_{primary+secondary}$ is the dose quantity of interest (exposure, absorbed dose, etc.) considering the primary and secondary photons
- B is the buildup factor appropriate for that dose quantity in the given scenario
- $DQ_{primary}$ is the calculated dose quantity of interest based solely on the primary photons

In Section 6.3.1.1, we were calculating $DQ_{primary}$, but the use of the buildup factor now gives us the ability to calculate the total dose $DQ_{primary+secondary}$.

The buildup factor for a given scenario understandably depends on several parameters as follows:

- The dose quantity itself
- The energy of the primary photons
- The material from which the shield is constructed
- The geometry of the source (point, slab, line, etc.)
- The thickness of the shield, which is expressed in units of mean free paths (mfp), which is μx if using linear thickness and the linear attenuation coefficient, or $(\mu/\rho)(\rho x)$ if using density thickness and the mass attenuation coefficient. Greater shield thickness gives more opportunity for primary photon interactions to occur.

The buildup factor is an experimentally measured quantity and thus is not derived from first physics principles. It can be obtained by laboratory measurement or by Monte Carlo simulation; the values are then placed into tables that can be used in practical calculations.

Table 6-2 shows values of B for air kerma for an isotropic point source in iron for various photon energies and varying shield thicknesses (expressed in mean free paths). A report with buildup factor for other materials can be found at https://technicalreports.ornl.gov/1988/3445605718328.pdf.

Table 6-2. Buildup factors for iron for air kerma from an isotropic point source.(Chilton et al. 1984)

Mean free paths μx	Buildup Factor (B) at the stated photon energy for air kerma				
	0.1 MeV	0.2 MeV	0.5 MeV	1 MeV	2 MeV
1	1.40	1.86	1.99	1.85	1.71
2	1.61	2.59	3.12	2.85	2.49
3	1.78	3.33	4.44	4.00	3.34
4	1.94	4.08	5.96	5.30	4.25
5	2.07	4.85	7.68	6.74	5.22
6	2.20	5.64	9.58	8.31	6.25
7	2.31	6.44	11.7	10.0	7.33
8	2.41	7.25	14.0	11.8	8.45

The values in Table 6-2 emphasize the importance of accounting for secondary photons when performing shielding calculations. In the example that we worked in the previous section, we calculated a shield thickness of 37.35 g/cm^2 (4.75 cm) of steel to reduce the absorbed dose rate $D_{primary}$ to 0.1 mGy/hour due to 0.662 MeV photons from a 0.8 Ci point source of Cs-137. This shield thickness represents roughly 2.76 mean free paths (multiplying the mass attenuation coefficient of 0.0739 cm^2/g by the density thickness of the shield). Using the table above (and assuming that B for air kerma and B for absorbed dose in tissue are similar in value), we can see that the buildup factor would be between 3 and 4. Thus, for our 4.75 cm thick shield, the total dose rate could be close to 0.3-0.4 mGy/hour, which we obtain by $B*D_{primary}$. This means that in practice our shield would need to be much thicker to account for the secondary photons.

There have been multiple efforts to perform curve fits to the buildup factors that have resulted in several mathematical forms. None of these are perfect formulations for B, but they do allow a method to quickly estimate B. Three common forms are the (a) Berger approximation; (b) Taylor approximation; and (c) Geometric progression approximation. Each of these approximations uses a different mathematical formulation with fitting parameters to estimate the buildup factor. The details of these methods are beyond the scope of this text, but each method has some advantages in allowing a rapid calculation of B in computer programs.

An additional complicating factor arises if multiple shield types are used. For example, if a room is constructed of concrete but also has a steel lining inside it for shielding purposes, what is the buildup factor for that combination? It is largely a judgment call, but one approach is to calculate the total number of mean free paths for all shielding materials and then determine the buildup factor based on the material that gives the most conservative value.

At this point, it should be obvious that solving a problem explicitly for a shield thickness including buildup is an impossibility. We will have to perform multiple iterations of trial and error to determine the proper shield thickness for a given shielding problem. Greater shield thicknesses reduce the dose contribution from the primary photons but increase the dose contribution from the secondary photons. For practical shielding calculations, the process might go as follows:

1. Select an initial shield thickness based on professional judgment. In some cases, a thickness is estimated based only on the primary beam as we did above to have a starting point.
2. Perform the dose calculation, including buildup to determine the effectiveness of the shielding in achieving the desired dose rate.
3. If the shield thickness is still insufficient, add another mean free path of shielding and perform the dose calculation again.
4. Continue with this process until the calculations indicate that the desired dose rate has been achieved; the final shield thickness may then be supplemented with an added thickness of 10% as a margin of safety.

6.3.1.3 Computational methods for photon shielding calculations

Thus far, we have emphasized the basics of shielding analysis using hand calculations. While sometimes tedious, hand calculations can nonetheless be useful for simple shielding scenarios involving point sources with single photon energies. One drawback of hand calculations is that the shielding problem can become complex quickly if multiple nuclides in a single source are involved, if multiple radioactive sources are arranged around the receptor location, or if the source is a distributed source (such as a planar or a volumetric source). It can be difficult to account for self-absorption in the source, and simplifying assumptions to make the hand calculations more manageable often come at the expense of being overly conservative.

To advance beyond hand calculations, there are some other alternatives that involve computer resources. The degree of sophistication of the method depends on the problem at hand; in some cases, a simpler approach will work better or be more affordable while in other situations, a more sophisticated approach will be necessary.

One free program addresses point sources in simple configurations. RadPro Calculator is available either to run in a web browser or as a downloadable file from www.radprocalculator.com. While not robust enough to handle complicated shielding scenarios, it is a useful and intuitive tool and illustrates many of the concepts that have been covered already. In addition to beta and photon shielding calculations (including buildup), it handles decay correction for radionuclides and other basic functions.

6.3.1.3.1 Point-kernel computer programs

The general approach to point-kernel computer programs is that the source/receptor geometry is represented by a predefined shape that is easy to break into individual mathematical elements based on trigonometry. Each element is then treated as a discrete radiation source (a point source), and the fluence from each element is calculated along with any reduction due to attenuation from any shields (including self-shielding by the source) and any increase due to buildup between the element and the receptor.

The total fluence and dose rate at the receptor is calculated as the sum of the fluence and dose rates from each of the mathematical elements in the source. Therefore, the program performs an approximation of an integral by summing up the contributions from the receptors much like a Riemann sum approximates a true integral.

Dividing the source into more elements is obviously going to increase accuracy, but the greater number of elements causes additional computational time and may not change the results drastically.

The concept of how this works can be visualized by referring to Figure 6-4 below, which shows a line source that has a shield between it and the receptor. The line is 200 cm long and is 30 cm from the receptor; the shield is 5 cm thick. For this case, the line could be divided into 200 individual elements that are 1 cm in length with 1/200th of the total activity located in each element. Further, in each element, the activity could be located in the center of the element as a point source. Using trigonometry, it is straightforward to determine the source-to-receptor distance for each element as well as the effective shielding thickness for each segment as shown in the figure for one segment.

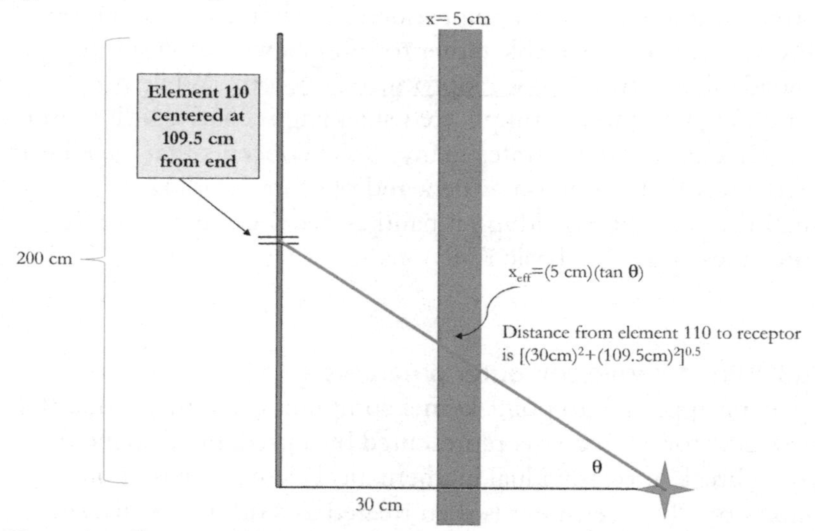

Figure 6-4. Diagram of how a line source can be subdivided into mathematical elements to facilitate a shielding calculation.

If we look just at the 110th element in the line, the distance to the source for that element can be calculated using either the Pythagorean theorem or using the angle. The effective shield thickness x_{eff} for that particular segment can likewise be calculated as the shield thickness multiplied by the tangent of the angle. For each element, a new angle would be calculated, along with a new distance between the element and receptor and a new effective shield thickness. As discussed earlier, the buildup factor B can be assessed for each individual element based on the shield material and the effective thickness for that element. Finally, the contribution to the fluence and dose rate at the receptor would be obtained by summing the contributions from each individual element.

While it is certainly possible to design a rudimentary program or even a spreadsheet that could perform such calculations, it is usually more defensible in the workplace to use a well-documented program that has gone through rigorous validation and verification (V&V), especially if the results of the shielding calculations are used for facility design or for potentially high-dose situations.

One commercial program that has been available for several decades is MicroShield®, designed by Grove Software (www.radiationsoftware.com). While it is not free, it is nonetheless recognized in the nuclear industry for its robustness and ease of use.

NOTE: For academic institutions, Grove Software can make MicroShield® available for use so long as it is not distributed. Contact Grove Software through their website (www.radiationsoftware.com) for more information on this opportunity. **All images in this section from MicroShield are provided courtesy of Grove Software.**

MicroShield employs a graphical user interface (GUI) that allows the user to select the source geometry and then proceed through a series of input screens to provide additional information about the receptor location, shield(s) materials, radioactive source specification, buildup, and integration parameters. Figure 6-5 below shows the listing of the source geometries that are included.

While the source geometries may not be all-encompassing to cover every possible source configuration in the real world, the selections are still sufficient to cover many scenarios that would be encountered. For irregularly shaped sources that are not included in MicroShield, it may be necessary to find a source geometry that can be sized to provide a bounding estimate of the fluence and dose rate.

As discussed earlier, non-point sources can be divided into individual elements whose contributions are then summed to approximate the integral over the source dimensions. In MicroShield, this is specified by the user as part of the integration parameters. Figure 6-6 shows how a three-dimensional right cylindrical source can be divided into elements radially, circumferentially, and vertically.

Welcome to MicroShield!

New Project

Create new geometry case(s):

Point
Line
Disk
Rectangular Area - Vertical
Rectangular Area - Horizontal
Sphere
Cylinder Volume - Side Shields
Cylinder Volume - End Shields
Cylinder Surface - Internal Dose Point
Cylinder Surface - External Dose Point
Annular Cylinder - Internal Dose Point
Annular Cylinder - External Dose Point
Rectangular Volume
Truncated Cone
Infinite Plane
Infinite Slab

Recent Cases

Recent

Use the control key to select more than one geometry.

Create Case(s)

Open Recent Case(s)

Open Existing Case

Figure 6-5. Source geometries included in MicroShield

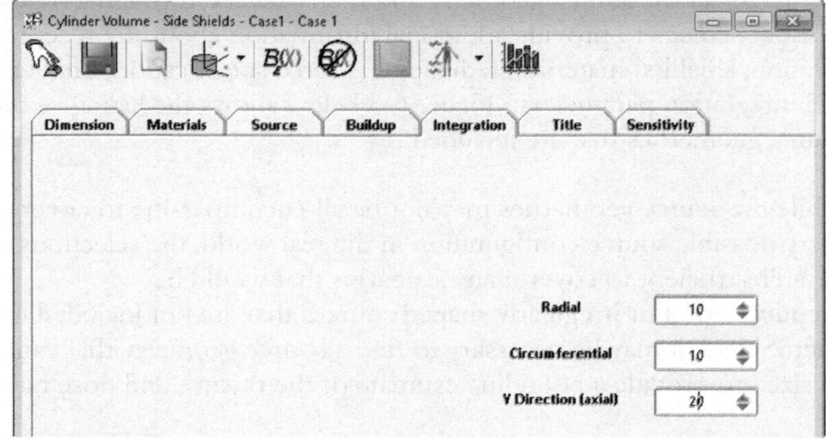

Figure 6-6. Specifying integration parameters for a right circular cylinder in MicroShield

Finally, there is one useful feature that helps in determining the shield thickness that would be necessary to achieve a particular dose rate. Because we use the buildup factor B, it is not practical to solve for a shield thickness analytically; thus, shielding analysis is often an iterative process. MicroShield contains a feature called "Sensitivity," which allows the user to specify a range for any of the variables that are in the scenario. Thus, the user can specify a minimum shield thickness and a maximum shield thickness with the number of steps in between those thicknesses as shown below in Figure 6-7.

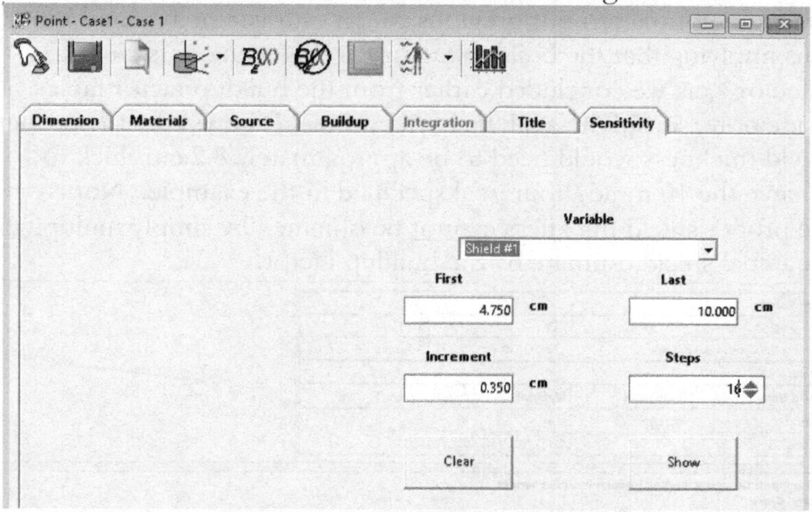

Figure 6-7. Using the sensitivity analysis tool in MicroShield to see the impact of shield thickness on the calculated absorbed dose rate.

Following the selection of the sensitivity parameters, the program executes repeatedly and displays the results of all the individual runs so that the user can see the effect of increasing or decreasing shield thickness. This keeps the user from having to perform the iterations manually with totally new cases in the program. Of course, other variables can be subjected to the same analysis, including the number of elements into which the source is divided, the receptor distance, etc.

Our earlier shielding example in Section 6.3.1.1 of the 0.8 Ci point source of Cs-137 provides a straightforward way for us to see how a shielding program can help analyze a scenario. Previously, we determined that a shield thickness of 4.75 cm (considering no buildup) would reduce the absorbed dose rate from the primary photons to approximately 10 mrad/hour. Placing those parameters into MicroShield now confirms that a 4.75 cm thick shield would be insufficient as shown in Figure 6-8 below. The output shows the dose contribution from multiple photon energies and buildup that our manual calculations ignored earlier and predicts an absorbed dose rate to tissue of approximately 41 mrad/hour (we recognize that the table value of 37.25 mrad/hr is ~0.87 of the exposure rate of 42.67 mR/hr, indicating this is the absorbed dose rate to air and not tissue as we saw with the example following Equation 4.11 in Chapter 4). This is in contrast to our previous rough estimate of 10 mrad/hour, thus implying that the buildup factor for this scenario is close to a value of 4, as we concluded earlier from the buildup factor tables. As a side note, sensitivity analysis in MicroShield shows that the proper shield thickness would need to be approximately 8.2 cm thick to achieve the 10 mrad/hour goal specified in the example. Notice that the proper shield thickness <u>cannot</u> be obtained by simply multiplying the initial shield estimate by the buildup factor.

Dose Points			
No.	X	Y	Z
#1	125.0 cm (4 ft 1.213 in)	0.0 cm (0 in)	0.0 cm (0 in)

Shields			
Shield Name	Dimension	Material	Density (g/cm³)
Shield 1	4.75 cm	Iron	7.86
Air Gap		Air	0.00122

Y Z X

Source Input: Grouping Method - Actual Photon Energies
Library: Grove

Nuclide	Ci	Bq
Ba-137m	7.5680e-001	2.8002e+010
Cs-137	8.0000e-001	2.9600e+010

Buildup

Buildup: The material reference is Shield 1.

Integration Parameters

Integration does not apply.

Results With Buildup

Energy (MeV)	Activity (Photons/sec)	Energy Flux (MeV/cm²/sec)	Photon Flux (Photons/cm²/sec)	Exposure Rate (mR/hr)	Absorbed Dose Rate (mrad/hr)	Absorbed Dose Rate (mGy/hr)
3.182e-02	5.797e+08	9.083e-25	2.855e-23	7.565e-27	6.605e-27	6.605e-29
3.219e-02	1.070e+09	1.703e-24	5.289e-23	1.370e-26	1.196e-26	1.196e-28
3.640e-02	3.892e+08	7.463e-25	2.050e-23	4.240e-27	3.702e-27	3.702e-29
6.616e-01	2.520e+10	2.201e+04	3.326e+04	4.267e+01	3.725e+01	3.725e-01
Total	**2.723e+10**	**2.201e+04**	**3.326e+04**	**4.267e+01**	**3.725e+01**	**3.725e-01**

Figure 6-8. Output results in MicroShield from shielding a 0.8 Ci point source of Cs-137 with 4.75 cm of iron with a receptor distance of 1.25 meters. Note: the reported absorbed dose rate in this table is for air, not tissue.

Clearly, point-kernel computational methods have versatility and can be implemented in a number of ways, whether in commercial programs, standalone programs written for a single project, or even rudimentary spreadsheet calculations that can carry out the calculations. Answers can be achieved relatively quickly, and the learning curve for these programs typically is not as steep as other methods. However, for facility-level calculations with multiple sources, multiple receptors, and complicated geometries, point-kernel methods may lack the necessary robustness. Thus, the more advanced Monte Carlo techniques may be helpful in these situations.

6.3.1.3.2 Monte Carlo computer programs/codes

Monte Carlo calculational methods have been discussed earlier in reference to modeling various radiation interaction processes and obtaining a variety of results. It should come as no surprise that Monte Carlo methods can be used to set up a virtual experiment to evaluate the effectiveness of shielding.

As with point-kernel codes, it is possible at the student level to write a Monte Carlo program for simple shielding scenarios for photons. It requires a thorough understanding of the physics involved in radiation interactions along with good computer programming skills to track the progress of each photon history as the photon moves through the virtual environment that has been programmed. A pseudo-random number generator is critical as well, since random numbers are used to simulate the random events that are inherent in radiation interactions in materials. A simplified flow of following a single photon history through the process of Monte Carlo simulation is as follows, with pseudo-random numbers used in each step to select the parameters for transport.

- For each new history, the energy of the photon is selected, along with its original position and orientation.
- The photon's travel distance before its interaction is selected, and the position of that interaction is tracked.
- The photon's interaction type is selected.
- Secondary photons and their orientation are tracked relative to the point of interaction.
- All photons and other radiations associated with the original history are tracked until they are terminated, interact within the

receptor region, are below an energy cutoff value, or reach a spatial location in the virtual environment that is outside the bounds of interest.

- Once all tracking for a history is completed, a new history is initiated, and the process begins again.

Advanced Monte Carlo programs for radiation transport are available that can be used for a variety of purposes. The Visual Monte Carlo (VMC) program was already referenced in Chapter 4 as an example of a program that could calculate the effective dose from photon sources. However, for shielding purposes, we need Monte Carlo programs that can be customized for a wide variety of scenarios. Besides just simulating the shield that may lie between the source and receptor, other structures in the area need to be simulated as well because they could provide scattering of radiation that then irradiates the receptor. Indeed, entire facilities can be designed as a virtual environment for Monte Carlo simulation. Granted, the more complex the scenario and environment, the more computational run time will be required, and there are statistical techniques needed to address improbable interactions that could still cause significant radiation dose.

While this is hardly an exhaustive list, below are some of the readily available programs that could be used for shielding analysis that can be obtained at no cost for academic use. There are also commercial programs that can be purchased, but they are not included in this list. Many of these programs can be used to simulate transport of more than just photons and can thus be used for shielding and dose calculations from a variety of radiation sources, including charged particles (even high-energy particles from accelerators or outer space), neutrons, and other elementary particles.

	GEANT4 – (https://geant4.web.cern.ch/) – Maintained by CERN, the European Organization for Nuclear Research, and home of the Large Hadron Collider, GEANT4 is freely available for direct download from their website. Like the other programs below, it has extensive interaction libraries and can be used for a wide variety of simulations.

	FLUKA (http://www.fluka.org/) – FLUKA has attracted a large following due to its ease of designing complex virtual environments and the ability to display the results. FLUKA runs only in a Linux environment, which may be a limitation for some users. The program is available for direct download through their website.
	PHITS (https://phits.jaea.go.jp/index.html) – This code is maintained and distributed by the Japan Atomic Energy Agency (JAEA). Interested users must submit an application to the JAEA; if approved, a download link is provided. The output files from PHITS can be displayed and animated via external programs to give a visual representation of the radiation fields and interactions.
	EGSnrc (https://nrc-cnrc.github.io/EGSnrc/) – egsNRC is maintained by the National Research Council Canada but is freely distributed to interested users. It is an extended version of EGS4, a popular Monte Carlo simulation program used in the 1980s and 1990s.
	MCNP (https://mcnp.lanl.gov/)– This program is maintained by Los Alamos National Laboratory but is distributed by the Radiation Safety Information Computational Center website (RSICC), based in Oak Ridge, Tennessee. Because it can be used for purposes other than radiation shielding, it is export-controlled. **NOTE: There is a distribution fee from RSICC that may not be waived for all academic institutions.**

The choice of which program/code to use will depend on a variety of factors. The user interface is different for each of the programs listed above. In general, the programs are designed for computational speed, which means that the user input screens may be rudimentary or may be text files that provide a shorthand input notation for the program. In some cases, the programs are already compiled and ready for use; for others, the user may need to compile the program after selecting the proper scenario inputs using a standalone compiler.

Regardless of which program is selected, there will be a learning curve. Unfortunately, familiarity with one of the programs will not necessarily guarantee ease of use with a different program. Tutorials, example programs, and training sessions are available with many of the programs, and online user groups are available as well to help deal with specific issues. Monte Carlo techniques as applied to radiation physics is a specialized area and is typically learned best by doing and working with others in it as opposed to reading about it; therefore, the best approach is to select one of the programs above that seems to have the most support and background material and dig in with example problems that can be tested against known solutions to be sure that the program is being used correctly.

6.3.2 Shielding for Beta Radiation

The shielding of high-energy electrons can be approached in a number of ways. If weight and size are no concern, then a heavy metallic shield will be more than sufficient for the task. However, if weight and size are a concern, such as what might be the case in a medical environment where a syringe might be filled with a radionuclide that is a beta emitter, then there are some further considerations that must prevail.

When high-energy electrons undergo rapid deceleration in a material, they give off bremsstrahlung radiation, as illustrated in Figure 6-9 below. Bremsstrahlung production is desirable in some situations; indeed, this is the operating principle behind medical x-ray machines which accelerate electrons to kinetic energies of hundreds of keV and then impinge them on a target, typically tungsten. The amount of bremsstrahlung produced depends strongly on the kinetic energy of the electron but also on the atomic number of the absorber such that high-Z materials cause greater amounts of bremsstrahlung to be produced.

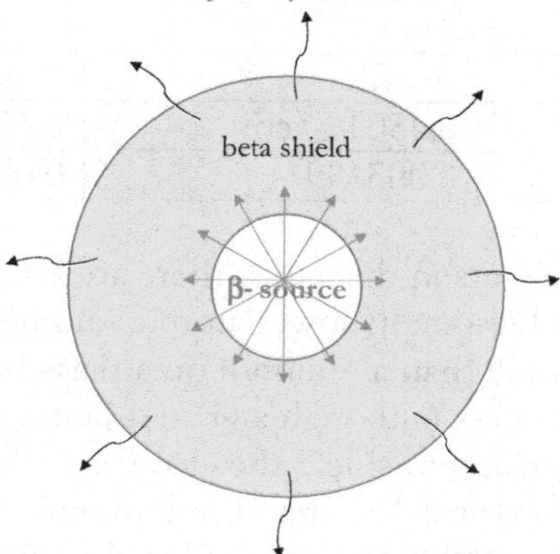

Figure 6-9. Illustration showing that shielding of beta radiation can produce a secondary photon radiation field due to bremsstrahlung emission.

When shielding beta radiation, however, bremsstrahlung production is undesirable. Therefore, a shield for beta emitters is typically a two-part problem; the beta particles must be shielded first and then any subsequent bremsstrahlung must be shielded separately. It is typical to use a low-Z material to shield the beta radiation and then supplement that with a metallic high-Z shield to absorb any bremsstrahlung radiation that is produced.

The amount of shielding for the beta particles obviously depends on the energy of the beta particles. Unlike photon radiation, though, electrons have a range in absorbers that has been tabulated. If an absorber with thickness equal to at least the range of the most energetic electrons is used, then the beta radiation will be totally stopped. As discussed in Chapter 4, the estar database maintained by NIST can be used to determine the range for common materials.

https://physics.nist.gov/PhysRefData/Star/Text/ESTAR.html

For a given absorber, the CSDA range and the radiation yield (the fraction of the beta energy that is radiated as bremsstrahlung) are included as a function of energy. To give some sense of this information, the CSDA range and radiation yield for 1 MeV electrons in aluminum and in acrylic (polymethyl methacrylate) are given below in Table 6-3.

Table 6-3. CSDA Range and radiation yield for 1 MeV electrons in aluminum and acrylic (from NIST estar database)

	Acrylic	Aluminum
CSDA Range	0.4504 g/cm^2	0.5546 g/cm^2
Radiation Yield	0.003199	0.007636

With the few data points in this table, there are already some interesting results. Despite its lower atomic number, acrylic provides better shielding overall than aluminum on a mass basis. It requires a smaller density thickness (although a greater linear thickness owing to the difference in density) to shield the electrons. Further, the radiation yield for acrylic is less than for aluminum, indicating that the amount of bremsstrahlung produced in the acrylic is less than in the aluminum.

For hand calculations of beta shielding for a point source, there are some differing approaches that could be used. Below is one such technique:

1. The primary shield for the beta particles should be thick enough to stop electrons with energy equal to the maximum beta energy from the radionuclide.
2. To estimate the potential photon absorbed dose from bremsstrahlung, the energy fluence rate ψ of MeV/cm^2/sec can be calculated using Equation 6.4 below and assuming that this energy fluence is comprised of photons with energy equal to the average beta energy (this is necessary to use the appropriate mass energy absorption coefficient).

$$\psi = \frac{AE_{\beta-(avg)}f}{4\pi r^2} \tag{6.4}$$

Where

- ψ is the energy fluence rate (MeV/cm^2/sec)
- A is the source activity (decays/sec or Bq)
- $E_{\beta\text{-}(avg)}$ is the average beta energy (MeV)
- f is the radiation yield evaluated at the average beta energy from the estar database (dimensionless)

- r is the distance from the center of the source, since a point source is assumed.

3. The absorbed dose rate due to the bremsstrahlung can be calculated as shown in Equation 6.5 below, which is a variation of Equation 4.6 where the energy fluence rate ψ replaces the product of photon fluence rate ϕ and photon energy E.

$$D = \psi(\frac{\mu_{en}}{\rho}) \tag{6.5}$$

Where
- D is the absorbed dose rate (MeV/g/sec)
- ψ is the energy fluence rate ($MeV/cm^2 sec$)
- (μ_{en}/ρ) is the mass energy absorption coefficient (cm^2/g) evaluated at the photon energy equal to the average beta energy

To convert the absorbed dose rate to Gy/sec, the results from Equation 6.5 would be multiplied by 1.6×10^{-10} (J/MeV)(kg/g). If desired, this result can then be multiplied by 3600 sec/hour to obtain a final result in Gy/hour.

Example: A 10 mCi point source of P-32 is to be shielded. What thickness of acrylic would be needed to totally shield the betas from this source, and what is a rough estimate of the photon absorbed dose rate to tissue at the outer edge of the beta shield due to the bremsstrahlung from the source, assuming point source geometry?

Solution: From the IAEA Isotope Database App, P-32 has a maximum beta energy of roughly 1711 keV and an average beta energy of roughly 695 keV. The activity is 3.7×10^8 decays/sec.

A quick inspection of the estar database table for polymethyl methacrylate (acrylic) shows that the range of 1.75 MeV electrons (only slightly larger than the 1.711 MeV maximum beta energy) is 0.8715 g/cm^2; thus, it is appropriate to shield the betas with that thickness of acrylic. Likewise, the radiation yield of 700 keV electrons (only slightly larger than the 695 keV average beta energy of P-32) is 0.00232. From the mass energy absorption coefficient table on the NIST website for soft tissue, we obtain the following values for (μ_{en}/ρ)

Energy (MeV)	μ_{en}/ρ (cm²/g
6.00E-01	3.25E-02
8.00E-01	3.18E-02

A bit of linear interpolation shows that at 0.695 MeV, the mass energy absorption coefficient should have a value of 0.0322 cm^2/g.

Thus, we have the following:

1. The primary beta shield should be a thickness of at least 0.8715 g/cm^2 of acrylic.
2. After that shield is in place, the energy fluence of bremsstrahlung will be:

$$\psi = \frac{AE_{\beta-(avg)}f}{4\pi r^2}$$

Where we assume that the receptor is located at the outside edge of the acrylic shield at a distance of 0.7324 cm, since the density of acrylic is 1.19 g/cm^3.

$$\psi = \frac{3.7 \times 10^8 \text{ decays/sec})(0.695 \text{ MeV/decay})(0.00232)}{4\pi(0.7324\ cm)^2}$$

$$=8.85\times10^4 \text{ MeV/cm}^2\text{/sec}$$

3. The absorbed dose rate is thus

$$D = \psi(\frac{\mu_{en}}{\rho})$$

$$=(8.85\times10^4 \text{ MeV/cm}^2\text{/sec})(0.0322 \text{ cm}^2\text{/g})$$
$$=2.85\times10^3 \text{ MeV/g/sec}$$
$$=1.64\times10^{-3} \text{ Gy/hour}$$

Obviously, the photons must be shielded using techniques described in the previous section while accounting for buildup. Either hand calculations or point-kernel software can be used to estimate the requisite photon shield thickness to arrive at a total shield package design. As mentioned earlier, the RadPro Calculator software can be used for point sources and includes bremsstrahlung photon production and shielding calculations.

In addition, the Monte Carlo programs discussed earlier for photon shielding can be used to design a total shield package for the beta particles and bremsstrahlung photons and arrive at a more robust solution since the Monte Carlo programs can simulate transport of both electrons and photons.

6.3.3 Shielding for Neutrons

The general design of shielding for neutrons follows some basic principles:

1. Fast neutrons must be thermalized, typically by using a material with a low atomic number so that fewer collisions are necessary to slow the neutrons. Typically, this is achieved by using materials rich in hydrogen, such as water, wax, concrete, or polymers.
2. Thermal neutrons must be absorbed in the shield, usually by an absorption reaction.
3. Any secondary radiation, usually photons, that are produced following absorption must be sufficiently shielded.

These principles are stylized below in Figure 6-10, which shows that a combination of more than one material is typically needed for neutron shielding. The capture gamma rays in the figure originate from the $^{1}H(n,\gamma)^{2}H$ reaction discussed earlier in our text, which results in the production of 2.2 MeV gamma rays that are very penetrating. To reduce the production rate of these photons, a neutron-absorbing material such as ^{10}B can be incorporated into the hydrogen-rich material, possibly using something as simple as boric acid. The boron has a much higher cross section for absorption than the $^{1}H(n,\gamma)^{2}H$ reaction and results in an alpha particle being released in an (n,α) reaction that is easily stopped by the neutron shield. Thus, the need for gamma shielding can be reduced greatly.

Shielding calculations for neutron sources are best done in the modern era by using Monte Carlo computer codes as discussed earlier. While the general principles of neutron shielding can help in choosing material types and arrangements, the necessary thicknesses of shielding material cannot be predicted readily by hand calculations without the possibility of extensive error; thus, computer codes provide the best tool for neutron shielding analysis.

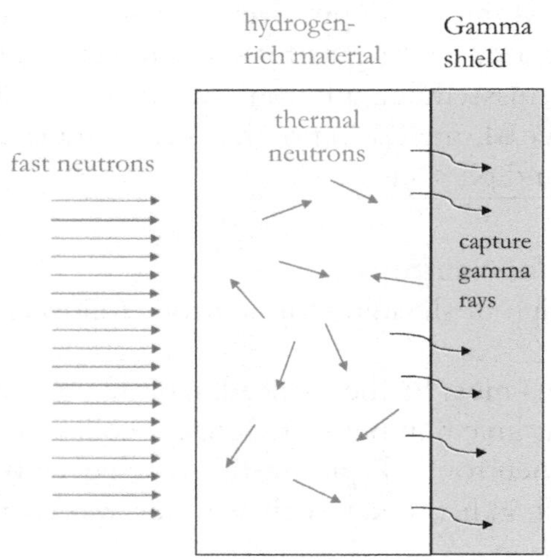

Figure 6-10. Conceptual principles of shielding neutrons, which results in the production of a secondary photon field that must also be shielded.

6.3.4 Some Final Notes on Shielding

While we will not cover these points in detail, there are other practical points to consider when designing shielding, particularly in the event of shielding large radiation sources or if the shield is room-sized.

1. Physical security of the source and shield must be maintained. If a radionuclide source is stored within a shield, then a locking mechanism (as simple as a padlock) should be used to secure the source so that personnel cannot be accidentally exposed. An example of a simple lock is shown in Figure 6-11. In addition, the source itself should be in a locked area that is not generally accessible.

Figure 6-11. Example of a simple lock mechanism on an isotopic calibrator source to prevent accidental exposure to personnel.

2. Penetrations through the shielding must be considered in the design. For small shields, the penetration could be for a handle to carry the source, or for hinges, or other material structure. For room-sized shielding, the penetration could be for utilities (electrical wiring, water, air conditioning) or doors/windows. These void spaces must be addressed to ensure that the shielding effectiveness is not compromised. A maze concept is often used for penetrations as shown below in Figure 6-12 where one or more additional walls are used to prevent a direct line of sight for radiation to penetrate through a door. Also, right angle or diagonal penetrations through walls can be used to reduce streaming of radiation through penetrations into the room for utilities.

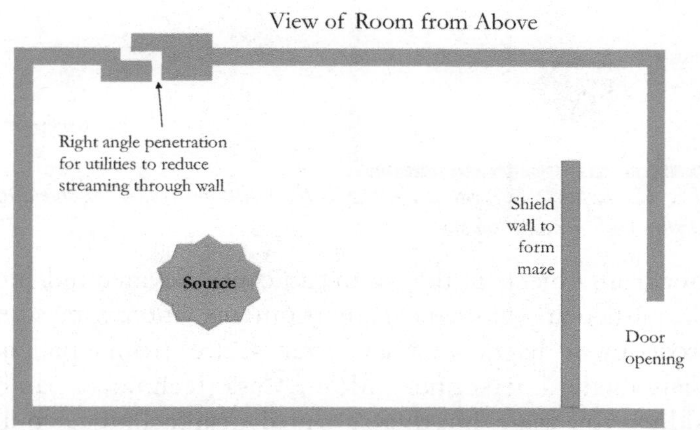

Figure 6-12. The use of maze entrances and offset penetrations to reduce undesired streaming of radiation from a room.

3. While a lot of emphasis is placed on direct shielding between the source and the receptor, it can be easy to overlook other shielding needs. One example of this is the phenomenon of skyshine. Often, when designing a room or building sized shield, the walls of the facility are designed to reduce dose rates to personnel outside the walls. However, if the ceiling/roof of the facility is left with too small a thickness for shielding purposes, radiation can penetrate through the roof, scatter off the atmosphere, and irradiate people at some distance from the facility as shown in Figure 6-13. In some cases, it may be possible to shield the roof with the same amount as the walls; however, this may not always be possible due to structural loading issues. Monte Carlo simulation can help with analysis of such scenarios. In addition, Grove Software has a point-kernel based package called MicroSkyshine® to address this unique shielding situation for gamma shielding. As with MicroShield®, academic institutions can inquire at www.radiationsoftware.com regarding making this software available for student use.

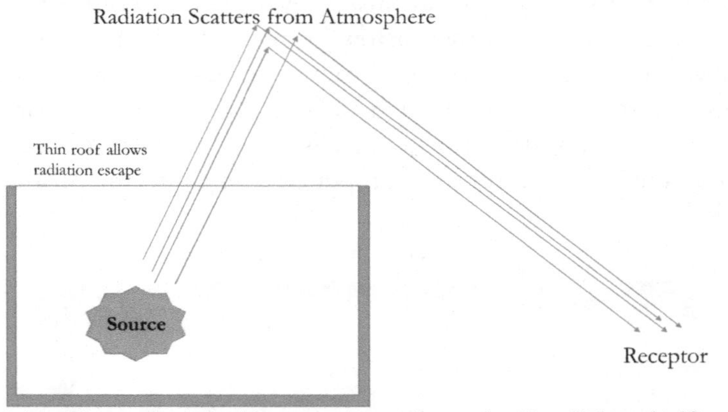

Figure 6-13. Example of skyshine occurring as the result of insufficient shielding in the roof of a facility with a radiation source.

4. Scatter from objects in the room can cause elevated radiation levels, particularly in medical environments where x-rays used for fluoroscopy or diagnostic images may scatter from a patient and irradiate medical personnel. Monte Carlo techniques can be used to address this issue, but there are analytical techniques published in documents such as NCRP Report 147 for medical facilities.

Free Resources for Chapter 6

- Questions and problems for this chapter can be found at www.hpfundamentals.com
- The RadPro Calculator is a free online resource for calculating a wide range of external dose quantities (including shielding with buildup) and is available at http://www.radprocalculator.com/
- A detailed set of point-kernel buildup factors for a variety of materials can be downloaded freely from https://technicalreports.ornl.gov/1988/3445605718328.pdf

Additional References

- Chilton, A.B., Shultis, J.K. and Faw, R.E. Principles of Radiation Shielding. Prentice Hall, Englewood Cliffs, NJ 1984.
- NCRP Report 147, *Structural Shielding Design for Medical X-Ray Imaging Facilities (2004),* available for purchase through ncrponline.com.
- Shultis, J.K. and Faw, R.E Radiation Shielding, American Nuclear Society, La Grange Park IL, 2000.

Chapter 7 Dosimetry Models for Internal Sources of Radiation

The previous three chapters dealt with the issue of calculating and measuring dose for radiation sources that are outside the body. However, in many facilities, a worker may be in proximity to radioactivity that could be inhaled or ingested such as:

- airborne radioactivity suspended in the air
- gaseous radionuclides
- contamination on a surface or an object that the person touches or that could be resuspended into the air

We must thus consider many factors that pertain not just to the radioactive material itself but also to how that material moves in the body and how tissues/organs will be irradiated.

This chapter discusses the methodology developed by the ICRP using data gained over decades of research and measurements. In particular, this chapter will focus on how the ICRP develops models based on the starting assumption that a worker has had an intake of a unit amount of activity (i.e., 1 Bq). An intake is defined as the amount of radioactivity that crosses one of the body's boundaries (i.e., skin, nose, or mouth).

Since the dose will scale proportionately with the activity in the intake (i.e., more activity in the intake directly causes a proportionate increase in dose), this allows the calculation of doses in advance to create a database of dose coefficients (dose conversion factors) so that in the workplace, we can use the results of those calculations to estimate the dose for a given intake: that is, we estimate the intake and multiply by a dose coefficient that has been previously determined).

Chapter 8 will focus on how to use the ICRP dose coefficients, coupled with measurements from the workers and workplace, to estimate the dose in a given scenario. While it is possible to use the results of the ICRP dose calculations without having a thorough understanding of all the nuances in the models, it is wise and judicious to be familiar with the main assumptions of the models so that proper interpretation of measurements and the results of the dose calculations can be made. Thus, this current chapter will discuss the models at a high level to give an overview of the main assumptions and their impact on dose assessment.

Most of the discussion in these chapters focuses on publications of the ICRP that are available as free downloads from www.icrp.org. In preparation for studying these chapters, these publications should be downloaded and referred to for additional details beyond what is practical to include in this textbook.

- ICRP Publication 130 – Occupational Intakes of Radionuclides: Part 1 (General discussion of assumptions and models) https://www.icrp.org/publication.asp?id=ICRP%20Publication%20130
- ICRP Publication 134 – Occupational Intakes of Radionuclides: Part 2 (chemical, metabolic, and dosimetric data on selected elements) https://www.icrp.org/publication.asp?id=ICRP%20Publication%20134
- ICRP Publication 137 – Occupational Intakes of Radionuclides: Part 3 (chemical, metabolic, and dosimetric data on selected elements) https://www.icrp.org/publication.asp?id=ICRP%20Publication%20137
- ICRP Publication 141 – Occupational Intakes of Radionuclides: Part 4 (chemical, metabolic, and dosimetric data on selected elements) https://www.icrp.org/publication.asp?id=ICRP%20Publication%20141

At the time of the writing of this book, the following ICRP Publication is available for purchase but is not free for download.

- ICRP Publication 151 – Occupational Intakes of Radionuclides Part 5 (chemical, metabolic, and dosimetric data on selected elements)

However, the numerical data from Publication 151 is available in the Electronic Annex that is discussed in Section 7.4 of this chapter.

There are a couple of general ideas that we can use as a starting point for internal dose assessment. First, there are four possible modes for how radioactive material can enter the body:

- Inhalation
- Ingestion
- Absorption through skin
- Injection/wounds (including intentional injection in nuclear medicine as well as wounds)

Based on these modes for entry, we can recognize that there will need to be the ability to model mathematically how material enters the body. Once it enters the body, though, we will also have to model how it moves around. Obviously, not every element on the periodic table has the same chemistry, and we should not expect that all radioactive material moves identically throughout the human body. Generally, radioisotopes behave chemically like the nonradioactive isotopes of the same element. For example, Na-24 and Na-22, though radioactive, behave chemically like stable Na-23. Since the chemistry of each element is expected to be different, though, this presents a conundrum for assessing the dose, and a new concept is needed that is explained in the next section.

7.1 Committed Dose

One major difference between dosimetry for external sources and dosimetry for internal sources is what occurs after a person is removed from the location where the radioactivity is present. An example of this is illustrated below in Figure 7-1. For an external source of radiation where the radioactivity is contained, removing the person from the location with the radiation source removes the fluence and the person receives no further radiation dose.

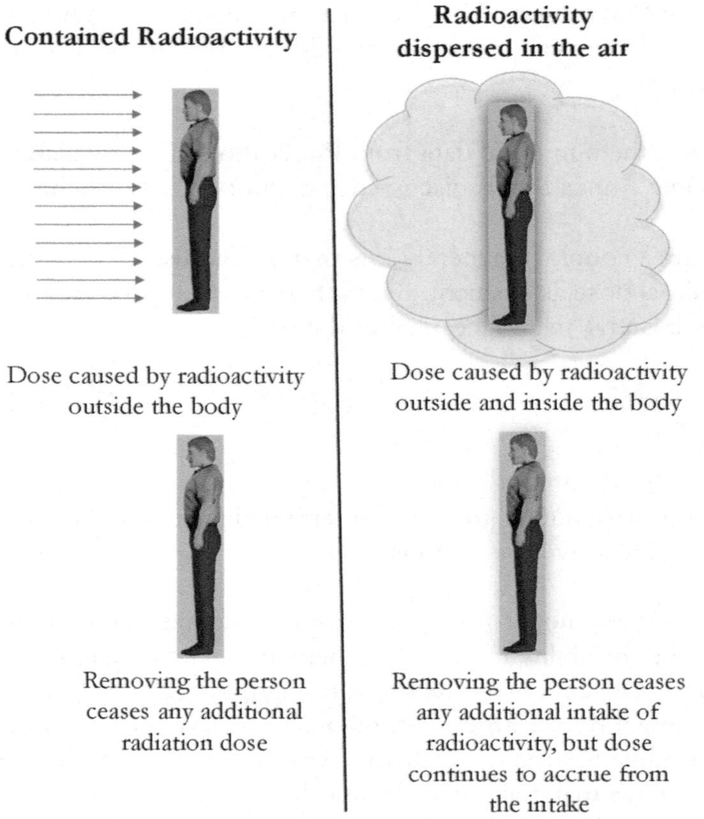

Contained Radioactivity

Radioactivity dispersed in the air

Dose caused by radioactivity outside the body

Dose caused by radioactivity outside and inside the body

Removing the person ceases any additional radiation dose

Removing the person ceases any additional intake of radioactivity, but dose continues to accrue from the intake

Figure 7-1. Contrast between radiation dose caused by external contained radioactivity and dispersed radioactivity that can enter the body.

In contrast, when a person is in a room or location where radioactivity can enter the body, removing the person from that location may stop additional radioactivity from entering the body but does not remove the radioactivity that has already entered the body. The presence of radioactivity in the body means that dose will continue to accrue even after the person physically leaves the area, leaves the work location, or even returns home. Once the radioactive material is in the person, a dose is committed to be received until the material either decays radiologically or is removed from the body. We thus introduce the concept of committed dose for purposes of internal dose assessment. Organs and tissues of the body will receive a committed equivalent dose; similarly, we can calculate a committed effective dose.

As might be expected, radioactive decay removes activity exponentially from the body following the pattern of Figure 1-1 earlier in this book; biological removal can follow a similar mathematical pattern (i.e., first-order kinetics). However, because of the mathematics involved, the dose to the person theoretically continues to infinity, although the dose rate decreases with time for most radionuclides. How quickly the dose rate declines depends on the radionuclide, its chemical form, its physical form (and particle size if inhaled), and biological parameters.

Figure 7-2 below shows how the relative dose rate from two different radionuclides (Cs-137 and Co-60) might decline over time based on the amount of activity expected to be present in the body. As shown, the dose rate from Co-60 declines drastically shortly after the material enters the body while the dose rate from Cs-137 remains relatively constant for the first 100 days or so. Eventually, though, the Cs-137 dose rate decreases to be less than that due to Co-60.

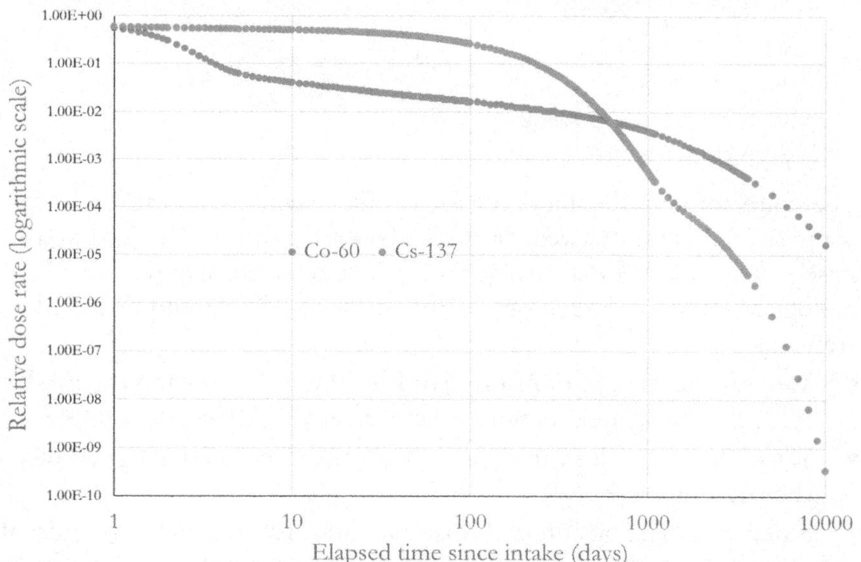

Figure 7-2. Relative dose rates from Cs-137 and Co-60 at times following an intake.

This raises the reasonable question of how the dose will be assessed. If the dose rate is changing over time, and the length of time that the material is in the body potentially is different for every single scenario, how does the health physicist assess and report the dose?

To provide some basis for comparison of intakes and dose consequences, it is logical that we assess the dose from an intake for a fixed time interval that is the same regardless of radionuclide. By historical convention and agreement in the dosimetry community, the dose from an intake of radioactivity is assessed over a 50-year period. This committed dose is then assigned for recordkeeping to the calendar year of intake. Thus, an intake that occurs today is assessed to calculate the total dose that accrues over the next 50 years, and that dose is assigned to the dose record for the current calendar year as shown below. Anecdotal evidence suggests that the rationale for a 50-year period was based on an expected lifespan of roughly 70 years and assuming a person could become a radiation worker at 20 years of age. Obviously, this is not the case for every person, but the 50-year assessment period has persisted, nonetheless.

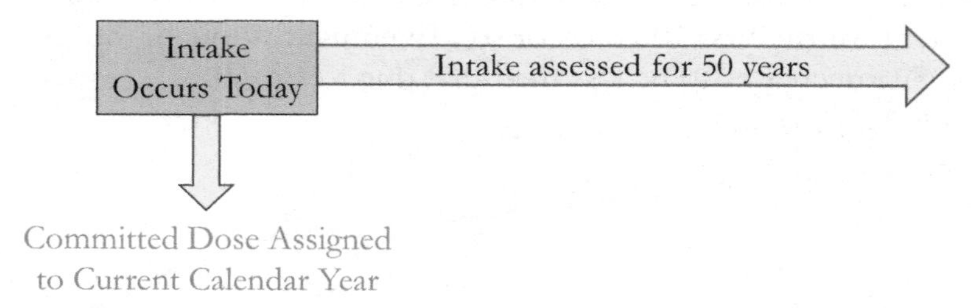

There are some understandable logical flaws with assigning a 50-year committed dose to the year of intake based on our understanding of dose assessment for external sources, dose quantities, and the biological effects of radiation. Some questions that could be asked are:

- What if the person does not live for 50 years following the intake (e.g., the intake occurs when the person is in their 50s or 60s)?
- Why is a future dose that has not yet been received assigned to the present year?
- Since external and internal doses are additive, why add an internal dose that <u>has not</u> yet been received to an external dose that <u>has</u> been received in the current year?
- What if the person's biological processes change over the course of 50 years and they remove either more or less of the activity than expected?
- Is there a lot of error with this sort of approach?

To address these questions and any others, there are some fundamental principles that must be remembered:

1. We are not calculating the dose to a real person. We will estimate the dose to a model that we call Reference Person. The dose to the actual worker may vary from what is calculated.
2. The bookkeeping of radiation doses assigned to people must be workable and practical. While it may not be scientifically correct to assign a committed dose to the year of intake, it simplifies recordkeeping while still remaining conservative. Otherwise, a new annual dose due to the intake would have to be assessed each year for the rest of the person's life. That would become the responsibility of each future employer of that person and would become unduly burdensome.
3. The fundamental goal is to provide adequate protection for the workers. The use of committed equivalent dose and committed effective dose and assigning these doses to the year of intake is conservative and helps workers and employers make decisions on future work activities that could cause further dose.
4. Because the health physics/radiation protection community has agreed on this approach, it promotes consistency in the dose estimation process, regardless of facility. While the exact dose estimate from various facilities may not be identical, they should be similar and able to be compared.

The ICRP defines the committed equivalent dose in a relatively straightforward equation as shown in Equation 7.1 below.

$$H_T(50) = \int_0^{50\,yrs} \dot{H}(r_T, t)dt \qquad (7.1)$$

Where

- $H_T(50)$ is the 50-year committed dose equivalent
- $\dot{H}(r_T, t)$ is the equivalent dose rate to the particular tissue T of interest at time t. The means of calculating this term will be discussed as we proceed through this chapter.

Equation 7.2 below shows how the equivalent dose is used to calculate the committed effective dose. It may seem a little puzzling at first but is designed to consider the difference in modeling males and females in the workplace. Thus, the effective dose uses an average dose to each tissue based on separate dose calculations for male and female as shown in Equation 7.2. The average dose is then multiplied by the appropriate tissue weighting factor, and these are summed in keeping with the definition of effective dose. Equation 7.2 is consistent with Equation 1.23 in Chapter 1, because the value of H_T in Equation 1.23 is intended to be the sex-averaged dose for the Reference Male and Reference Female.

$$E_{50} = \sum_T w_T \left[\frac{H_T^M(50) + H_T^F(50)}{2} \right] \quad (7.2)$$

Where

- E_{50} is the 50-year committed effective dose
- w_T is the appropriate tissue weighting factor for the target tissue r_T
- $H_T^M(50)$ is the committed equivalent dose to tissue T based on the male model
- $H_T^F(50)$ is the committed equivalent dose to Tissue T based on the female model

Obviously, Equations 7.1 and 7.2 will become more complex as it applies to internal dose calculations, but they form the starting point for how dose assessments are performed.

7.2 Summary of Models from ICRP Publication 130

As might be expected, it is not a trivial feat to develop tools for how to model the human body and how to assess the dose once an intake has occurred. Logically, the committed equivalent dose in any target region (r_T) of the body from radioactivity in a source region (r_S) of the body is directly proportional to:

- The number of decays that occur in the source region over the 50-year commitment (the integrated activity that was discussed in Chapter 1)
- The equivalent dose in the target region due to each decay that occurs in the source region.

The ICRP has developed mathematical models to determine both of these parameters, and this section will break down the models into three main areas:

- Intake models – how radioactive material enters the body and moves from the point of entry to the bloodstream.
- Systemic models – how the intake moves from the bloodstream to other organs of the body, including excretion.
- Dosimetric models – the anatomical structure of the body and the physics of how to calculate the energy deposition to the various tissues once the location(s) of the radioactive material is known.

The intake models and systemic models will help determine the number of decays that occur in each source region over the 50-year period while the dosimetric models will then help determine the absorbed dose and equivalent dose in each target region per decay in the source region.

Before digging into specifics, Figure 7-3 below shows an overview of the ICRP intake and systemic models and how they relate to each other.

Radioactive material enters the body through one of the four routes listed earlier; once in the body, mathematical models are used to predict the amount and time-dependent rate of movement between the various compartments. Obviously, Figure 7-3 does not address every single organ and tissue of the body. In particular, the box labeled "Other organs" must of necessity incorporate much of the body. However, the figure gives a general impression of the task ahead in developing the specific models.

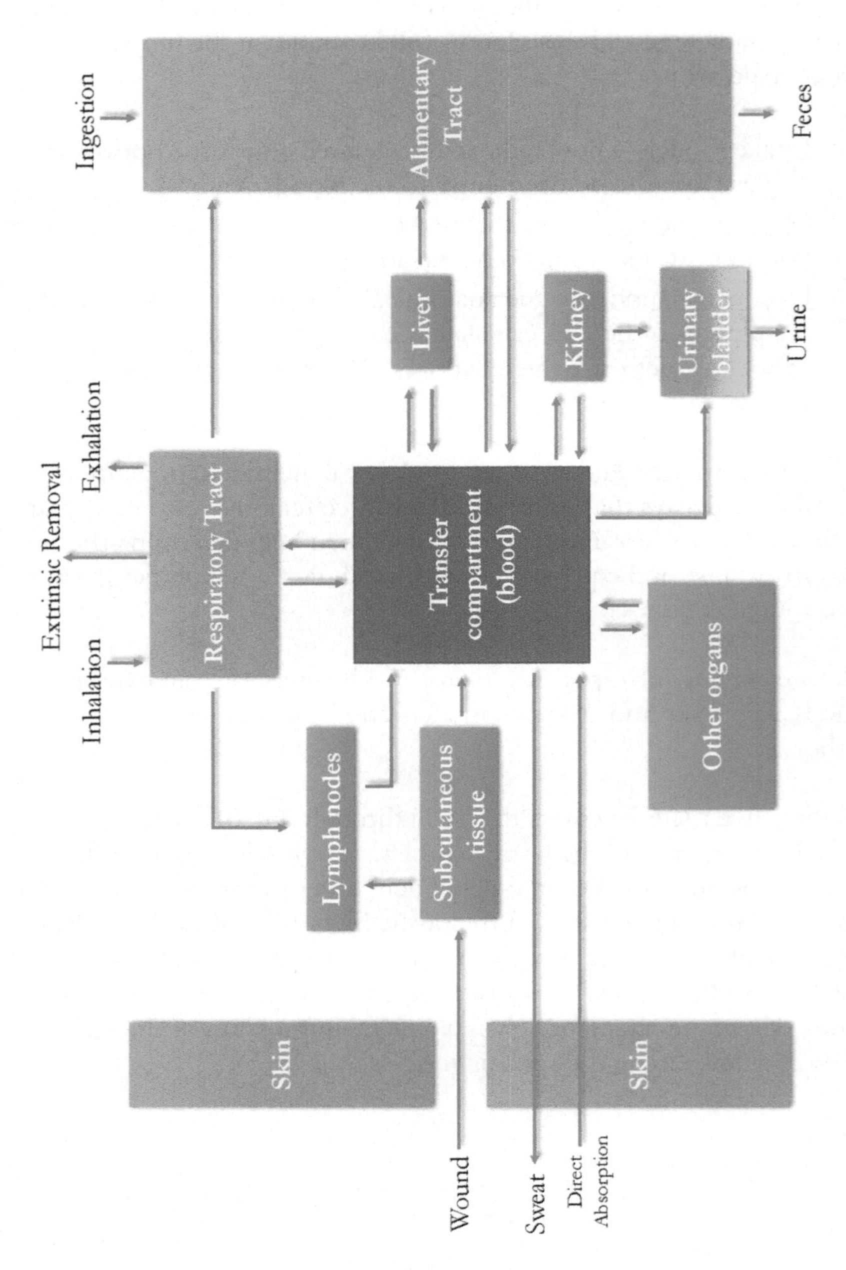

Figure 7-3. Schematic of modes of intake and movement of radioactive material in the body as used in ICRP Publication 130.

There are some general assumptions that go into the intake and systemic models that should be understood at the outset:

1. First order kinetics is assumed to govern the movement of material, even though the human body is much more complex than this would suggest. This means that first order linear differential equations are used as the basis for the time-dependent movement.
2. As part of the kinetics, partitioning factors and biological transfer coefficients are assumed as shown below in Figure 7-4. For illustrative simplicity, movement in only one direction is shown.

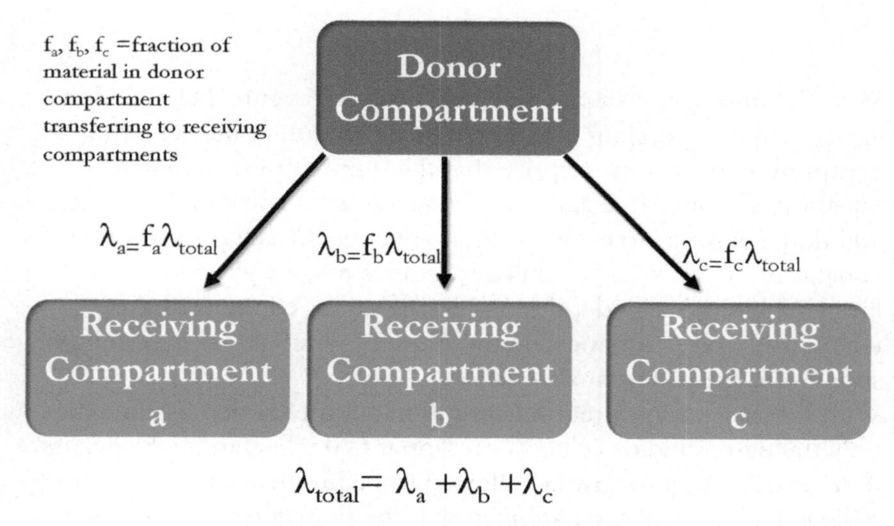

Figure 7-4. Simplified illustration of first order kinetics with no recycling and with partial biological transfer coefficients.

Because the material may have multiple destinations from any given donor compartment, the partitioning factor (represented as f_a, f_b, and f_c in the figure) gives the fraction of material in a particular donor compartment that is assumed to go to the next receiving compartment along its route. Much like partial decay constants in radioactive decay, partial biological transfer coefficients for each compartment can be calculated based on the partitioning factor and the total biological transfer coefficient as shown in Figure 7-4. A biological half-life can also be calculated for each transfer coefficient and is the time required for half the material to be removed from a compartment assuming only biological elimination as shown below in Equation 7.3.

$$T_{bio} = \frac{\ln(2)}{\lambda_{bio}} \tag{7.3}$$

Since the material will be decaying radiologically at the same time as it physically moves between compartments, this means that there is an effective decay constant that encompasses both biological removal and radioactive decay as shown in Equation 7.4 along with an effective half-life.

$$\lambda_{eff} = \lambda_{rad} + \lambda_{bio}$$

$$T_{eff} = \frac{\ln(2)}{\lambda_{eff}} \tag{7.4}$$

With the number of steps between intake and eventual elimination, along with the recycling of material between donor and receiving compartments, it is no surprise that the number of differential equations is quite large and would be tedious to solve on paper. In addition, solutions to each equation are coupled as inputs to other equations. Thus, while overall generalities may be gleaned by knowing biological and radiological half-times, computer-based techniques using numerical methods are required to get a meaningful solution to the differential equations.

3. With the exception of radon isotopes in air, the models assume that only the parent radionuclide is present at t=0 when an intake occurs. After intake, the progeny are allowed to be produced as the result of radioactive decay of the parent so that the final dose includes the dose contribution from ingrowth of radioactive progeny from the original parent in addition to the parent radionuclide. In the case of radon isotopes, the progeny that are present at the time of intake are treated as intakes simultaneous with the radon.

With all of this as an introduction, we are ready to delve into more details of the various models.

7.2.1 ICRP Intake Models

Because all the dose quantities will scale with the amount of activity that serves as the intake, the ICRP takes the approach of assuming a 1 Bq intake and then calculating the dose based on that unit activity. This makes tabulating dose coefficients possible so that in the workplace, the dose for a given scenario can be calculated as the intake amount multiplied by a dose coefficient per unit intake.

7.2.1.1 The Respiratory Tract Model

Figure 7-5 illustrates the human respiratory tract model adopted by the ICRP and updated in Publication 130. The respiratory tract is divided into two main tissue groups: the extrathoracic (ET) and the thoracic (TH) airways. These main tissues are then subdivided further as outlined in Table 7-1 below.

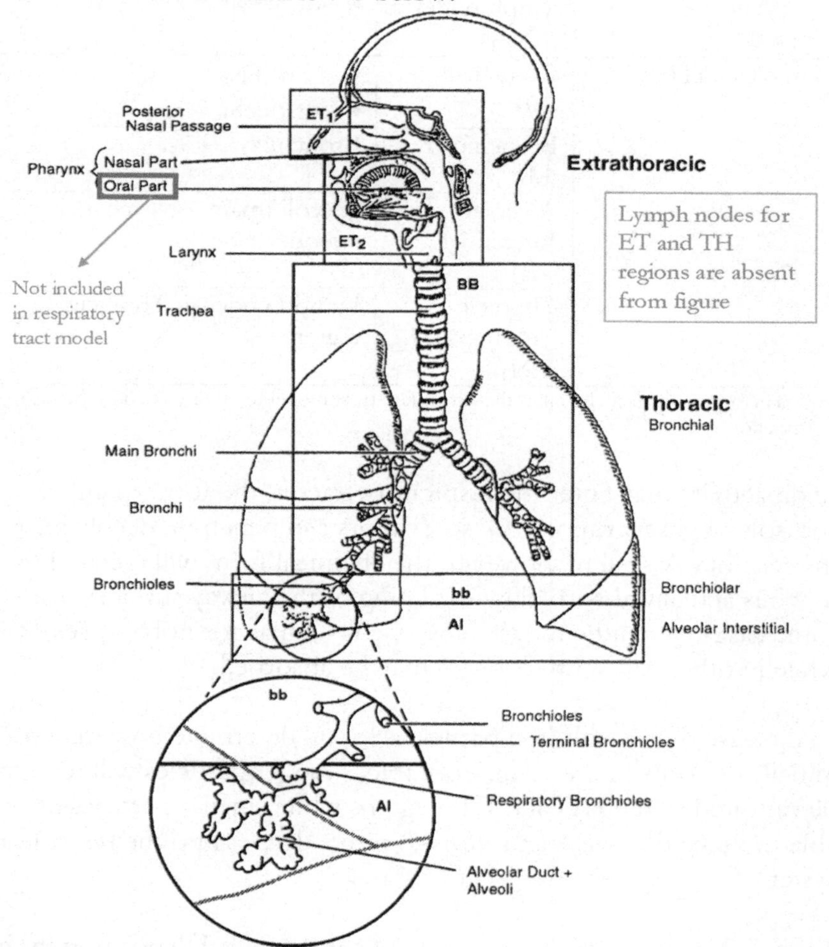

Figure 7-5. The ICRP human respiratory tract model as used in ICRP Publication 130 (Image courtesy of ICRP).

Table 7-1. Main tissue regions of the respiratory tract model with the included respiratory regions.

Main Tissue	Region	Included airways
Extrathoracic (ET)	ET_1	Anterior nasal passages
	ET_2	• Posterior nasal passage • Nasal part of pharynx (oral part[a] of pharynx is <u>not</u> included) • Larynx
	Extrathoracic lymph nodes (LN_{ET})	Lymph nodes in extrathoracic region
Thoracic (TH)	Bronchial (BB)	• Trachea • Bronchi
	Bronchiolar (bb)	Bronchioles
	Alveolar-Interstitial (AI)	Alveoli (main gas exchange region)
	Thoracic lymph nodes (LN_{TH})	Lymph nodes in Thoracic region

a. The oral cavity is included in the alimentary tract model to be discussed in the next section.

Radioactivity may enter the respiratory tract in the form of either aerosols or gases/vapors. Gases/vapors can penetrate deeply into the respiratory system; however, the chemical form will dictate how much is actually absorbed by the lining of the airway surfaces. In some cases, very little may be absorbed (such as for noble gases) while in other cases, up to 100% may be absorbed.

In the case of aerosols, the particle size will determine how material initially deposits in the respiratory tract. Larger particles will tend to be captured in the ET_1 and ET_2 regions while smaller particles are able to make the twists and turns to move deeper into the respiratory system.

Table 7-2 below shows an example of the modelled deposition in the various regions from an inhalation of 5 μm AMAD (activity median aerodynamic diameter) particles. These are not particles with a single size but represent a distribution of particles where 50% of the activity is attached to particles whose settling velocity is greater than a 5 μm sphere and 50% of the activity is attached to particles whose settling velocity is less than a 5 μm sphere.

Table 7-2. Deposition percentages for inhaled aerosols of 5 μm AMAD in the ICRP Publication 130 respiratory tract model.

Region	Deposition percentage (%)[a]
ET_1	47.94
ET_2	25.82
BB	1.78
bb	1.10
AI	5.32
Total deposited	81.96
Amount exhaled	18.04

a. These deposition fractions are for a worker who is assumed to be breathing through the nose and mouth while working.

As shown in Table 7-2, a percentage of what is inhaled is promptly exhaled back to the environment. However, the total amount inhaled is considered to be the intake, including the material that may be exhaled. Thus, the ICRP models account for this exhaled material that does not contribute to the dose, and calculations are always carried out per unit intake.

By default, the ICRP assumes that all intakes are 5 μm intakes. If the particle size is known to be different, that will affect the deposition percentages as presented in Table 7-2 and eventually the calculated dose. Methods for accounting for this are discussed in Section 7.4 and Chapter 8.

7.2.1.1.1 ICRP Inhalation Classes

While it would be ideal to know the exact rate at which every element and chemical form clears from the respiratory tract (thus affecting the integrated activity in the respiratory system), insufficient information is available for that level of complexity. Thus, the ICRP has designated four general inhalation classes that can be applied to a given intake as follows:

- Type S (Slow): intakes in this class are removed slowly from the respiratory tract and are relatively insoluble in the respiratory system (roughly 30% of material deposited in the AI region eventually reaches the blood)
- Type M (Moderate): intakes in this class are moderately soluble in the respiratory system (roughly 80% of material deposited in the AI region eventually reaches the blood)
- Type F (Fast): intakes in this class are readily soluble in the respiratory system (essentially all material deposited in the AI region reaches the blood in a short period of time)
- Type V (Very Fast): intakes in this class are assumed to be instantly and totally absorbed into the bloodstream from the respiratory system. These would typically be certain gases/vapors; one of the most common examples is tritiated water (HTO).

These four types are general groupings based on the chemical form of the radionuclide and are intended to be applied consistently regardless of the element. For example, all Class F materials, regardless of the element or chemical form, are assumed to move identically in the respiratory tract based solely on chemistry. However, the ICRP accounts for radioactive decay in transit that causes reduction of the material that is available to move in the various regions in the final dose calculations.

For each element, the ICRP has designated which chemical forms should be assigned to a given inhalation class; these are included in the elemental data in Publications 134, 137, 141, and 151. The concept of inhalation class highlights a major distinction between external dosimetry and internal dosimetry. In the case of external dosimetry, the chemical form of the radionuclide is seldom of interest for performing calculations; usually, only the identity of the radionuclide is required to perform a dose calculation. In internal dosimetry, the chemical form and the particle size must be known or assumed in order to perform a meaningful dose evaluation for inhalation intakes. This requires knowledge of workplace conditions that may not always be known in advance of an intake and requires informed professional judgment.

As an example, Table 7-3 presents the inhalation class assignments for the chemical forms of cobalt as outlined in ICRP Publication 134.

Table 7-3. Example of inhalation classes assigned to chemical compounds of cobalt from ICRP 134.

Chemical Form	Inhalation class
Cobalt nitrate, cobalt chloride	F
Unspecified, unknown	M
Cobalt oxide, fused aluminosilicate particles, polystyrene	S

If the chemical form of the cobalt is known, then Table 7-3 shows the assigned inhalation class; however, in some cases the chemical form may not be known. In that case, the ICRP usually has a default assumption (Class M in the case of cobalt) of inhalation class as a starting point for the calculations.

7.2.1.1.2 Movement from the Respiratory Tract

Once material is in the respiratory tract, it has a few different paths in which to move, as shown earlier in Figure 7-5. In general, these options are as follows:

- A portion of material (roughly 30%) in ET_1 can be removed by what is called "extrinsic removal," (for example, blowing the nose). The remainder moves to ET_2 where the vast majority of it is swallowed along with material already deposited in ET_2. A small fraction is physically moved to the extrathoracic lymph nodes.
- Absorption - Material in the TH area can dissolve to some degree and move into the bloodstream, where it is able to move to other organs and tissues of the body (absorption)
- Particle transport - Material in the TH area can be moved physically by mucociliary action either to the lymph nodes or upwards in the respiratory system, eventually entering the esophagus and being swallowed.

The amount of material that dissolves into the bloodstream versus what is removed by the cilia understandably depends on its solubility. Therefore, in general, we would expect that Class S materials (being less soluble) would tend to have a greater fraction of the material remain in the respiratory tract longer and would be removed by the mucociliary action to the throat and swallowed, while a smaller fraction would dissolve into the bloodstream. Class F materials, on the other hand, would have a larger fraction of the material to dissolve into the bloodstream compared to the fraction that would be eventually swallowed. As stated earlier, Class V materials are assumed to dissolve instantly to the bloodstream and thus have no component that is eventually swallowed. Hence, the inhalation class is key to predicting the movement of material from the respiratory tract.

As discussed in the introduction of this section, the movement of material within the respiratory tract and from the respiratory tract to other locations is modeled by first order kinetics with biological half-times assigned to each compartment and subcompartment within the model. The details of the exact values for the various inhalation classes are beyond the scope of this text but can be found in the ICRP publications listed earlier in this chapter.

The further movement of materials that are swallowed is covered in Section 7.2.1.2 while the further movement of materials in the bloodstream is covered in Section 7.2.2.

7.2.1.1.3 Radioactive Progeny in the Respiratory Tract Model

In the respiratory tract model, the ICRP recognized that some radionuclides would decay to produce progeny while in the tract. The ICRP assumes in Publication 130 that the progeny formed in the tract will have shared kinetics with the parent radionuclide: that is, the progeny will have the same absorption parameters as the parent to reach the bloodstream. While this may seem counterintuitive, the ICRP performed calculations that indicated that treating the progeny with element-specific absorption parameters did not greatly affect the final dose results and instead added complexity that was not warranted. The one exception to this assumption is the case of noble gases, where clearly the noble gas would not be readily absorbed into the bloodstream. Once radionuclides leave the respiratory tract and enter the bloodstream, the systemic model takes over; the assumptions on radioactive progeny in that model are discussed in Section 7.2.2.

7.2.1.2 The Alimentary Tract (Digestive Tract) Model

Radioactive material can enter the alimentary tract via ingestion either by consuming liquids or solids that contain radioactive material or by swallowing material that was initially inhaled that has worked its way into the esophagus as discussed in the previous section on the respiratory tract model.

Figure 7-6 presents a schematic illustration of the ICRP alimentary tract model. As noted earlier, the oral cavity (mouth) is not included in the respiratory system model and is modeled in the alimentary tract instead. Material that is ingested via food or liquids travels from the mouth subsequently through the downward path illustrated in the figure until some amount is excreted in the feces. Material from the respiratory tract enters below the mouth at the esophagus but follows the same downward path.

As the contents of the tract move through each section, radioactive material already present in the bloodstream and in secretory organs/glands can be injected into various parts of the tract due to the normal digestion process. In the oral cavity, material can be partly absorbed into the teeth and oral mucosa. In the rest of the tract, the radioactive material in the contents of each section can absorb into the wall of that section and into the bloodstream/transfer compartment. Although Figure 7-6 includes arrows that indicate possible radionuclide movement from the walls back into the contents, the ICRP states in Publication 130 that the default assumption is that there is no recycling from the wall back to the contents of the alimentary tract.

One item to note is the presence of two virtual compartments from the esophagus to the stomach. While the esophagus is not physically divided into two pathways to the stomach, it is understandable that the swallowing of material does not happen all at once. Rather, some material quickly moves to the stomach while multiple swallowing actions may be required to remove the remainder of the material.

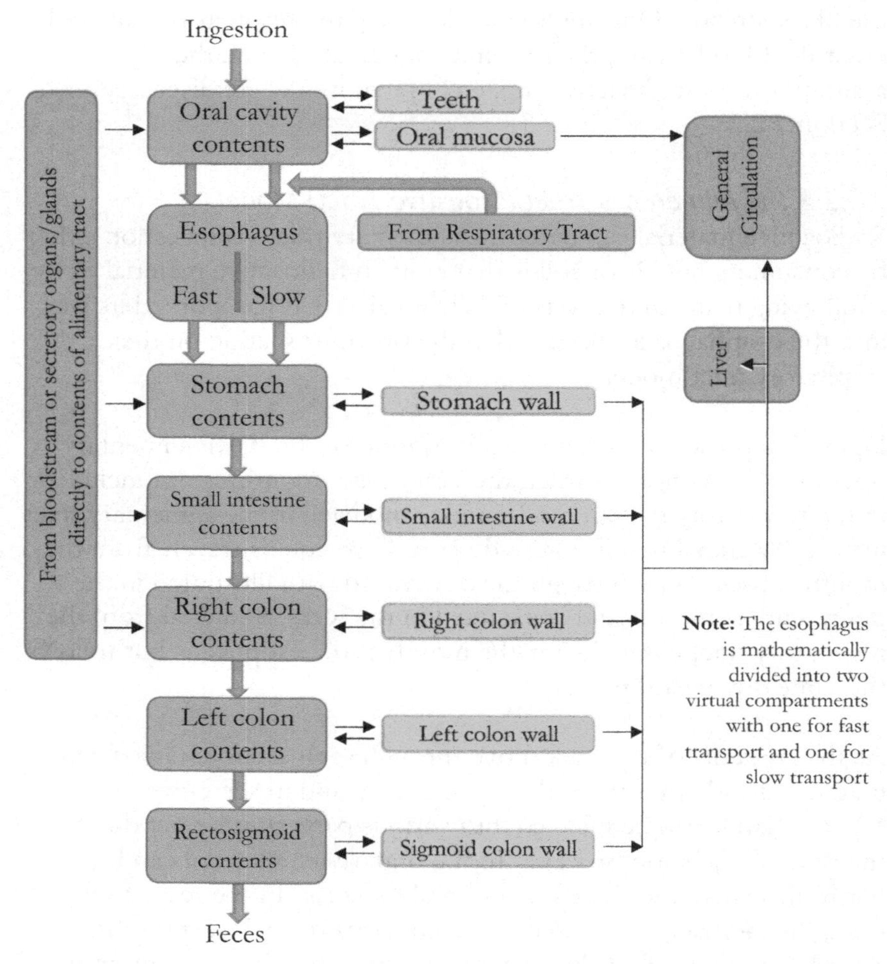

Figure 7-6. Schematic view of the alimentary tract model for the Reference Worker as used in ICRP Publication 130. The relationships to the respiratory tract model and the systemic models are also shown in the figure even though they are not officially part of the alimentary tract model.

7.2.1.2.1 Movement from the Alimentary Tract

Inspection of Figure 7-6 gives a rough indication of the number of first order differential equations to be derived and solved. For each movement indicated by an arrow above, a biological half-time and an estimate of the fraction of intake absorbed into the bloodstream is needed. Table 7-4 below shows the biological transfer coefficients and biological half-life associated with each compartment.

Table 7-4. Transfer coefficients from ICRP 130 and associated biological half-times for each part of the alimentary tract in the Reference Worker.

From	To	ICRP 130 transfer coefficient $\lambda_{from,to}$ (d^{-1})	Approximate biological half-time (minutes)
Oral cavity	Esophagus (fast transport)	6480	0.154
	Esophagus (slow transport)	720	1.4
Esophagus (fast transport)	Stomach contents	12,343	0.081
Esophagus (slow transport)	Stomach contents	2160	0.46
Stomach contents	Small intestine contents	20.57	48.5
Small intestine contents	Right colon contents	6	166
Right colon contents	Left colon contents	2	500
Left colon contents	Rectosigmoid contents	2	500
Rectosigmoid contents	Feces	2	500

As might be surmised, the fraction of material absorbed into the bloodstream/transfer compartment is affected by the solubility of the radionuclide. Radionuclides with higher solubility are more likely to be absorbed into the bloodstream while less soluble radionuclides are more likely to pass through the alimentary tract and be excreted via the feces.

The ICRP has defined the total absorption f_A as the fraction of the amount entering the alimentary tract that is absorbed. These are defined for each element and for the chemical form within the element. More soluble compounds will have higher values of f_A while less soluble compounds will have lower values.

ICRP 130 assumes that for most cases, the transfer of material from the alimentary tract to the bloodstream occurs via the small intestine contents. The biological transfer coefficient for this process is derived based on the following equations so that the final equation below shows how to calculate the transfer coefficient when the value of f_A is known and using the values of λ from Table 7-4.

$$\lambda_{SI,total} = \lambda_{SI,blood} + \lambda_{SI,RightColon}$$

$$\lambda_{SI,blood} = f_A \lambda_{SI,total}$$

$$\lambda_{SI,RightColon} = (1 - f_A)\lambda_{SI,total}$$

$$\lambda_{SI,blood} = \frac{f_A \lambda_{SI,RightColon}}{(1 - f_A)}$$

Logically, with larger values of f_A, the transfer coefficient to the blood will also be larger, indicating a shorter biological half-life for the transfer. One complicating factor is that the f_A value will also depend on whether the material was directly ingested via food/liquid or if it entered the alimentary tract via the respiratory tract. However, the ICRP has accounted for these differences in its calculations, and the data tables in Publications 134, 137, 141, and 151 for each element include the differing f_A values for these routes of entry.

Table 7-5 presents some values of f_A for several elements assuming that the material was consumed in food/liquids.

Table 7-5. Designated values of f_A for several elements based on information in ICRP Publications 134, 137, 141, and 151 after consuming materials with the element.

Element	Chemical Form	f_A
Iron	All forms	0.1
Strontium	Strontium titanate	0.01
	All other chemical forms	0.25
Iodine	All forms	0.99[a]
Cesium	Relatively insoluble forms (e.g., fuel fragments)	0.1
	All other forms	0.99[a]
Plutonium	Oxides and hydroxides	1E-5
	Nitrates, chloride, bicarbonates, all other unidentified chemical forms	5E-4

a. The ICRP designates a value of 1 for f_A, but 0.99 is used in practice to avoid division by zero in the equation above to calculate $\lambda_{SI,blood}$.

7.2.1.2.2 Radioactive Progeny in the Alimentary Tract Model

For the alimentary tract model, the ICRP takes a different approach than that in the respiratory tract model with regard to radioactive progeny. For the alimentary tract model, the general position is that the progeny are assigned f_A values based on their own unique elemental/chemical form and not that of the parent radionuclide. Thus, radioactive progeny have independent kinetics from the parent radionuclide in the alimentary tract model.

7.2.1.3 Intakes through intact skin

When the human skin is intact, it serves as an effective barrier to particulate forms of radionuclides; nevertheless, a few compounds such as tritiated water (HTO) are known to move through the intact skin. Historically, the ICRP included tritium absorption through the skin as an additional intake in addition to respiration (i.e., a person in an area containing airborne tritiated water was assumed to have an intake via both respiration and direct absorption through the skin). In its newest recommendations in Publication 130, the ICRP no longer takes this approach, and the model dose calculations assume no intake through the skin. However, if it is suspected that a significant intake has occurred through the skin for a particular scenario, measurements taken after the intake can be made to help determine the resultant dose (these methods are discussed in Chapter 8).

7.2.1.4 Intakes through wounds

The ICRP has no official model for intakes through wounds that may occur, such as what might happen if a worker injures himself with a contaminated tool. While wounds with contamination are not an everyday occurrence, they do happen. The ICRP suggests referring to a report of the National Council on Radiation Protection and Measurements (NCRP) entitled "Development of a Biokinetic Model for Radionuclide-Contaminated Wounds for Their Assessment, Dosimetry and Treatment" (Publication 156).

The ICRP does include dose coefficients for injection (i.e., direct entry into the bloodstream) in their reports; these values are provided as additional information to aid in the assessment of radiation doses from wounds; however, these coefficients do <u>not</u> address the complex movement of material from the wound site into the bloodstream.

7.2.2 ICRP Systemic Models

Once radioactive material has entered the bloodstream where it can be transferred to other organs and tissues of the body from its original intake location, the ICRP set of systemic models takes over. The general overview of how systemic models are used can be visualized by referring again to Figure 7-3 earlier in this chapter. Following the intake, some portion of the radioactive material enters the transfer compartment in a time-dependent fashion depending on the absorption fraction and the biological half-life of the material.

How the material distributes in the body depends understandably on the chemistry of the radionuclide. The human body has an intricate set of chemistry controls that can direct higher concentrations of some elements to certain tissues while having lower concentrations elsewhere. Consequently, there is no single systemic model that will work for all radionuclides other than the general understanding that the radionuclides will travel to various tissues of the body in a time-dependent fashion and will be eventually excreted commonly through urine, feces, and sweat.

The ICRP has tabulated systemic models for many elements in Publications 134, 137, 141 and 151. As an example, Figure 7-7 below shows the assumed systemic model for inorganic sulfur.

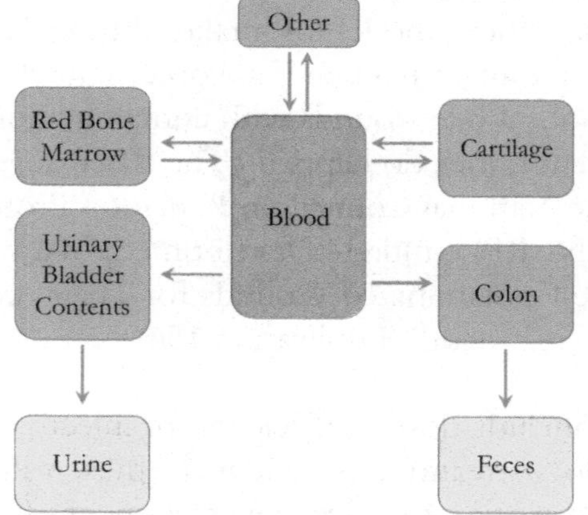

Figure 7-7. ICRP systemic model for inorganic sulfur from ICRP 134.

As shown in the figure, the bloodstream/transfer compartment is the beginning point. Following that, various compartments are identified based on studies that show how the chemical compound moves in the body. In some cases, the compartments represent an entire tissue or organ; in other cases, the compartment represents only part of a tissue or organ. As shown in Figure 7-7, there may be a compartment that is not associated with a single tissue or organ (labeled as "other").

Once the compartments are identified, they are used for solving the differential equations. Table 7-6 gives the biological transfer coefficients and the partitioning fractions of material for inorganic sulfur that are used with the model in Figure 7-7. As discussed earlier and shown in Figure 7-4, the partitioning fraction is the ratio of λ_{bio} for a particular donor compartment to receiving compartment divided by the total λ_{bio} for that donor compartment.

Table 7-6. Mathematical data for the systemic model of inorganic sulfur from ICRP 134.

From (donor compartment)	To (receiving compartment)	Biological transfer coefficient λ_{bio} (d^{-1})	Partitioning fraction (fraction from blood to compartment)
Blood	Red marrow	0.075	0.03
Blood	Cartilage	0.25	0.10
Blood	Other	0.175	0.07
Blood	Urinary bladder contents	1.8	0.72
Blood	Right colon contents	0.2	0.08
Red marrow	Blood	0.3	
Cartilage	Blood	0.1	
Other	Blood	3.5	

Inspection of the table above allows us to make a few observations. Most notably, inorganic sulfur travels to several different tissues as shown; however, the largest fraction in the bloodstream heads for the urinary bladder and ultimately to excretion by urine with a biological half-life of 0.38 days ($0.693/\lambda_{bio}$), indicating relatively fast removal. The longest biological half-life is 9.24 days for any of the compartments, which is for the transfer from blood to red marrow. Thus, we might expect that after inorganic sulfur enters the bloodstream, it is metabolized relatively quickly. Note, however, that this conclusion only applies to the systemic model and not to the movement of inorganic sulfur from the respiratory tract or the alimentary tract into the systemic model.

Some elements and compounds tend to have greater concentration in certain tissues and organs of the body due to their chemistry. For example, iodine is used as an essential element in the thyroid gland so that radioactive iodine tends to have a greater concentration in the thyroid compared to the rest of the body. As a side note, some people mistakenly think that all iodine ends up in the thyroid, but that is not the case; iodine is present throughout the body but does have a greater concentration in the thyroid gland.

There are also some elements and compounds that have a higher concentration in the skeleton of the body. We commonly refer to these as bone-seekers. As an example, strontium, due to its chemical similarity to calcium, is one such element that has a higher concentration in the bone.

Radioactive progeny in the systemic models are generally modeled with independent kinetics. Thus, as radionuclides decay, the chemistry of the progeny takes over in the calculations at the location(s) where the progeny are produced.

7.2.3 ICRP Dosimetric Models

Having now dealt with the models that allow the calculation of the integrated activity following an intake, we turn our attention to an overview of the models for calculating the dose to target regions of the body from radioactive material that is located in source regions of the body.

In its Publication 110 (Adult Reference Computational Phantoms), the ICRP defined the anatomical structure of the reference male and female used for its calculations in Publication 130. The phantoms are anthropomorphic and can be used in Monte Carlo calculations to simulate radiation transport (this was discussed for external dose calculations in Chapter 4). Details on organ size and placement are included in Publication 110. In addition, a downloadable file with electronic data files is available so that information can be imported into external programs.

Figure 7-8 below is taken from Publication 110 and shows a sample of a cross section of the computational phantoms to demonstrate the level of detail that is contained in the description. However, these are not perfect phantoms. For example, the phantoms are static and do not reflect the different postures of real people throughout the day. In addition, there are assumptions of exact organ sizes and placement even though there may be some variation in those variables in the real world. As stated earlier in this chapter, our dose calculations are intended to be for these reference phantoms, and the phantoms do not necessarily represent absolute reality; however, the consistency with which the phantom and its traits are applied give strength to the dose calculation methodology.

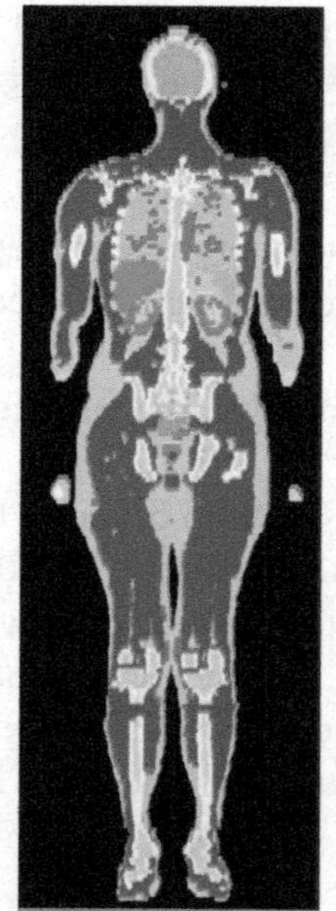

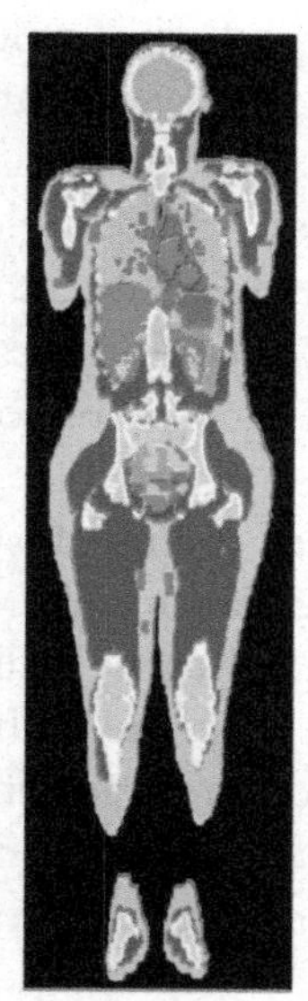

Figure 7-8. Coronal images of the ICRP Publication 110 Reference male phantom (left) and female phantom (right) (Images courtesy of ICRP)

With the anatomical structure of the phantom settled, it is now time to discuss the physics of how to calculate the absorbed dose. From our earlier discussion in Section 7.2, we need to calculate the absorbed dose in a target region per decay in a source region. Thus, all the possible combinations of source and target regions must be established. As might be imagined, radioactivity in a single organ of the body could potentially irradiate all other organs of the body. Likewise, a single organ of the body could be the target for multiple source organs. With the number of organs and tissues in the body, the number of possible combinations is quite large.

Ideally, the dose calculation could be broken down into the following general steps:

1. Identify all possible source/target combinations.
2. Select a single target region.
3. Select a single source region.
4. For a given radionuclide in the source region, identify each radiation type and energy emitted per decay along with the yield.
5. Determine for each decay the amount of energy absorbed in the target region based on the radiation type, energy, and yield.
6. For each radiation type, divide the average energy absorbed in the target region by the target's mass to obtain the absorbed dose due to that radiation type.
7. Pick a new source region (Step 3) and then repeat Steps 4-6. Continue this process until all source regions have been accounted for in the calculations.
8. Pick a new target region (Step 2) and begin the process again.

The physics of the radiation interactions in Step 5 above lends itself to the Monte Carlo method to determine the energy deposited in the target region, whereby individual radiation types and energies are simulated to originate within the source region and then allowed to be transported in three-dimensional space throughout the body. Tallies of where the radiation deposits its energy and the location of that deposition are used to calculate the dose.

To facilitate computations, the ICRP introduced the concept of the absorbed fraction $\phi(r_T \leftarrow r_S)$, which is defined as shown in Equation 7.5.

$$\phi(r_T \leftarrow r_S) = \frac{energy\ absorbed\ in\ target\ region}{energy\ emitted\ in\ source\ region} \tag{7.5}$$

Though absorbed fraction uses the same symbol as fluence rate discussed in Chapter 4, these are not the same concept and are not related to each other. The absorbed fraction is conceptually the fraction of energy emitted from r_S that is absorbed in r_T.

The absorbed fraction can then be used to calculate the specific absorbed fraction $\Phi(r_T \leftarrow r_S)$, which is defined as in Equation 7.6.

$$\Phi(r_T \leftarrow r_S) = \frac{\phi(r_T \leftarrow r_S)}{m_T} \tag{7.6}$$

Where

- $\Phi(r_T \leftarrow r_S)$ is the specific absorbed fraction
- $\phi(r_T \leftarrow r_S)$ is the absorbed fraction as defined above
- m_T is the mass of the target region (kg)

The advantage of using the absorbed fraction and specific absorbed fraction concepts is that the Monte Carlo simulations do not have to be performed for every individual radionuclide. Instead, the specific absorbed fractions for all radiation types at discrete energies for source/target combinations can be determined using the Monte Carlo method. The results from these calculations can then be applied to the particular decay scheme of a radionuclide based on the yield and energy of radiations from the radionuclide.

The absorbed fraction and specific absorbed fraction depend on the geometry between the source region and the target region in terms of proximity and shape. Some generalities can be made based on considering four common arrangements of a source region (r_S) and target region (r_T)as shown below in Figure 7-9.

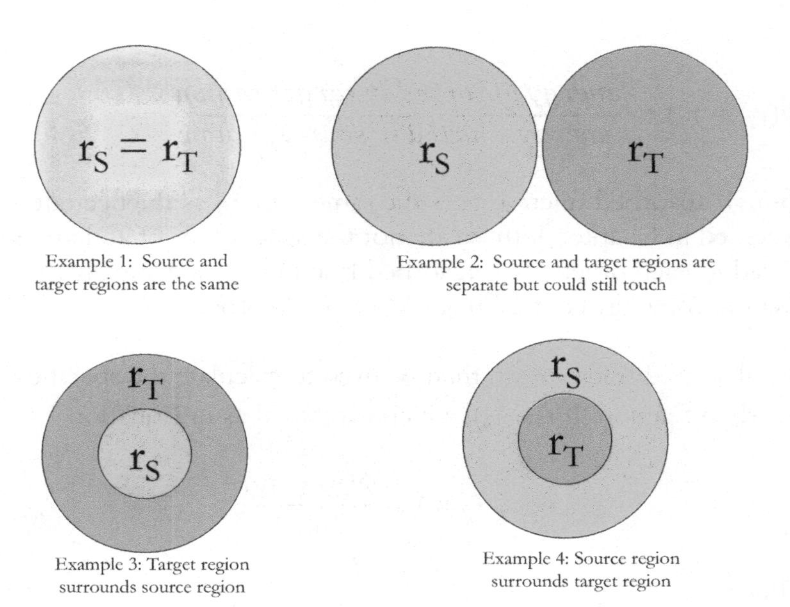

Figure 7-9. Common arrangements of source region and target region for internal dose assessment.

Example 1 above is very common in that most source regions are also target regions and the source thus irradiates itself. For alpha particles that are emitted in this source, essentially 100% of the energy will be absorbed due to the short range of alpha particles; therefore, $\phi(r_T \leftarrow r_S) \approx 1$, and $\Phi(r_T \leftarrow r_S) \approx 1/m_T$. By comparison, beta particles, gamma rays, and neutrons are more penetrating and it is possible that not all the emitted energy would be absorbed in the same region; therefore, the absorbed fraction and specific absorbed fraction would be energy-dependent and size-dependent. With higher energies and smaller region sizes, the absorbed fraction would be relatively small, while lower energies and larger regions would lead to higher absorbed fractions. As a side note, not all source regions are target regions. For example, the urinary bladder contents are a source of radiation but is not a target region, since radiation dose to the liquid contents of the urinary bladder poses no health risk. The same is true of the alimentary tract contents.

Example 2 in Figure 7-9 is also common in that one source region may irradiate a target region that is physically distinct from it. In some cases, the regions may be separated by relatively great distances; in other cases, they might be next to each other and touch at a portion of their boundaries. In general because of their short range, alpha particles from the source region would not be able to penetrate to the target region with any appreciable number, and thus $\phi(r_T \leftarrow r_S)=0$, which also means that $\Phi(r_T \leftarrow r_S)=0$. For beta radiation, the distance between the source region and target region will play a role. At distances beyond the maximum range of the beta particles, the absorbed fraction and specific absorbed fraction would be zero, while shorter distances would result in nonzero values. For photons and neutrons, the distance and size of both the source and target regions will affect the absorbed fraction and specific absorbed fraction.

Example 3 has the target region surrounding the source region, such as what might be expected with the alimentary tract where the tract walls surround the tract contents. The intimate contact between the source region and target region will dictate that the absorbed fraction and specific absorbed fraction will be nonzero for all radiation types, even for alpha particles.

Example 4 has the source region surrounding the target region, such as what might be expected with bone marrow being surrounded by the bone structure. As with Example 3, the intimate contact between the source and target regions (especially if the target region is small with respect to the source region) will dictate nonzero values for the absorbed fraction and specific absorbed fraction for all radiation types.

Once the specific absorbed fraction is determined for a given source-target configuration, the absorbed dose in the target region per decay in the source region can be determined and then the radiation weighting factor can be applied to arrive at the equivalent dose in the target region per decay in the source region using Equation 7.7

$$S_W = \sum_R w_R \sum_i E_{R,i} Y_{R,i} \; \Phi(r_T \leftarrow r_S, E_{R,i}) \qquad (7.7)$$

Where

- S_W is the equivalent dose in the target region per decay in the source region for a given radionuclide
- w_R is the radiation weighting factor for radiation type R

- $E_{R,i}$ is the energy of the ith radiation of type R emitted in the source region
- $Y_{R,i}$ is the yield of the ith radiation of type R emitted in the source region
- $\Phi(r_T \leftarrow r_S, E_{R,i})$ is the specific absorbed fraction for the ith radiation of type R for the given source-target configuration

While the symbology in Equation 7.7 may seem intensive, the conceptual meaning is straightforward: for each radiation emitted in the source region, the energy of that radiation is multiplied by its yield, specific absorbed fraction, and radiation weighting factor to obtain the equivalent dose. The summation is carried out for all radiations emitted in the source region as part of the decay scheme for a given radionuclide.

7.3 Calculation of Equivalent Dose and Effective Dose

The rest of the ICRP dose calculation methodology is now a matter of multiplying the S-factor from Equation 7.7 for each source/target configuration by the appropriate integrated activity for that source region and repeating that process for each source region that irradiates the same target region. This will give the appropriate committed equivalent dose coefficient for that target region, as shown below in Equation 7.8.

$$h_T = \sum_j \sum_{r_S} \tilde{a}_j(r_S)\, S_w(r_T \leftarrow r_S)_j \qquad (7.8)$$

Where

- h_T = the committed equivalent dose to tissue T following a 1 Bq intake
- j = each radionuclide member of a given chain following intake (a different radionuclide for the parent and then for each progeny, if applicable)
- r_S = source region
- r_T = target region
- $\tilde{a}_j(r_S)$ = the number of decays over 50 years occurring in the source region r_S for radionuclide j following a 1 Bq intake
- $S_W(r_T \leftarrow r_S)_j$ = the S-factor defined in Equation 7.7 for radionuclide j that is part of the decay chain

Equation 7.8 thus sums the equivalent dose to a target tissue from all source tissues and for all radionuclides that are part of the decay chain. As mentioned earlier, the ICRP tabulated dose coefficients that assume an intake of pure parent but include the dose contribution from radioactive progeny that are produced through radioactive decay over the 50-year period. If there are no radioactive progeny, then "j" in Equation 7.8 would represent only one radionuclide (common for radionuclides such as Co-60, H-3, etc.).

The process of calculating h_T values is repeated for both male and female reference models for all other target regions. The ICRP then used these values of h_T to determine the dose coefficients to calculate the committed effective dose. The dose coefficients are written in **lower-case** and are designated as follows:

- The committed effective dose per unit intake (Sv/Bq) is designated as e(50)
- The committed equivalent dose per unit intake (Sv/Bq) for each tissue is designated as $h_T(50)$

The value for e(50) is calculated as a sex-averaged value based on the $h_T(50)$ values using the logic as described earlier for Equation 7.2. Thus, e(50) is represented as shown below in Equation 7.9

$$e(50) = \sum_T w_T \left[\frac{h_T^M(50) + h_T^F(50)}{2} \right] \tag{7.9}$$

Where

- e(50) is the committed effective dose per unit intake
- $h_T^M(50)$ is the committed equivalent dose per unit intake for tissue T for the male worker
- $h_T^F(50)$ is the committed equivalent dose per unit intake for tissue T for the female worker
- w_T is the tissue weighting factor for tissue T

Table 7-7 below shows an example of how the $h_T(50)$ values can be used to determine the e(50) value for Co-60 Type M. In the table, the $h_T(50)$ values for both male and female are taken from the ICRP Electronic Annex, which can be downloaded directly from https://www.icrp.org/docs/ICRP%202022%20OIR%20Electronic%20Annex%20Distribution%20Set.zip.

Table 7-7. Calculation of e(50) for inhalation of 5 μm Type M Co-60 based on the average $h_T(50)$ values for male and female.

	$h_T^M(50)$ (male) Sv/Bq	$h_T^F(50)$ (female) Sv/Bq	Average $h_T(50)$ Sv/Bq	w_T	$w_T*h_T(50)$
Bone marrow	3.60E-09	4.70E-09	4.15E-09	0.12	4.98E-10
Colon	3.80E-09	4.00E-09	3.90E-09	0.12	4.68E-10
Lung	1.90E-08	2.20E-08	2.05E-08	0.12	2.46E-09
Stomach	4.60E-09	4.40E-09	4.50E-09	0.12	5.40E-10
Breast	3.70E-09	4.80E-09	4.25E-09	0.12	5.10E-10
Ovaries	0.00E+00	2.30E-09	1.15E-09	0.08	9.20E-11
Testes	6.80E-10	0.00E+00	3.40E-10	0.08	2.72E-11
Urinary bladder	1.50E-09	1.90E-09	1.70E-09	0.04	6.80E-11
Esophagus	6.80E-09	8.60E-09	7.70E-09	0.04	3.08E-10
Liver	6.70E-09	7.50E-09	7.10E-09	0.04	2.84E-10
Thyroid	4.10E-09	4.70E-09	4.40E-09	0.04	1.76E-10
Bone Surface	2.40E-09	3.20E-09	2.80E-09	0.01	2.80E-11
Brain	1.20E-09	1.60E-09	1.40E-09	0.01	1.40E-11
Salivary glands	1.60E-09	2.40E-09	2.00E-09	0.01	2.00E-11
Skin	1.10E-09	1.40E-09	1.25E-09	0.01	1.25E-11
Remainder Tissues			5.43E-09[a]	0.12	6.51E-10
Adrenals	5.00E-09	5.90E-09			
ET of HRTM	1.80E-08	2.10E-08			
Gall bladder	5.70E-09	5.90E-09			
Heart	8.00E-09	9.70E-09			
Kidneys	5.30E-09	6.10E-09			
Lymphatic nodes	4.10E-09	4.90E-09			
Muscle	1.70E-09	2.20E-09			
Oral mucosa	2.30E-09	3.30E-09			
Pancreas	4.30E-09	4.10E-09			
Prostate	1.20E-09	0.00E+00			
Small intestine	2.50E-09	3.10E-09			
Spleen	5.00E-09	5.10E-09			
Thymus	6.70E-09	8.30E-09			
Uterus	0.00E+00	2.60E-09			
					e(50) (Sv/Bq) =6.16E-09

a. The average remainder tissue dose is the average equivalent dose for both male and female for all tissues listed under the Remainder Tissue heading.

The tissue weighting factors are taken from Chapter 1 of this book (Table 1-2). As explained in footnote c of Table 1-2, the equivalent dose to the remainder tissues is averaged and then multiplied by the remainder w_T value before being added to other tissue-weighted doses to obtain the committed effective dose.

Since the ICRP publications contain all the necessary e(50) values, we do not have to calculate e(50) in everyday dosimetry; thus, we will use the results of the ICRP calculations of e(50) from this point forward.

7.4 Dosimetric data contained in ICRP 134, 137, 141,and 151

ICRP Publications 134, 137, 141, and 151 present data by element and include the following information:

- Chemical forms in the workplace
- Routes of intake, including biological data on studies of inhalation and ingestion absorption to form the basis for inhalation class and f_A values. In addition, the systemic movement of the material is summarized along with the systemic model that is assumed for the element.
- Individual monitoring methods for determining the intake amount based on the individual isotopes of the element.
- Dosimetric data for each radionuclide with the committed equivalent dose coefficients $h_T(50)$ for each sex along with the effective dose coefficients e(50). There is also information on the time-dependent retention and excretion of the element. Many radionuclides are included in the actual hard copy report; additional radionuclides are presented in the electronic annex to the report (discussed below)

The ICRP electronic annex has more detailed information on each radionuclide, including the ability to adjust the dose coefficients for inhalation intakes if the particle size of the intake is known. For example, Figure 7-10 provides the Fe-59 dose coefficients obtained from the electronic annex for 1 µm, 5 µm (default ICRP AMAD particle size), and 10 µm particle sizes.

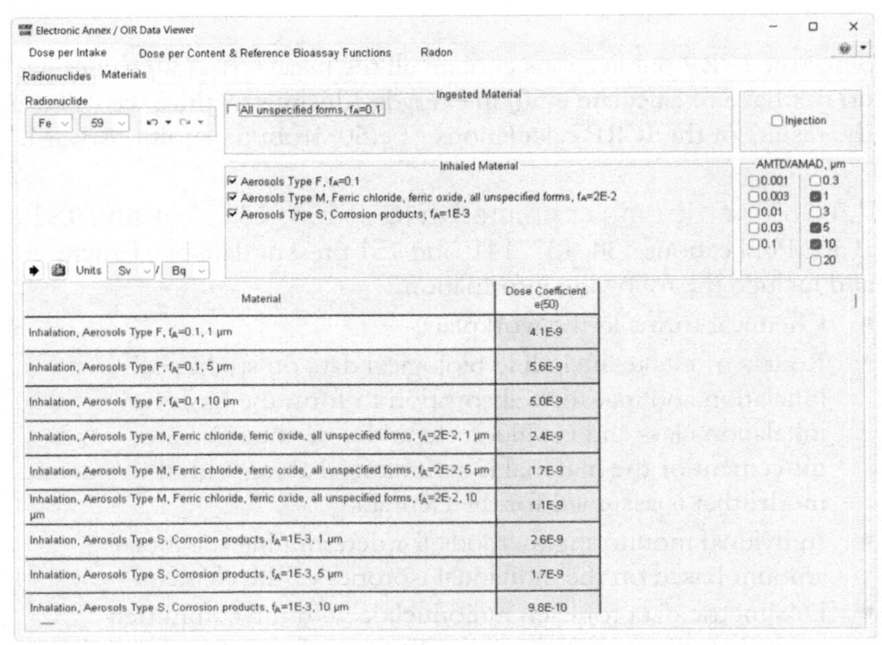

Figure 7-10. Fe-59 dose coefficients for 1 μm, 5 μm, and 10 μm particle sizes as obtained from the ICRP Electronic Annex. Within the annex, the dose coefficients can be selected based on method of intake as well as particle size in the case of inhalation intakes (Image courtesy of ICRP).

Figure 7-10 illustrates the danger of assuming the magnitude of a dose coefficient based on its inhalation class. Logically, it would be tempting to assume that the most soluble inhalation class (Type F) would quickly move through the body and be removed, thus causing the least amount of dose. However, Figure 7-10 shows that for Fe-59, this is not the case; in fact, Type F has the highest dose coefficient. Thus, it is always important to review the dose coefficients rather than making judgment calls without the numerical data.

Figure 7-11 shows the graphical output of the electronic annex for the Fe-59 activity in various compartments of the body with respect to time following an inhalation intake of Type F Fe-59. The y-axis of this graph shows the fraction of the intake that is present at time t, again based on an assumed intake of 1 Bq so that the values can be scaled accordingly. The electronic annex also provides tables with the values used to produce the figure so that the values can be copied/imported into an external program or into a spreadsheet for further use. More on how this data can be used is discussed in Chapter 8.

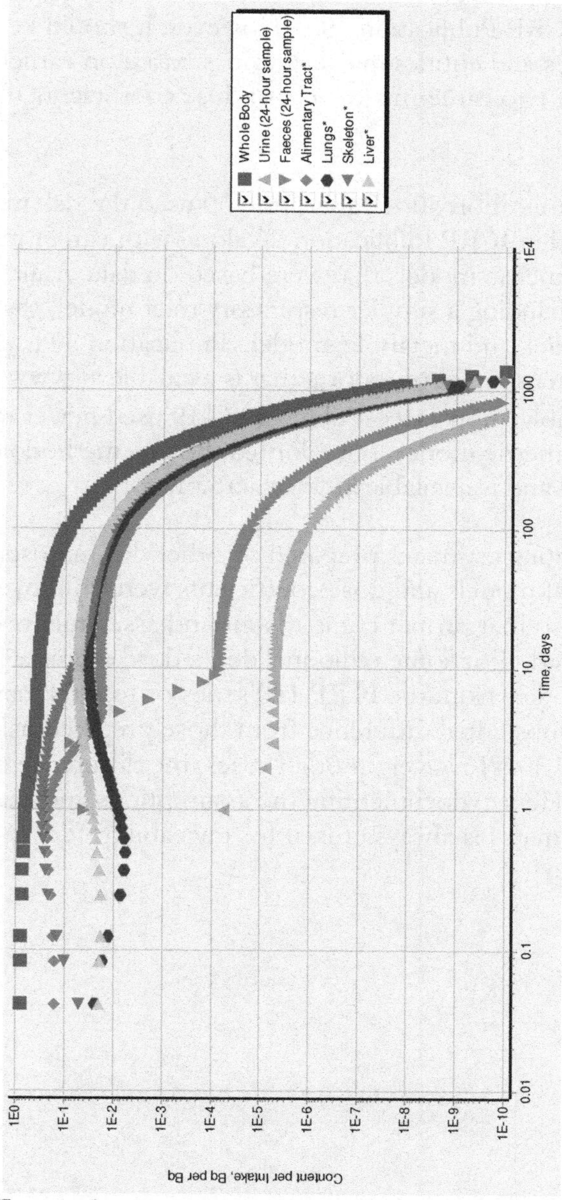

Figure 7-11. Content of various tissue compartments as a function of time for Type F Fe-59 based on the ICRP intake and systemic models (Image courtesy of ICRP).

7.5 Dosimetric data based on previous ICRP Reports

This chapter focuses on the newest ICRP models and dose coefficients that are used in conjunction with risk methodology outlined in ICRP Publication 103. However, it should be noted that some facilities and entities may use values based on earlier ICRP models. The two earlier major sets of dose coefficients that may still be in use are:

- ICRP Publication 30 – Publication 30 used the risk methodology described in ICRP Publication 26 along with earlier metabolic and dosimetric models that were based on data available in the 1970s, including a simpler respiratory tract model, gastrointestinal tract model, and metabolic model. Publication 30 is a comprehensive set of reports that is available at www.icrp.org.
- ICRP Publication 119 – Publication 119 used newer metabolic and dosimetric models but adopted the risk methodology of ICRP 60 and is available at www.icrp.org.

When evaluating estimates prepared by other dosimetrists, it is crucial to know what models and dose coefficients were employed in the calculations so that proper comparisons and assessments of the doses can be enacted. For some radionuclides, a dose estimate prepared using the models from the ICRP 130 series of reports may not represent a substantial difference from those prepared using ICRP 119 or ICRP 30. However, in other cases, the difference could be significant. Hence, documenting the assumptions and scientific basis for dose estimates is always crucial for traceability.

Free Resources for Chapter 7

- Questions and problems for this chapter can be found at www.hpfundamentals.com
- ICRP Publication 130, *Occupational Intakes of Radionuclides: Part 1.* Ann. ICRP 44(2), 2015. Available from www.icrp.org.
- ICRP Publication 134, *Occupational Intakes of Radionuclides: Part 2.* Ann. ICRP 45(3/4), 1–352, 2016. Available from www.icrp.org.
- ICRP Publication 137, *Occupational Intakes of Radionuclides: Part 3.* Ann. ICRP 46(3/4), 2017. Available from www.icrp.org.
- ICRP Publication 141, *Occupational Intakes of Radionuclides: Part 4.* Ann. ICRP 48(2/3), 2019. Available from www.icrp.org.
- The electronic annex developed by the ICRP that includes the relevant dose coefficients and other information can be downloaded freely from the ICRP web site at https://www.icrp.org/docs/ICRP%202022%20OIR%20Electronic%20Annex%20Distribution%20Set.zip (this link can be accessed via the web page for ICRP Publication 151.

Additional References

- ICRP Publication 151, *Occupational Intakes of Radionuclides: Part 5.* Ann. ICRP 51(1-2), 2022. Available from www.icrp.org.
- NCRP Report 156, *Development of a Biokinetic Model for Radionuclide-Contaminated Wounds for Their Assessment, Dosimetry and Treatment (2006).* Available from www.ncrponline.org.

Chapter 8 Measurements and Calculation of Dose from Internal Sources of Radiation

The last chapter dealt with the models for how the ICRP calculated the dose per unit intake for both inhalation and ingestion. This chapter will deal with the more practical issue of calculating doses based on monitoring and sampling. Strictly speaking, monitoring usually refers to real-time measurements of the total activity while sampling refers to measuring the activity of a representative portion of the total activity. As an example from environmental health physics (covered in Chapters 11 and 12), it is not practical to determine the total activity of a lake by direct monitoring, but a representative sample of the water in the lake can be obtained to estimate the total activity in the lake. Granted, it is not always clear nor is it necessary to draw such a distinction between monitoring and sampling. As an example, we will refer to obtaining a urine sample for internal dose assessment, but that is part of the internal dosimetry monitoring program.

Measurements related to internal dosimetry can be broken down into prospective monitoring (monitoring that can occur before the person has an intake) and retrospective monitoring (monitoring after the intake has occurred) as follows:

- Prospective monitoring
    - Air concentration measurements
    - Water/foodstuff concentration measurements
- Retrospective monitoring
    - Direct measurement of radioactivity in the body (in-vivo)
    - Measurement of excreta (urine, feces, sweat) and inferring the activity in the body (in-vitro)

Prospective monitoring has the advantage that decisions can be made in some cases before a person is exposed to the air/water/foodstuffs. Thus, limits on concentrations in these media can be enacted and facilities can employ protective measures to reduce potential doses.

Retrospective monitoring uses the measurement of activity that is directly related to an individual. For in-vivo monitoring, a detector system is placed in such a way to measure real-time activity in the person's body. For in-vitro monitoring, a sample of urine or feces or other excreta is gathered and then analyzed in a laboratory. Both of these measurements are insensitive to the breathing rate or ingestion rate of the person, and the intake can be inferred without the variability of intake rate.

8.1 Prospective Monitoring and Dose Calculations

Like most topics in this text, the monitoring of airborne, waterborne, and foodborne concentrations of radioactivity could take an entire book on its own. While it is not possible to cover all aspects of prospective monitoring, there are some general principles to be considered as outlined in the following sections.

8.1.1 Monitoring for Airborne Concentrations of Radioactivity

There are generally three scenarios for which measurements of airborne activity are made:

- Measuring the amount of airborne radioactivity leaving a facility through a ventilation stack
- Measuring the amount of airborne radioactivity inside a facility where workers will be conducting activities.
- Measuring the amount of airborne radioactivity in the environment outside a facility where the public could be exposed.

In the first scenario, a gamma ray detector can be placed where the air in the ventilation stack flows by the detector on the way to the exit point of the stack. Thus, the detector system is able to identify specific radionuclides and determine the activity based on the known flow rate and the detection efficiency of the system so that a concentration of each radionuclide can be calculated. Obviously, this represents only the concentration as it exits the stack; the concentration at the location where an individual is located relies on environmental transport modeling, which is covered in Chapter 11.

For the second and third scenarios, a lone gamma detector placed in the room usually is not capable of measuring the activity in the ambient air with any accuracy. For radionuclides that are particulate in nature, it is possible to collect a sample of the airborne particles and measure the activity; this has the advantage of concentrating the activity to make it easier to detect. A vacuum pump with a known flow rate can be attached to a filter head that has filter paper inside it with a known collection efficiency. After the pump/filter has run for a known amount of time (thus sampling a fixed volume of air that is the product of flow rate and operating time), the filter paper can be removed and taken to a lab for analysis. The lab can determine the activity on the filter; this number is used in conjunction with the filter collection efficiency and the total air volume through the filter to determine the airborne concentration. More information on air sampling/monitoring in the workplace is presented in Chapter 10, while air sampling in the environment is presented in Chapter 12.

Sampling/monitoring for gaseous radionuclides or vapors is a specialized topic because of the difficulty in collecting the sample since these will not collect on a paper filter as particulates do. It is necessary to collect the vapors/gases in a different medium; in some cases, activated charcoal is a possibility. In other cases, condensation of vapors or bubbling through liquids to collect the sample may be necessary. While this topic will not be covered further, it is critical for the health physicist to be aware of the challenges in sampling this type of airborne radioactivity.

8.1.2 Monitoring for Waterborne and Foodborne Concentrations

For water and foodstuff monitoring, the process is somewhat simpler than air monitoring and involves obtaining a suitably sized sample of the material (a liter or more of water and a few hundred grams of solid foodstuff) to allow detection of small amounts of activity. Once the activity is known, then the concentration in the medium can be determined.

8.1.3 Calculation of Dose from Prospective Monitoring

Regardless of the mode of intake, the potential dose based on prospective monitoring follows the logic of Equation 8.1 below.

$$E_{(50)} = I \cdot e(50)$$

Or (8.1)

$$H_T(50) = I \cdot h_T(50)$$

Where

- $E_{(50)}$ is the committed effective dose
- I is the intake amount
- e(50) is the committed effective dose coefficient specific to the mode of intake of interest
- $H_T(50)$ is the committed equivalent dose to a given target tissue
- $h_T(50)$ is the committed equivalent dose coefficient specific to the mode of intake of interest

As discussed in Chapter 7, many values for $h_T(50)$ and e(50) are tabulated in ICRP Publications 134, 137, 141 and 151 with a more comprehensive listing in the Electronic Annex.

The usual challenge in internal dosimetry is to determine the intake amount to use in Equation 8.1. For prospective monitoring based on air concentration measurements, though, the process is relatively straightforward. The intake amount for inhalation can be calculated using Equation 8.2 below:

$$Intake = (Conc_{air})(BR)(time) \qquad (8.2)$$

Where

- $Conc_{air}$ = the air concentration (Bq/m^3) of the radionuclide
- BR = breathing rate (m^3/hr)
- time = exposure period (hours)

Obviously, the breathing rate is an assumed value for the individual and will vary depending on the physical size of the person and the level of activity. For a reference male worker, the ICRP assumes a breathing rate of 1.2 m^3/hr. A sex-averaged breathing rate is assumed by the ICRP to be 1.1 m^3/hr.

For waterborne and foodborne concentrations of radionuclides, the intake amount can be calculated using Equation 8.3 below:

$$Intake = (Conc_{medium})(Ingestion\ amount) \quad (8.3)$$

Where

- $Conc_{medium}$ is the concentration in either the water or the foodstuff (Bq/l for water, Bq/kg for foodstuff)
- Ingestion amount is expressed in liters (for water) and kg (for foodstuff)

For a short-term acute intake, the amount of water or foodstuff ingested could be estimated based on real-time information. For calculations involving chronic intakes due to environmental releases from facilities, the methodology and assumptions are covered in Chapter 11.

Example: A worker is exposed to an airborne concentration of 6000 Bq/m^3 of Co-60 for 4 hours. What is an estimate of the committed effective dose that the worker could receive?

Solution: As with many problems, there are some assumptions to be made in the absence of more detailed information. Since it is an inhalation intake, we assume the following:

- The particle size of the airborne Co-60 is 5 μm AMAD, which is the ICRP default size.
- With no information on the chemical form, we must assume the inhalation class associated with unspecified forms of Co-60, which in this case is Type M.
- The breathing rate is 1.1 m^3/hr

Based on the first two assumptions, the appropriate committed effective dose coefficient e(50) from the ICRP Electronic Annex is 6.2×10^{-9} Sv/Bq.

Using Equation 8.2, the intake amount is:
(6000 Bq/m^3)(1.1 m^3/hr)(4 hr) = 26,400 Bq

Using Equation 8.1 $E_{(50)}$ is therefore (26,400 Bq)(6.2×10^{-9} Sv/Bq), which is equal to 1.6×10^{-4} Sv (0.16 mSv or 16 mrem).

Example: An adult ingests 1.5 liters of water contaminated with Cs-137 at a concentration of 4500 Bq/l. What is an estimate of the committed effective dose for this scenario?

Solution: As in the previous example, there are assumptions to be made since the form of Cs-137 is not known. The ICRP Electronic Annex gives a committed effective dose coefficient e(50) of 1.4×10^{-8} Sv/Bq for ingestion of Cs-137 in unspecified forms (this form also has an f_A value of 0.99, indicating it is highly absorbed).

Using Equation 8.3, the intake is estimated to be (4500 Bq/l)(1.5 l) = 6750 Bq. Using Equation 8.1, the committed effective dose is therefore (6750 Bq)(1.4×10^{-8} Sv/Bq) = 9.5×10^{-5} Sv (9.5×10^{-3} rem or 9.5 mrem)

In both these examples, the final dose is reported to only two significant figures.

8.2 Retrospective Monitoring and Dose Calculations

Retrospective monitoring of individuals to determine the intake and dose is usually divided into the broad categories of in-vivo monitoring/bioassay and in-vitro monitoring/bioassay. In-vivo monitoring involves making a direct measurement of the radiation emitted by the individual using a detection system to determine the amount of activity present at some particular time to estimate the intake and the dose. In-vitro monitoring, on the other hand, involves collecting a sample from the individual, usually urine or feces, and determining the activity in that sample to estimate the intake amount and dose.

Facilities that have a likelihood for workers to have intakes will set up a program that is designed to monitor workers under several scenarios:

- Routine bioassay, where the worker is monitored at a set frequency depending on the degree of risk of an intake and the difficulty of detecting the radionuclide.

- Non-routine/special bioassay, where the worker is monitored in response to a potential or suspected intake. Examples include a situation where a worker is known to be performing special work in an area with a high potential for intake or a situation where the worker becomes unexpectedly contaminated while performing routine work.
- Baseline/Termination bioassay, where a worker is monitored when beginning employment at a facility (to establish a baseline) and when the worker leaves employment at the facility (to ensure no intakes have been missed prior to termination).

The following sections discuss some of the considerations for both in-vivo and in-vitro monitoring and how dose calculations are performed using the ICRP 130 methodology and information in the ICRP Publications.

8.2.1 In-Vivo Monitoring

In-vivo monitoring equipment varies depending on the manufacturer and the particular application. While it is common to purchase ready-made equipment from a company, some facilities construct their own in-vivo system using radiation detectors, electronics, and shielding (to reduce the ambient background radiation).

By far, the most common application of in-vivo monitoring is for what is termed "whole-body counting," where the detection system is used to measure the activity in the total body based on gamma emission from the person; the system used for this is called a whole-body counter (WBC). However, the WBC is not appropriate for all circumstances. For suspected intakes of iodine radioisotopes, it may be desirable to determine the activity in the thyroid gland, since the thyroid uses iodine in its function and hence a higher concentration of iodine would be in the thyroid compared to other body tissues. Therefore, a standalone thyroid counter may be used, or a thyroid detector may be incorporated into a WBC.

Figure 8-1 shows several common examples of in-vivo counting systems. Within the figure, Counters (a), (b), and (c) are illustrative of what would be used as a WBC. In Counter (a), the person would stand in the booth area facing the front column where the detectors are located. Shielding in the walls and around the detectors in the column are used to reduce spurious background counts. The detectors can be either stationary or may move vertically to scan the activity of the person from head to toe. This system is used when the total count time is of short duration so that a person can stand comfortably (usually no more than five minutes).

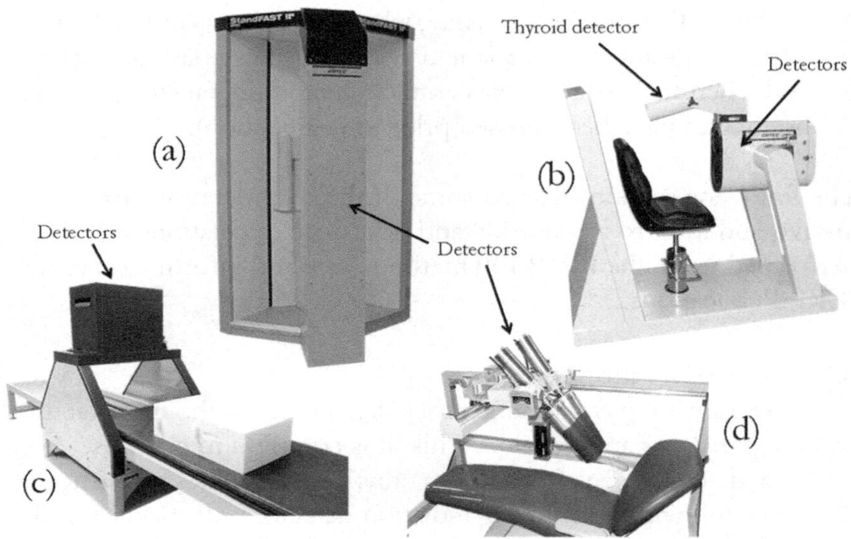

Figure 8-1. Examples of several in-vivo counting systems, including (a) standup counter; (b) chair counter with supplemental thyroid detector; (c) scanning bed counter; and (d) chest/ lung counter. (Photos courtesy of Ortec, www.ortec-online.com)

Counter (b) in the figure shows an example of a chair counter where the person would sit in the chair and the detectors would be brought up to the person in close proximity to the body. Again, there could be some variety in the design; in some cases, one detector might be behind the person to determine the lung activity while a second detector would be in the front to determine the activity in the alimentary tract. In Counter (b), there is also a thyroid detector.

Counter (c) in the figure is a scanning bed counter. The person would lie prone on the padded platform and would be scanned from head to toe. In some configurations, the detectors would move along a track while the person remains stationary. In other cases, as shown in the figure, the detectors remain stationary while the person would be slowly moved in their field of view.

Counter (d) in the figure is illustrative of a case where a traditional WBC is not appropriate. Typical WBC systems work well when the activity in the person is comprised of gamma-emitting radionuclides where the gamma energy is several hundreds of keV in magnitude. However, when attempting to measure the activity for low-energy/low-yield photons, the usual WBC system would be inadequate. Transuranic nuclides in particular tend to have low-energy photon emissions (usually x-rays following radioactive decay) that are easily absorbed by the shielding of the human body. Thus, a different approach is taken for these situations with three design features:

1. The detectors themselves are special detectors designed for low-energy photons. A thin entrance window for the detector is used to minimize shielding by the detector.
2. The person is placed in a stationary position (a reclining position of some sort), and the detectors are placed in intimate contact with the chest area to put them as close to the lobes of the lungs as possible.
3. Because of the need to reduce ambient background, the entire apparatus is housed in a heavily shielded room that can appear to be vault-like in appearance.

For Counters (b), (c), and (d) in Figure 8-1, the count times would typically be tens of minutes up to a half hour or so in order to achieve the desired sensitivity.

One caution when performing in-vivo analysis is the possibility that the individual being counted could have external contamination on their skin or clothing, which would be detected and mistakenly analyzed as being part of an intake. If contamination is suspected, the individual could be asked to shower and change clothing prior to counting. Additionally, for the standup counter [Counter (a) in Figure 8-1], the person might be counted one time facing the detectors and one time with their back to the detectors. If localized contamination on the skin is present, then a significant difference in the recorded activity from these two positions would be present.

Despite the issue of possible external contamination, there are several advantages to in-vivo monitoring. First is that it is less invasive and is usually preferred by individuals. Second is that it is relatively straightforward, and quantitative results can be obtained quickly within a matter of minutes. In addition, recounts of the individual can be performed daily, if necessary, without much inconvenience to the individual. Therefore, most facilities will try to use in-vivo monitoring as much as possible. However, when attempting to detect radionuclides that emit few photons or that emit low-energy photons, in-vivo monitoring may be ineffective, in which case in-vitro monitoring will be the preferred method for bioassay.

8.2.2 In-vitro Monitoring

As discussed earlier, in-vitro bioassay relies on collecting an excreta sample from the individual and using analysis of that bioassay sample to determine the intake and the dose. While a saliva or nasal swab sample might be useful qualitatively to indicate if a person had an intake, these types of samples are highly variable and are not usually useful for quantitative determination of an intake or the dose (the health physics literature does show some research on how this could change). Thus, urine and feces are normally the selected sample type. These samples are collected and then analyzed for radioactivity by counting the alpha or beta radiation emissions from the sample. In some cases, however, radioanalysis lacks the needed sensitivity, and some form of mass spectrometry may be employed to achieve even better results.

There are some issues to consider when using in-vitro bioassay:

- Due to the normal human variation in dietary and excretion habits, using a single sample from an individual is not sufficient for dose assessment (tritium in urine is an exception because it mixes completely with the body water relatively fast). Therefore, the individual normally would submit 24-hour samples or longer.
- The first void or defecation following an intake is typically not useful because the contents of the bladder or the colon have had insufficient time to be affected by the intake.
- The unpleasantness for the individual of collecting the sample cannot be ignored. Every void or defecation must be collected for the specified time period and stored until the sample is submitted.

- The samples are biohazards and must be stored and treated appropriately.
- The samples require laboratory preparation, usually consisting of acid digestion and other steps to concentrate any radioactivity to increase counting efficiency. The time for these steps is not trivial and contributes to a delay in obtaining results. It is common for the turnaround time for an in-vitro sample to be a few weeks.
- It is usually necessary to collect additional samples from the individual before the results of the first sample are known due to the turnaround time for analysis. This helps to ensure that multiple data points are available to be used for any eventual dose assessment. Thus, the unpleasantness factor for the individual is multiplied by having to give multiple samples without knowing the results of the early samples.

Facilities with large numbers of samples may choose to perform the actual analysis of the samples in their own onsite laboratory. However, smaller facilities may choose to have an offsite laboratory perform the analysis and report the results back for dose assessment.

8.2.3 Calculation of Dose from Retrospective Monitoring

8.2.3.1 General Approach

Dose assessment based on retrospective monitoring relies on the ability to determine the intake based on the bioassay measurement. This requires that the intake and biokinetic models discussed in the last chapter be used to interpret bioassay measurements so that we know what amount of activity should be present in a given bioassay measurement at a given time post-intake.

ICRP Publication 130 uses the concept of the "reference bioassay function" with the symbol m(t) to assist in estimating the intake. Conceptually, m(t) is the fraction of the intake expected to be in a respective bioassay compartment. If in-vivo monitoring is employed, then m(t) represents a retention function and is the fraction present specifically at time t in that compartment. If in-vitro monitoring is employed, then m(t) represents the amount expected to be present in a 24-hour sample that ends at time t.

The electronic annex to Publications 134, 137, 141 and 151 provides tables of m(t) values listed as "Content per intake (Reference Bioassay Function)." Figure 8-2 below shows a screenshot of the data for 5 μm AMAD Type M Co-60 assuming an inhalation intake.

The exact compartments for which m(t) values are provided depends on the element and radionuclide. Examples of compartments include the following:

- Whole body
- Urine (24-hour sample)
- Feces (24-hour sample)
- Alimentary tract
- Lungs
- Thyroid
- Liver
- Skeleton

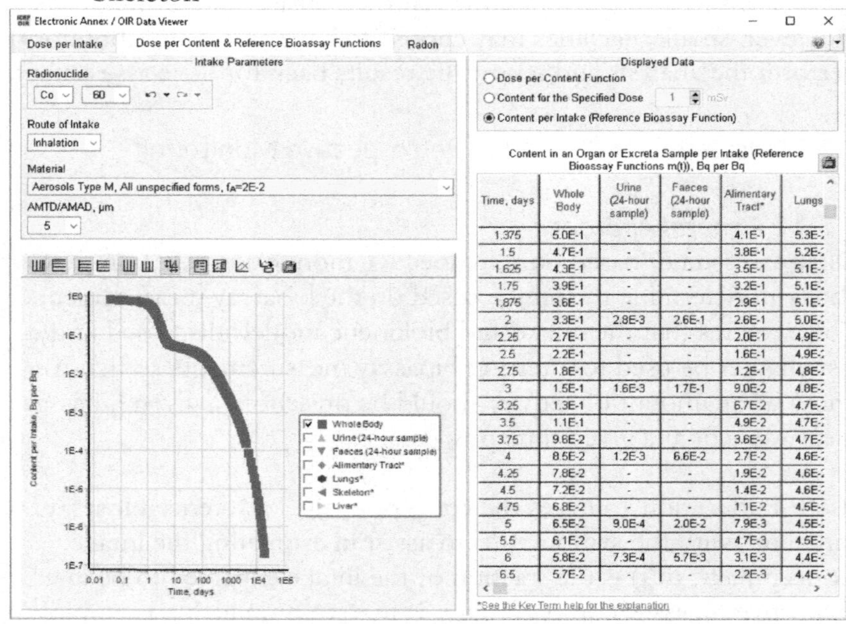

Figure 8-2. Reference Bioassay Function m(t) information for 5 μm Type M Co-60 following an inhalation intake (Image courtesy of ICRP).

The data for m(t) can be copied and pasted into a spreadsheet or other external program so that automated calculations can be performed. From inspection of the data table, m(t) values are provided at discrete points in time as a matter of convenience. For time periods that fall in between the points in the table, logarithmic interpolation can be used to estimate the proper value of m(t). If the values of m(t) are not changing drastically over the time interval, then linear interpolation may be used for expediency without sacrificing accuracy.

The simplest use of m(t) to determine intake is represented below in Equation 8.4 (taken from ICRP 130):

$$I = \frac{M}{m(t)} \tag{8.4}$$

Where

- I is the calculated intake
- M is the bioassay measurement
- m(t) is the reference bioassay function appropriate for the radionuclide, mode of intake, and assumed time of intake based on the explanation of m(t) given above.

This equation can be used regardless of the bioassay method; however, it is necessary to ensure that the selected m(t) value is appropriate to be used with the particular bioassay data that is provided, as shown in the following examples. These are highly idealized examples but illustrate the basic premise of how the reference bioassay functions are to be used.

Example: A worker is suspected to have inhaled Co-60. She is sent to have a whole-body count performed at 1.75 days post-intake. The result of the whole-body count indicates an activity of 6.2 nCi. What is the best estimate of intake based on that data point? What is an estimate of the committed effective dose based on this data point?

Solution: Using Equation 8.4 above, we need to be able to assign a value of m(t) for the given intake. Absent any other information, as a first estimate, we assume that the Co-60 is the most generic form, which according to the ICRP Electronic Annex is Type M (the Annex assigns Type M to Co-60 when the chemical form is not known). Further, with no specific information on particle size, we use the default size of 5 μm. From Figure 8-2 above, the Reference Bioassay Function m(t) for the Whole Body at t=1.75 days is equal to 0.39. Thus, the intake is:

$$I = \frac{M}{m(t)} = \frac{6.2\, nCi}{0.39} = 15.9\, nCi = 588\, Bq$$

The committed effective dose is the intake multiplied by the committed effective dose coefficient, again obtained from the ICRP Electronic Annex. To be consistent with the assumptions used in the intake estimate, we use the dose coefficient for Type M Co-60 of 5 μm particle size, which is 6.2×10^{-9} Sv/Bq. Thus, the committed effective dose is 3.6 μSv, or 0.36 mrem.

Example: A worker is suspected to have inhaled Sr-90. He is put on a urine bioassay program and submits a 24-hour urine sample that ends on day 5 following the suspected intake. After radiochemical analysis, the urine sample is measured to have an activity of 2.3 pCi. What is the best estimate of intake based on that data point? What is an estimate of the committed effective dose based on this data point?

Solution: As with the previous example, we need to be able to assign a value of m(t) for the given intake. Without any information on the chemical form of the Sr-90, the ICRP Electronic Annex assigns Type M to Sr-90. As before, we use the default size of 5 μm. From the Electronic Annex, we must determine the Reference Bioassay Function for a 24-hour urine sample that ends at Day 5 (Reminder: For in-vitro bioassay, the value of m(t) is at the end of the sample collection period. From the figure below, the appropriate value of m(t) is 1.5×10^{-3}. As a side note, the plot of how the activity in 24-hour urine samples is expected to decline is shown on the left of the figure.

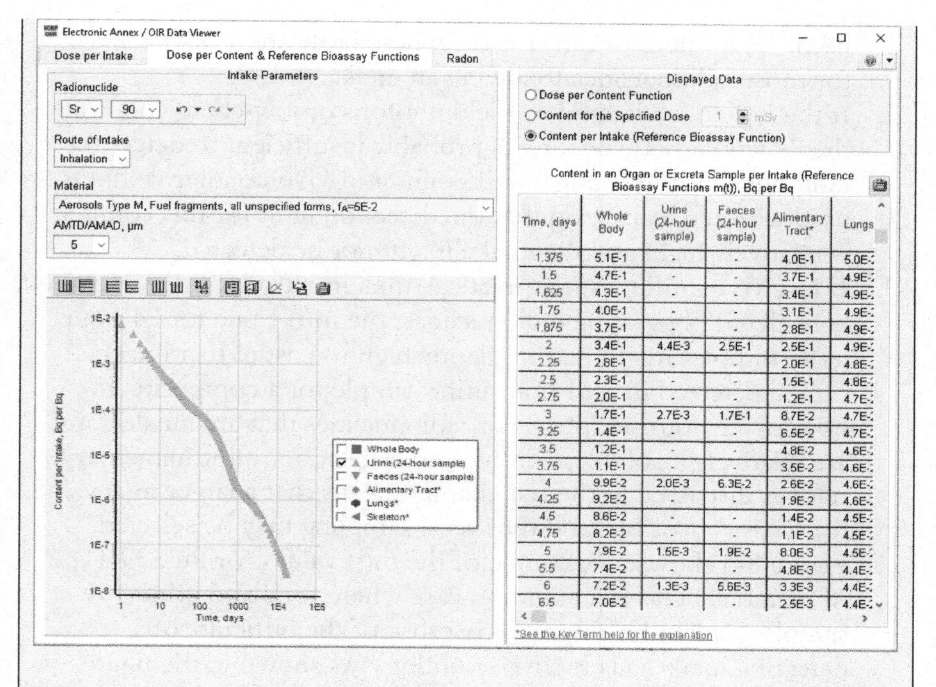

Time, days	Whole Body	Urine (24-hour sample)	Faeces (24-hour sample)	Alimentary Tract*	Lungs
1.375	5.1E-1	-	-	4.0E-1	5.0E-
1.5	4.7E-1	-	-	3.7E-1	4.9E-
1.625	4.3E-1	-	-	3.4E-1	4.9E-
1.75	4.0E-1	-	-	3.1E-1	4.9E-
1.875	3.7E-1	-	-	2.8E-1	4.9E-
2	3.4E-1	4.4E-3	2.5E-1	2.5E-1	4.9E-
2.25	2.8E-1	-	-	2.0E-1	4.8E-
2.5	2.3E-1	-	-	1.5E-1	4.8E-
2.75	2.0E-1	-	-	1.2E-1	4.7E-
3	1.7E-1	2.7E-3	1.7E-1	8.7E-2	4.7E-
3.25	1.4E-1	-	-	6.5E-2	4.7E-
3.5	1.2E-1	-	-	4.8E-2	4.6E-
3.75	1.1E-1	-	-	3.5E-2	4.6E-
4	9.9E-2	2.0E-3	6.3E-2	2.6E-2	4.6E-
4.25	9.2E-2	-	-	1.9E-2	4.6E-
4.5	8.6E-2	-	-	1.4E-2	4.5E-
4.75	8.2E-2	-	-	1.1E-2	4.5E-
5	7.9E-2	1.5E-3	1.9E-2	8.0E-3	4.5E-
5.5	7.4E-2	-	-	4.8E-3	4.4E-
6	7.2E-2	1.3E-3	5.6E-3	3.3E-3	4.4E-
6.5	7.0E-2	-	-	2.5E-3	4.4E-

Therefore,

$$I = \frac{M}{m(t)} = \frac{2.3\,pCi}{1.5 \times 10^{-3}} = 1533\,pCi = 56.7\,Bq$$

The committed effective dose is the intake multiplied by the committed effective dose coefficient, again obtained from the ICRP Electronic Annex. To be consistent with the assumptions used in the intake estimate, we use the dose coefficient for Type M Sr-90 of 5 µm particle size, which is 1.8×10^{-8} Sv/Bq. Thus, the committed effective dose is 1.0 µSv, or 0.1 mrem.

The value of m(t) for the various compartments can help decide the type of bioassay measurement that is performed in a given scenario. The decision process might follow a set of steps like those listed below:

1. If the radionuclide has a high-yield gamma ray that is emitted and is easily detected, then in-vivo counting is preferred; in all probability, using a traditional whole-body counter would be

used. The values of m(t) for the whole body are relatively large for most radionuclides following an intake.

2. If the radionuclide has low-yield photons or low-energy photons, then a whole-body counter is probably insufficient to detect the radionuclide, and a lung/chest counter in-vivo counter would be used with the realization that the detection limit for this counter is relatively high, and the intake might not be detected.
3. If in-vivo counting is not practical, then in-vitro bioassay must be considered. For some radionuclides, the m(t) value for 24-hour urine samples might be sufficiently high to ensure that intakes can be detected in a 24-hour urine sample (or a composite 48-hour or 72-hour sample). For radionuclides that are inhaled and are relatively insoluble, however, translocation to the alimentary tract as discussed in the last chapter means that a larger m(t) value for feces is possible, and thus fecal sampling may be selected. Figure 8-3 shows an example of the m(t) values for Pu-239 Type M material. This is a common case where fecal and urine sampling might be employed because of the difficulty of detecting intakes via in-vivo counting. As shown in the figure, m(t) values for fecal samples are several orders of magnitude higher than for urine samples, thus indicating greater probability of detection via fecal samples.

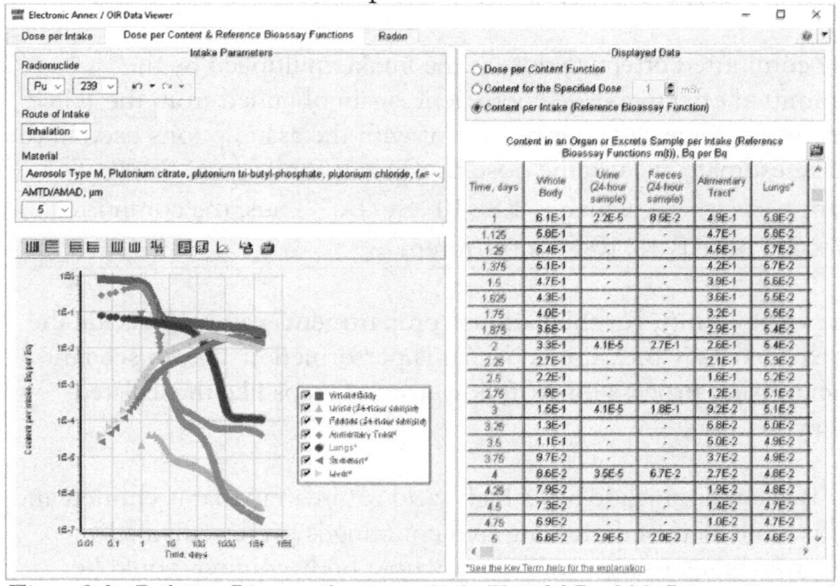

Time, days	Whole Body	Urine (24-hour sample)	Faeces (24-hour sample)	Alimentary Tract*	Lungs*
1	6.1E-1	2.2E-5	8.5E-2	4.9E-1	5.8E-2
1.125	5.8E-1	-	-	4.7E-1	5.8E-2
1.25	5.4E-1	-	-	4.5E-1	5.7E-2
1.375	5.1E-1	-	-	4.2E-1	5.7E-2
1.5	4.7E-1	-	-	3.9E-1	5.6E-2
1.625	4.3E-1	-	-	3.6E-1	5.5E-2
1.75	4.0E-1	-	-	3.2E-1	5.5E-2
1.875	3.6E-1	-	-	2.9E-1	5.4E-2
2	3.3E-1	4.1E-5	2.7E-1	2.6E-1	5.4E-2
2.25	2.7E-1	-	-	2.1E-1	5.3E-2
2.5	2.2E-1	-	-	1.6E-1	5.2E-2
2.75	1.9E-1	-	-	1.2E-1	5.1E-2
3	1.5E-1	4.1E-5	1.8E-1	9.2E-2	5.1E-2
3.25	1.3E-1	-	-	6.8E-2	5.0E-2
3.5	1.1E-1	-	-	5.0E-2	4.9E-2
3.75	9.7E-2	-	-	3.7E-2	4.9E-2
4	8.6E-2	3.5E-5	6.7E-2	2.7E-2	4.8E-2
4.25	7.9E-2	-	-	1.9E-2	4.8E-2
4.5	7.3E-2	-	-	1.4E-2	4.7E-2
4.75	6.9E-2	-	-	1.0E-2	4.7E-2
5	6.6E-2	2.9E-5	2.0E-2	7.6E-3	4.6E-2

Figure 8-3. Reference Bioassay function m(t) for Type M Pu-239 (Image courtesy of ICRP)

8.2.3.2 Practical Considerations in Bioassay Interpretation

As should be evidenced by now, the relation of measurements to dose for internal dose assessment differs from that for external dose assessment due to the nature of the measurements and variability that comes from several origins. While it is not possible to cover all the possible confounding factors when interpreting bioassay measurements, there are nonetheless some considerations to recognize as discussed below.

1. **Time of intake** – As demonstrated in the tables of reference bioassay functions, the bioassay measurement is not meaningful if a time of intake cannot be associated with the measurement. If a known incident has occurred, then the time of intake typically can be narrowed down to within a few hours. However, if the bioassay measurement was a routine sample for which no known incident had occurred, then it is necessary to assume a time of intake. The most conservative assumption would be that the intake occurred immediately after the most recent negative bioassay measurement. However, that may be needlessly conservative and is certainly not realistic. Thus, it is also defensible to assume a time of intake at the midpoint between the current positive bioassay measurement and the previous negative bioassay measurement.
2. **Sample collection time relative to intake** – It is common to assume incorrectly that following an intake, a bioassay measurement should be made as quickly as possible to have the best opportunity to detect the intake. However, it takes time for the intake to be incorporated into the body such that the mathematical models can be viewed with confidence. It may take several days (especially for feces) for the excreta measurement to be reflective of the intake with no dilution from excreta already present prior to the intake. This principle is true for in-vivo counting as well because of the time it may take for material to move to the predicted compartments for which the counting equipment is calibrated. Thus, it is usually advisable to treat early measurements (within the first day or two) circumspectly as they may not be representative of the models.
3. **Intervention/Chelation** – For some radionuclides, decorporation therapy/chelation therapy may be employed. A decision on treatment is made by a physician, who may seek

information on the amount of the intake or dose prior to making the decision. In particular, Ca-DTPA or Zn-DTPA (calcium- or zinc- diethylenetriamine pentaacetate) are approved for chelation of isotopes of plutonium, americium, and curium. More information on chelation therapy can be found at https://www.fda.gov/drugs/bioterrorism-and-drug-preparedness/questions-and-answers-calcium-dtpa-and-zinc-dtpa-updated. The goal of this therapy is to bind the radionuclide and enhance its removal with a shorter biological half-time than would normally be experienced. Therefore, we must recognize that the models in ICRP 130 will not be applicable to any measurements made while the chelating agent is influencing the removal because immediately after chelation, large "flushes" of activity will be excreted, giving rise to abrupt changes in any bioassay measurements. While some dosimetrists model intakes using bioassay data influenced by chelation with some success, another common approach is to use only bioassay data that were gathered after the influence of chelation has ceased in order to determine the intake. Jech et al. (1972 HPJ) determined that urine excretion rates return to near predicted levels within 15-25 days after the final chelation treatment; however, urinary excretion may still be elevated for another 100 days or longer.

4. **Mixtures of radionuclides** – When multiple radionuclides are involved in an intake, determining the amount for each radionuclide can be a challenge depending on the bioassay measurement. In particular, if in-vitro bioassay is performed, chemical separations are often needed to remove all but one element since alpha/beta counting of samples may not be able to distinguish between radionuclides. If multiple isotopes of the same element are present, chemical separations are not possible, and enhanced attention to the counting of the sample must be performed to estimate the contributions from the individual isotopes.
5. **Particle size** – Knowing the exact particle size of an inhaled radionuclide is not always possible in advance. Some facilities may use cascade impactors, laser diffraction sampling, or other equipment to determine the particle size in a given facility under operational conditions; however, during an unplanned event, there is no guarantee that the predetermined particle size is representative of what was inhaled.

6. **Chemical form** – Much like particle size, the chemical form of the radionuclide may not be known exactly. Based on processes occurring in a facility, an educated guess can be made. For instance, chemical processing of a particular radionuclide may mean it is in the form of an oxide or a nitrate. In addition, laboratory use of radionuclides in research generally means that the radionuclide is acquired in a specific chemical form. However, as with particle size, the chemical form of a radionuclide can change in the case of an unexpected event.
7. **Multiple bioassay measurements** – As mentioned in Section 8.2.3.1, Equation 8.4 for determining an intake applies to a single data point. Greater confidence in the intake and dose estimate can be obtained by making multiple measurements, and statistical techniques must be used to interpret these measurements. Equation 8.5 in an example later in this chapter shows one method for dealing with multiple measurements. However, more robust methods are usually implemented using computational tools. Some internal dosimetrists have written their own software, and there are also commercial software packages available.
8. **Bioassay measurements in conjunction with air sample measurements** – While air sample measurements may be used to provide a rough estimate of an intake as demonstrated in Section 8.1.3, it is not uncommon for intake estimates from air samples to be in conflict with intake estimates from bioassay measurements. This is not surprising, since air sample measurements are not always representative: that is, they may not be in the breathing zone of the individual or they may have been made prior to work beginning when the air was not disturbed. For that reason, bioassay measurements are usually preferred to air sample measurements. However, in the case of personal air sampler measurements (also called lapel air sampler measurements), the air sample results may provide a quick estimate of the potential intake if bioassay measurements are not available or if the detection limits for bioassay are insufficient to determine compliance with dose limits.
9. **Multiple bioassay methods** – For some intakes, multiple bioassay methods may contribute to the dose assessment, particularly when both urine and fecal bioassays are collected. Trying to resolve any disparities between urine and fecal samples

can be challenging at times; however, consideration of the possible errors in the data and adjustment of models to fit the data can often lead to a reasonable estimate. One caution should be emphasized in the use of nasal smears. Following a suspected intake, some facilities will use a swab to wipe inside the nostrils of the worker to determine if any activity is present in that area. The presence of activity on the swab may well indicate that an intake has occurred, but a quantitative dose assessment based on that measurement may be questionable. Further, a negative nasal swab is inconclusive because an intake could still have occurred, especially if the worker breathed primarily through the mouth.

10. **Timeliness of Bioassay Measurements** – Following an intake, particularly one that is unexpected, there is understandable attention given to obtaining an intake and dose estimate as quickly as possible. When in-vivo bioassay is sufficient to detect the radionuclide, an intake estimate may be obtained shortly after the intake. However, for in-vitro samples, it could be weeks to months before a defensible dose assessment can be performed, due to time required for sample collection, chemical preparation, and counting time.

8.2.3.3 Approaches to Analyzing Bioassay Data

Like many aspects of radiation protection, interpreting and analyzing bioassay measurements is a matter of professional judgment that may vary somewhat depending on the dosimetrist. However, any intake/dose assessment must be technically defensible and documented, including all assumptions that are made. This helps to ensure that there is a technical basis that can be used to reconstruct and review all steps that were taken to arrive at the dose estimate.

When reviewing the literature of internal dosimetry, there is understandably a wide variety of approaches ranging from simple basic assumptions to more detailed statistical methods to assess data. Sometimes a "sliding scale" approach is used in dose assessments, whereby the degree of sophistication and statistical analysis is related to the potential dose consequences. If the dose estimate is relatively small (which is subjective depending on the facility), then conservative simplifying assumptions may be used to arrive quickly at a dose estimate. However, if the dose consequences are large (again, a subjective term), then additional techniques may be employed to arrive at an estimate that employs more realistic assumptions that can still be justified and defended. The greater complexity and sophistication typically require more time to complete and would necessitate a more rigorous formal written report/defense of the process that is followed; however, it is necessary at times to avoid an overly conservative answer.

What constitutes a large dose or a small dose that may warrant more in-depth analysis? There are several factors that can be evaluated to arrive at the answer for a given facility.

- Would the initial dose estimate approach a regulatory dose limit?
- Would the initial dose estimate approach a facility-specific administrative limit for dose control?
- Would the initial dose estimate imply an unreasonably larger dose than would be expected for the work that was performed?
- Would the initial dose estimate cause unnecessary concern for the affected worker?
- Would the initial dose estimate cause unnecessary concern for the general public who is made aware of the intake and its dose consequence?

One downside to the sliding scale approach is a perceived inconsistency in analyzing cases. For those unfamiliar with the limitations of internal dose models and the inherent uncertainty in any dose estimate, it can appear that more rigorous approaches to data analysis are employed as an attempt to cover up significant intakes when the reality is that the dose consequences (and subsequent risk) of smaller intakes can be treated more conservatively to conclude that the risk from the intake is not significant.

Figure 8-4 below shows some possible approaches to addressing positive bioassay measurements. Fundamentally, whenever an intake occurs, a declaration must be made of the radionuclide identity, chemical form (which affects class for inhalation intakes or f_A value for ingestion intakes), particle size, and time of intake. In addition, the mathematical approach to determining the intake based on multiple bioassay data must be decided. Again, there may be some variation among dosimetrists as to what assumptions are best, but the assumptions should be selected based on reasonableness and technical defensibility.

The ICRP Electronic Annex provides easy access to data for making some adjustments in assumptions. For example, the particle size influence on the values of m(t) and on the corresponding dose coefficient can be determined quickly by selecting the alternate particle size. In some cases, using a different particle size helps the model to better fit the bioassay data.

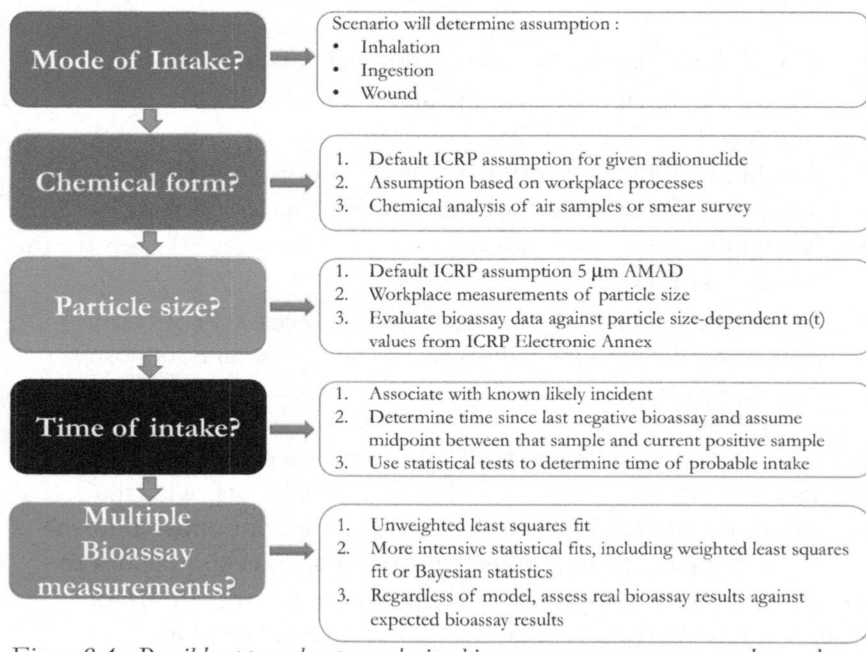

Figure 8-4. Possible approaches to analyzing bioassay measurements to conduct a dose assessment.

The following example serves to illustrate how the data after an intake might be handled when multiple bioassay data are available.

Example: A worker in a facility is exposed to airborne Co-60. Following the intake, the following bioassay results were obtained using a whole-body counter (WBC). What is a reasonable estimate of the intake and the committed effective dose?

Time following intake (day)	WBC Result (Bq)
1	9584
3	1753
6	1082
10	950

Solution: First, it is important to note that WBC for Co-60 is the appropriate bioassay method and is a very sensitive technique. Therefore, absent any other information, there is no reason to indicate that alternative bioassay methods (urine, feces) should be considered in this case. In addition, the scenario details that it is airborne Co-60, and thus we do not need to assume an ingestion or wound intake.

It is time to make some assumptions that can be used going forward in the calculations. With no information on the chemical form at the moment, the ICRP Electronic Annex identifies Type M as the default inhalation class for Co-60 (all unspecified forms). Further, with no information on particle size, the initial assumption will be 5 μm. If further data become available, these assumptions can be modified.

Figure 8-2 shows a screenshot of the partial information available for 5 μm Type M Co-60 from the ICRP Electronic Annex. By copying and pasting the full information into a table, it is possible to see all the various values of m(t) for the first 20 days following the intake.

At this point, the question rightly arises of how to deal with the multiple bioassay measurements. The table below shows the m(t) values along with an estimate of intake if each data point were used independently based on Equation 8.4 (i.e., I=M/m(t))

Time following intake (day)	WBC Result (Bq) M	m(t) for Whole Body	Intake estimate based on single data point
1	9584	0.61	15711.48
3	1753	0.15	11686.67
6	1082	0.058	18655.17
10	950	0.052	18269.23

Noticeably, there is some variation in the estimate of intake for the four data points. It is not difficult to imagine that if there were many data points, the variation of intake estimates based on each data point could be even greater. A simple arithmetic mean or even a geometric mean of the individual values could be considered as a suitable method to determine an average value of the intake if the variation is not too great. However, a more elegant technique that is often used when fitting data to a mathematical model is an unweighted least-squares fit to the data. When applied to multiple bioassay data, the best estimate of intake from this method is given below in Equation 8.5, which is taken from Report 37 of the International Atomic Energy Agency (IAEA). However, a form of this equation has been used for several decades in other publications.

$$I = \frac{\sum_i M_i \times m(t_i)}{\sum_i [m(t_i)]^2} \quad (8.5)$$

Where

- I is the estimate of intake (Bq).
- M_i is the ith bioassay measurement.
- $m(t_i)$ is the Reference Bioassay Function corresponding to the same time t as the ith bioassay measurement.

The summation carried out in the numerator as well as the denominator lends itself to a spreadsheet or other computer program to handle the arithmetic. Using this approach to the data for this example yields the following:

Time following intake (day)	WBC Result (Bq) M_i	$m(t_i)$ for Whole Body	$M_i \times m(t_i)$	$[m(t_i)]^2$
1	9584	0.61	5846.24	0.3721
3	1753	0.15	262.95	0.0225
6	1082	0.058	62.756	0.003364
10	950	0.052	49.4	0.002704
		Sum	6221.346	0.400668
		Intake	**15527.43 Bq**	

The next step is to see how well our estimate of intake fits the observed bioassay data. This is accomplished by plotting both the observed bioassay results and the expected bioassay curve (I×m(t)) on the same graph as shown below.

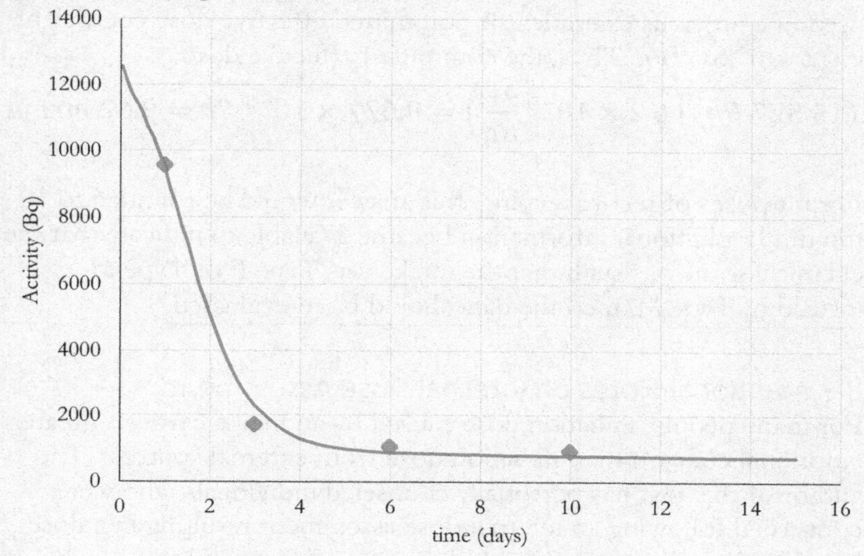

While not perfect, the expected bioassay curve does fit reasonably well to the measured data, thus giving us confidence that we could predict the results of future bioassay measurements. Hence, we have confidence in our estimate of the intake as well as our selection of the biokinetic model. NOTE: It is important to always plot the expected bioassay curve against observed measurements.

If the expected bioassay curve did not model the data correctly, then a different model would be needed. This could involve using a different inhalation type, changing the particle size, or possibly combining two different inhalation types. Additional statistical analyses could also be performed to estimate the intake rather than least squares. These modifications are not introductory in nature, however, and are beyond the scope of this text.

It may have been noticed that the expected bioassay result at t=0 is not equal to the intake. We must remember that the intake includes not only what remains in the body but also the amount that is exhaled. Thus, the projected bioassay result at t=0 for the inhalation pathway is less than the intake.

Now that the intake has been estimated, the calculation of the committed effective dose is accomplished using Equation 8.1. As used in a previous example, the committed effective dose coefficient is 6.2×10^{-9} Sv/Bq. Thus, the committed effective dose is:

$$(15{,}527\ Bq)\left(6.2\times10^{-9}\frac{Sv}{Bq}\right) = 9.627\times10^{-5}\ Sv = 9.62\ mrem$$

For purposes of recordkeeping, this answer would be rounded to 10 mrem. If additional information became available to indicate that the chemical form was such that the intake was Type F or Type M instead of Type M, then the data should be re-evaluated.

8.3 Additional Notes on Internal Dose Assessment

For many people, radiation dose caused by an intake carries a greater emotional concern than radiation dose from external sources. The author of this text has personally counseled individuals who were concerned following an internal dose assessment resulting in a dose less than 20 mrem even though the same workers might routinely receive tens or hundreds of millirem of dose from external sources. Besides calculating the dose, radiation protection personnel may be called upon to put the radiation dose into perspective; therefore, communication is critically important.

Being dismissive of the dose is not an acceptable approach; rather, the concern must be acknowledged. There are some points that may help put the internal dose in perspective:

- Every person receives internal dose from natural sources on a regular basis, as discussed in Chapter 3.
- Doses from intakes are evaluated over a long time period, and the dose is not acute as it might be for external doses.
- There is no enhanced biological damage from internal dose compared to external dose.

Based on questions or comments from the individual, it may be possible to narrow down the concern of the individual further and provide more information. While we must provide factual data, we must also address the reality that we are dealing with people who have legitimate concerns about their radiation dose.

An additional consideration for the person performing the dose assessment is the need to be reasonable in the dose estimation procedure. At times, it is tempting to provide an upper bound in the calculation and incorporate overly conservative assumptions. While this may expedite the calculation and help give an order of magnitude result to guide with follow-up actions, presenting that result as a preliminary dose estimate to the public, management, or the individual can have some unintended consequences. Particularly in the case of actinide intakes, where a dose estimate could take months to complete, preliminary estimates are prone to be revised as more bioassay data become available, especially if chelation therapy has been employed. Regardless of the number of caveats that are attached to the preliminary dose estimate, the number itself will be remembered, which could consequently lead to non-dosimetrists arriving at one of two incorrect conclusions when the final dose estimate is reported.

1. If the final dose estimate is substantially higher than the preliminary estimate, then the uninformed may conclude that the dosimetrist overlooked important information in the original dose estimate.
2. If the final dose estimate is substantially lower than the preliminary estimate, then the uninformed may conclude that the dosimetrist is intentionally hiding or discounting information to achieve a more favorable result.

Thus, it is important that dose estimates not be reported, even informally, until the data and dose assessments have been reviewed thoroughly. The usual process is for the dosimetrist to prepare the initial dose estimate paperwork with supporting documentation to a competent technical reviewer who can perform an independent assessment to confirm the original evaluation and resolve any discrepancies in assumptions or calculations.

In this chapter, our focus has been on calculating the committed effective dose. However, for some radionuclides, the equivalent dose to a given organ may be of interest, especially for those organs/tissues where an equivalent dose regulatory limit may be exceeded even though the effective dose may not exceed its respective regulatory limit. To aid in assessing organ equivalent doses, the ICRP Electronic Annex has a feature that provides the $h_T(50)$ dose coefficients. Figure 8-5 shows how the values are tabulated for Sr-90 once the "Dose per intake" and "Materials" tab are selected. A scroll bar at the bottom of the window allows viewing all the dose coefficients for both female and male reference individuals.

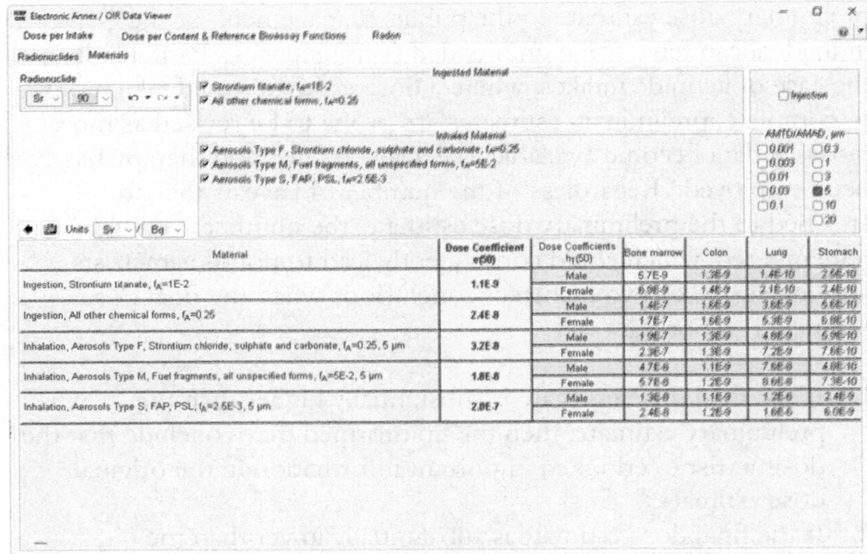

Material	Dose Coefficient e(50)	Dose Coefficients $h_T(50)$	Bone marrow	Colon	Lung	Stomach
Ingestion, Strontium titanate, f_A=1E-2	1.1E-9	Male	5.7E-9	1.3E-9	1.4E-10	2.5E-10
		Female	6.9E-9	1.4E-9	2.1E-10	2.4E-10
Ingestion, All other chemical forms, f_A=0.25	2.4E-8	Male	1.4E-7	1.6E-9	3.8E-9	5.6E-10
		Female	1.7E-7	1.6E-9	6.3E-9	6.8E-10
Inhalation, Aerosols Type F, Strontium chloride, sulphate and carbonate, f_A=0.25, 5 µm	3.2E-8	Male	1.9E-7	1.3E-9	4.8E-9	5.9E-10
		Female	2.3E-7	1.3E-9	7.2E-9	7.6E-10
Inhalation, Aerosols Type M, Fuel fragments, all unspecified forms, f_A=5E-2, 5 µm	1.8E-8	Male	4.7E-8	1.1E-9	7.6E-8	4.8E-10
		Female	5.7E-8	1.2E-9	8.6E-8	7.3E-10
Inhalation, Aerosols Type S, FAP, PSL, f_A=2.5E-3, 5 µm	2.0E-7	Male	1.3E-8	1.1E-9	1.2E-6	2.4E-9
		Female	2.4E-8	1.2E-9	1.6E-6	6.0E-9

Figure 8-5. Equivalent dose coefficients for Sr-90 for ingestion and inhalation intakes from the ICRP OIR Data Viewer; the scroll bar at the bottom allows viewing of dose coefficients for additional organs/tissues (Image courtesy of ICRP).

Free Resources for Chapter 8

- Questions and problems for this chapter can be found at www.hpfundamentals.com
- A journal writeup of plutonium intake cases showing the dose assessment process when chelation is involved following a wound intake is available at: https://www.researchgate.net/publication/46256021_Three_plutonium_chelation_cases_at_Los_Alamos_National_Laboratory
- ICRP Publication 130, *Occupational Intakes of Radionuclides: Part 1*. Ann. ICRP 44(2), 2015. Available from www.icrp.org.
- ICRP Publication 134, *Occupational Intakes of Radionuclides: Part 2*. Ann. ICRP 45(3/4), 1–352, 2016. Available from www.icrp.org.
- ICRP Publication 137, *Occupational Intakes of Radionuclides: Part 3*. Ann. ICRP 46(3/4), 2017. Available from www.icrp.org.
- ICRP Publication 141, *Occupational Intakes of Radionuclides: Part 4*. Ann. ICRP 48(2/3), 2019. Available from www.icrp.org.
- The electronic annex developed by the ICRP that includes the relevant dose coefficients and other information can be downloaded freely from the ICRP web site at https://www.icrp.org/docs/ICRP%202022%20OIR%20Electronic%20Annex%20Distribution%20Set.zip (this link can be accessed via the web page for ICRP Publication 151.
- International Atomic Energy Agency, *Methods for Assessing Occupational Radiation Doses due to Intakes of Radionuclides*, Safety Reports Series No. 37, IAEA, Vienna (2004)

Additional References

- ICRP Publication 151, *Occupational Intakes of Radionuclides: Part 5*. Ann. ICRP 51(1-2), 2022. Available from www.icrp.org.
- Jech, J.J., Andersen, B.V., and Heid, K.R. *Interpretation of Human Excretion of Plutonium for Cases Treated with DTPA*, Health Physics Journal, 22:787-792, 1972.

Chapter 9 Controlling Internal Exposures in the Workplace

As with external radiation exposures, there are some techniques and practices that can reduce the exposure of workers to internal sources of radiation. In fact, planning and use of technology can reduce the internal exposure drastically for many scenarios.

Most intakes in the workplace are due to airborne radioactivity. This does not mean that all intakes are purely inhalation intakes. As discussed in the ICRP models for intakes, a worker may often breathe through the mouth while performing strenuous actions such that the intake may be due in part to material being deposited in the oral cavity and swallowed in addition to material being inhaled.

Airborne radioactivity in the workplace often accompanies loose contamination that may be present in the same location. In some cases, the airborne radioactivity is the result of surface contamination that is resuspended in the air. In particular, contamination on floor surfaces may be resuspended as people walk about in an area, much like walking about in a dusty room causes increases in airborne dust in the room. The amount of loose material that can be resuspended depends on many factors such as humidity and particle size. This is characterized by a quantity known as the resuspension factor as defined in Equation 9.1 below.

$$f_r = \frac{airborne\ concentration\ (\frac{Bq}{m^3})}{surface\ contamination\ (\frac{Bq}{m^2})} \qquad (9.1)$$

The resuspension factor is often reported in units of m^{-1}, but Equation 9.1 shows that this is a mathematical simplification of the actual physical units (i.e., airborne concentration per unit surface area contamination).

In the absence of specific data for a given scenario, a commonly assumed value of the resuspension factor is 10^{-6} m^{-1}; one reference for this value is NUREG/CR-5512, which reviewed literature from 1964 to 1990. Physically, this means that for a given 1 m^2 of floor/ground area, one-millionth of the contamination in the area is assumed to be distributed in 1 m^3 of the air above that area. The resuspension factor is intended for crude calculations for planning purposes; actual measurements of airborne radioactivity in the workplace would always take precedence in real-world scenarios. However, when planning prior to monitoring, the resuspension factor can help in making decisions on monitoring, the degree of necessary decontamination efforts, etc.

Clearly, measures to reduce internal exposures must address both the issue of loose contamination as well as airborne radioactivity that may arise from other sources (leaks of gaseous radionuclides from systems, evaporation of liquid sources, etc.). The following sections are intended to present basic techniques and issues to consider when designing a program to control internal exposures.

9.1 Annual Limit on Intake and Derived Air Concentration

The concepts of the Annual Limit on Intake (ALI) and the Derived Air Concentration (DAC) were first introduced in ICRP Publication 30 as useful quantities that could be enacted in the workplace to control internal exposures.

ICRP 30 defined the ALI as follows:

> *Annual Limit on Intake (ALI)*
> *The activity of a radionuclide which taken alone would irradiate a person, represented by* Reference Man, *to the limit set by the ICRP for each year of occupational exposure.*

One complicating factor to this definition was that there were two separate ICRP dose limits to consider: a 50 mSv committed effective dose equivalent limit as well as a 0.5 Sv committed dose equivalent limit (more information on the ICRP dose limits is in Chapter 10).

Due to biokinetics and dosimetry, some radionuclides would cause the 0.5 Sv committed dose equivalent limit to be exceeded before the 50 mSv committed effective dose equivalent limit would be reached (one example of this was I-131, where an intake could cause the dose to the thyroid to exceed 0.5 Sv while the committed effective dose equivalent would still be less than 50 mSv). Thus, the ALI might be set by the committed effective dose equivalent limit for some radionuclides while being set by the committed dose equivalent to a tissue for other radionuclides.

In Publication 60 and also in Publication 103, the ICRP modified the definition of the ALI so that it is based only on the committed effective dose as shown below in Equation 9.2; the ICRP also changed the value of the dose limit that should be used in the calculation.

$$ALI = \frac{E_{limit,w}}{e(50)} \qquad (9.2)$$

Where

- ALI is the Annual Limit on Intake for a particular radionuclide
- $E_{limit,w}$ is the committed effective dose limit (Sv). In Publication 60 and in Publication 130, this is specified as 0.02 Sv for purposes of calculating the ALI.
- e(50) is the committed effective dose coefficient discussed previously (Sv/Bq)

The ALI thus gives dosimetrists a rough idea of the magnitude of dose consequences following an intake by quickly comparing the intake to the ALI. However, to provide another tool to prospectively be able to protect workers, ICRP 30 initially defined the Derived Air Concentration (DAC) as follows:

> *Derived Air Concentration (DAC)*
> *equals the ALI (of a radionuclide) divided by the volume of air inhaled by Reference Man in a working year (i.e., 2.4×10^3 m^3). The unit of DAC is Bq m^{-3}*

As discussed in Chapter 7, the current assumption in ICRP 130 of breathing rate is that a worker would breathe in 1.1 m^3/hour, which is different from the assumed breathing rate of ICRP 30. When multiplied by 2000 working hours in a year (40 hours per week for 50 weeks), the annual volume of air inhaled at work is 2200 $m^{3.}$ Thus, the DAC under the newest ICRP recommendations is calculated mathematically as shown in Equation 9.3.

$$DAC = \frac{ALI\ (Bq)}{2200\ m^3} \tag{9.3}$$

The DAC is a natural extension of the ALI to help manage the workplace. By definition, the DAC is the concentration that if breathed for 2000 hours per year would lead to an intake of 1 ALI. This is not to say that a worker should be exposed to 1 DAC of airborne activity continually. It merely gives radiation protection personnel the basis for evaluating airborne concentrations of radioactive material. In the workplace, air concentrations are measured as part of characterizing work areas. However, the concentration value itself does not indicate the potential dose consequence from inhalation because each radionuclide has its own dose coefficient. By comparing the measured air concentration against the DAC for the radionuclide of concern, decisions can be made about posting the area (posting means putting signs in place that indicate the radiological hazard) and identifying protective measures to minimize intakes before they occur.

The other advantage of the DAC concept is that airborne concentrations for radionuclides can be expressed as multiples of the DAC; for instance, a concentration of two times the DAC is expressed as "2 DACs." This is similar to how velocities of objects in air are expressed as multiples of the speed of sound (e.g., Mach 2, Mach 3, etc.). Thus, rather than having to know the exact DAC value for every nuclide, the concentrations can be quickly expressed in relation to the DAC to communicate the degree of radiological risk. This also allows the evaluation of airborne environments in which multiple radionuclides are present. Just as doses are additive, DACs are additive; for example, if multiple radionuclides are present in a room where one radionuclide is 1 DAC, a second radionuclide is 4 DAC, and a third radionuclide is 0.5 DAC, the health physicist can quickly recognize that the room has an effective concentration of 5.5 DAC.

Example: Calculate the ALI and the DAC for Cs-137 and tritiated water (HTO) vapor using the most current ICRP definitions.

Solution: The most current ICRP definitions of the ALI and DAC rely on using the effective dose coefficient as shown above in Equation 9.2. From the Electronic Annex/OIR Data Viewer discussed in Chapter 7, the following effective dose coefficients are appropriate to use:

Cs-137 (5 μm), Type M (unspecified forms): 5.6×10^{-9} Sv/Bq
H-3 (tritiated water vapor) Type V: 2.0×10^{-11} Sv/Bq

Using Equation 9.2 above, the ALI for Cs-137 is

$$ALI = \frac{E_{limit,w}}{e(50)} = \frac{0.02\ Sv}{5.6\times10^{-9}\ \frac{\text{Sv}}{\text{Bq}}}$$
$$= 3{,}571{,}428\ Bq\ (3.57\ MBq)$$

Using Equation 9.3, the DAC for Cs-137 is

$$DAC = \frac{ALI\ (Bq)}{2200\ m^3} = \frac{3.57\ MBq}{2200\ m^3} = 1623\ \frac{Bq}{m^3} = 1600\ \frac{Bq}{m^3}$$

The same calculations for H-3 in the form of tritiated water vapor yield an ALI of 1 GBq and a DAC of 4.5×10^5 Bq/m^3.

Example: A room in a facility is measured to have an airborne concentration of Cs-137 of 6000 Bq/m^3. How would this concentration be expressed in DACs?

Solution: As derived above, the DAC value for Cs-137 is 1600 Bq/m^3. Thus, the calculation would be based on unit analysis as follows:

$$\left(6000\frac{Bq}{m^3}\right)\left(\frac{1\ DAC}{1600\frac{Bq}{m^3}}\right) = 3.75\ DAC$$

This example shows the versatility of using the DAC to quickly assess radiological hazard in a particular area. The raw concentration value of 6000 Bq/m^3 does not carry meaning by itself without a knowledge of the dose consequences. However, by expressing the concentration as a multiple of the appropriate DAC for Cs-137, the proper perspective is obtained of the potential dose consequences if a worker should be present in that area.

From the definitions of ALI and DAC, it should be clear that for a Reference Person who is exposed to a concentration of 1 DAC for a period of 2000 hours, we would expect an intake of 1 ALI by inhalation to occur, leading to a committed effective dose of 0.02 Sv. Another way to view this is that the product of the number of DACs to which a person is exposed and the exposure time leads to an operational unit known as the DAC-hour, such that 2000 DAC-hours leads to an intake of 1 ALI and a potential committed effective dose of 0.02 Sv.

To take this one step further, an inhalation of 1 DAC-hour leads to a potential intake of 1/2000 ALI, which would in turn lead to a potential committed effective dose of 1/2000 (0.02 Sv) or 10 μSv. This is not to say that 1 DAC-hour equals 10 μSv; as defined, the inhalation has the potential for Reference Person to receive that dose. Thus, the number of DAC-hours evaluated for potential work gives the ability to prospectively assess the potential dose consequences and decide what actions may be necessary prior to work being performed.

Example: A room in a facility has a concentration of 8 DAC, and a worker needs to perform work in the room that will last 2.5 hours. What is the potential dose consequence of this exposure?

Solution: We assess this scenario using the product of the number of DAC and the exposure time in hours, which is (8 DAC)(2.5 hours) = 20 DAC-hours.

The potential dose consequence is then calculated as:

$$20\ DAC - hours \frac{0.02\ Sv}{2000\ DAC - hours} = 0.2\ mSv$$

Within the framework of regulations in the United States, it is permissible in some situations to assign an official internal dose based on the DAC-hours to which a person is exposed if bioassay measurements are too insensitive to detect a known intake. While not necessarily preferable, it is sometimes necessary to use this method to compensate for technology shortfalls in detection capabilities for bioassay measurements.

NOTE: In the United States, regulations promulgated in 10CFR20 (Nuclear Regulatory Commission licensees) and 10CFR835 (Department of Energy facilities) are based on the definitions of ALI and DAC as provided in ICRP 26/30 Recommendations. Therefore, the published DAC values in these regulations were calculated using higher dose limits than those defined in ICRP 60 and ICRP 130. Therefore, 1 DAC-hour using the published DAC values in these regulations does not potentially lead to 10 μSv committed effective dose. Instead, 1 DAC-hour potentially leads to either 0.05 Sv committed effective dose or to 0.5 Sv committed equivalent dose to a given tissue.

In addition, the concept of DAC-hours can be used as a trigger point for other actions, such as requiring a specific type of bioassay. Thus, as part of planning radiological work, a requirement could be put into place that if a worker is projected to exceed a threshold value of DAC-hours, a bioassay would be required.

More discussion on the use of DAC-hours and how it relates to the overall radiation protection program is discussed in Chapter 10 where we will examine the necessary decisions when considering workplace information on both external and internal sources of radiation.

9.2 Protective Measures to Reduce Internal Exposure

9.2.1 Source Term Reduction and Containment

By far, the best way to reduce internal exposures is to ensure that the environment itself has as little airborne radioactivity as possible. Two of the ways to reduce the activity are discussed below.

9.2.1.1 Area decontamination

As discussed earlier in this chapter, one source of airborne radioactivity is the resuspension of surface contamination. Thus, it should be expected that reducing surface contamination in work areas should reduce intakes on two fronts. First, decontamination reduces the amount of available radioactivity that could be directly transferred to a worker's skin when the worker is in the area and might brush up against a contaminated surface (contamination on the skin, especially the hands, could lead to an ingestion intake). Secondly, decontamination reduces the amount of available radioactivity that can be resuspended in the area that could lead to an inhalation intake.

The process of decontamination requires persistence and work but is not necessarily complicated. Surface contamination could be comprised of residue from evaporated liquids or could be attached to dust particles that are on the surface. Either way, the method to remove contamination is the same as if it were nonradioactive: the use of detergents in water and a rag or mop to wipe it away. To determine the success of the effort, a survey of the surface(s) using portable instrumentation would be performed prior to the cleaning and then after the cleaning is complete.

If cleaning efforts are unsuccessful in removing the contamination, then the contamination is said to be "fixed" as opposed to "loose"; to further ensure that it does not become loose in the future, the surface can be painted or treated with some other coating that will immobilize it.

Because of its mobility in nature, tritium may appear to have been removed successfully from surfaces by decontamination efforts; however, there are many instances where a surface has been decontaminated only to have tritium that was apparently deep within the finish coat of the surface migrate back to the surface a few hours later to detectable levels. Thus, facilities that use tritium have special considerations with regard to source term reduction.

9.2.1.2 Area ventilation

The design of ventilation in industrial facilities is a discipline unto itself that cannot be covered in a brief introductory text like this book. However, there are some fundamentals with which radiation protection personnel should be familiar.

Unlike residential ventilation design that largely relies on recirculated air, areas with potential airborne contamination are designed to have air pass through the area but not be recirculated. This reduces the risk of continual buildup of airborne radioactivity in the area.

In addition, the flow of air from room to room is important. As air moves from one room to the next, the ventilation system must be designed so that contamination does not move to an uncontaminated area, especially when doors are opened (opening doors causes pressure changes in rooms). Airflow should always move from uncontaminated areas to contaminated areas. If many rooms are involved in the ventilation design, then airflow should move from uncontaminated areas to progressively more contaminated areas so that the last room in the ventilation plan is the most contaminated area.

As a further consideration, the air from the facility is usually exhausted via a ventilation stack (which to the general public may mistakenly appear to be a smokestack because it may protrude above the building). However, before the air is discharged, it must be filtered to remove as much contamination as possible. This is accomplished usually by the use of a high efficiency particulate air (HEPA) filter that is designed to remove particulates while still allowing adequate airflow.

These concepts are not restricted to radiological facilities; indeed, facilities that have chemical, biological, or other possible airborne hazards must use these same concepts. Therefore, there is much information in the literature across multiple disciplines on designing ventilations systems to handle airborne concerns.

Figures 9-1 and 9-2 below show a simplified view of a scenario where a contaminated area is next to a clean area. As shown in the figures, the air flows from the clean area into the contaminated area before being filtered and ventilated through the stack. Because the contaminated area is nearest the exhaust fan, it is under negative pressure with respect to the surrounding areas. Thus, air will flow into the room through any openings so that the spread of contamination is minimized. This highlights the need for good ventilation design and the need to address any changes in airflow in a facility. Actions as simple as propping a door open can change the airflow patterns and cause a good design to be invalidated with the potential spread of contamination (this is why many facilities have rules against propping doors open).

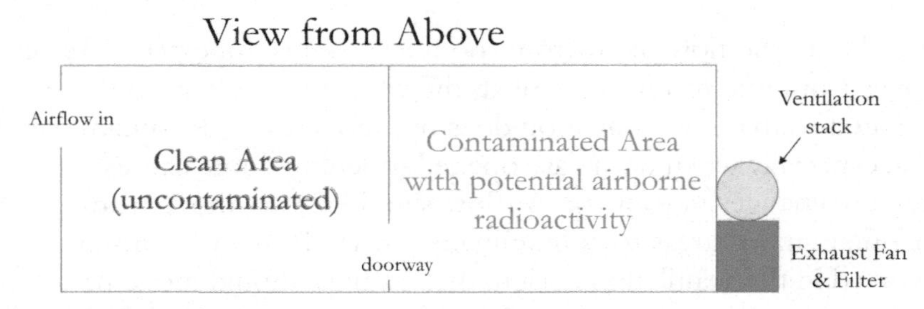

Figure 9-1. Ventilation design for two rooms where a clean room is adjacent to a contaminated area (overhead view).

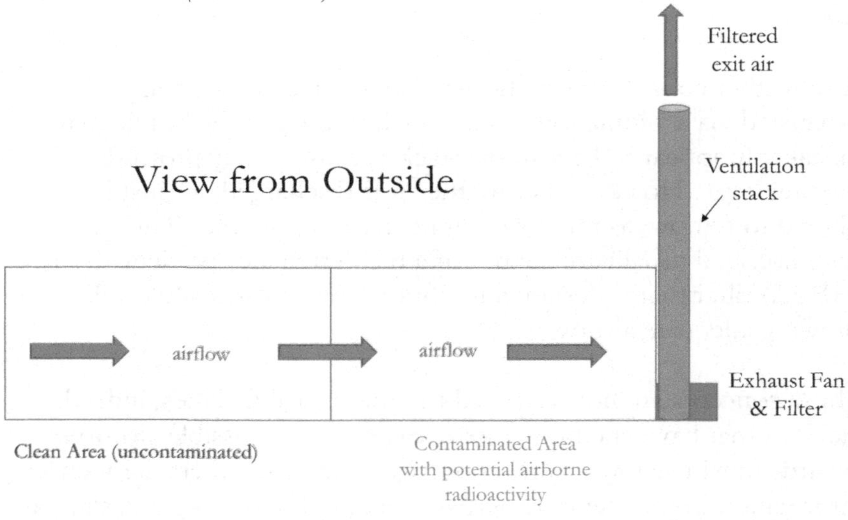

Figure 9-2. The same scenario from Figure 9-1 but from a view outside the facility to show the protruding ventilation stack above the facility.

While the ventilation design is often based on the airflow needed to maintain proper pressures between rooms or to maintain temperature and humidity, there is the added benefit that the inflow of clean air to a contaminated area is able to reduce the airborne concentration in the area.

Much like with radioactive decay, first order kinetics is often assumed for the sake of simplicity when calculating time-dependent airborne concentrations. If the concentration in the room would be constant without ventilation (i.e., no new airborne activity is being produced) and radioactive decay is negligible for the timeframe of interest, then the concentration in the room can be represented by Equation 9.4.

$$C = C_0 e^{-Jt} \tag{9.4}$$

Where

- C is the concentration at some time t
- C_0 is the concentration with no ventilation
- J is the ventilation removal rate constant calculated as the ventilation rate (m^3/hour) divided by the room volume (m^3) such that J has units of $hour^{-1}$. Other time units can be used so long as J and t have reciprocal units.
- t is the elapsed time in hours or an alternative unit of time consistent with how J is defined

More sophisticated differential equations are necessary if:

- Airborne radioactivity is being continually produced in the room.
- The airflow into the room is not clean but comes from one or more rooms that are also contaminated.
- The room does not have well-mixed air.

Example: With no ventilation, a room that measures 5 m x 6 m x 4 m has an airborne concentration of 2500 Bq/m^3 of Cs-137. If the room is ventilated at a rate of 300 cfm (ft^3/min), what would the expected airborne concentration be after one hour of ventilation?

Solution: Ventilation is often rated in cfm; so the first step is to convert this to a more suitable unit. Since the premise asks for the concentration after one hour, we can opt to use hours as our time unit. The room dimensions (and hence room volume) will be in SI units; therefore, it seems logical to express the ventilation rate in m^3/hour. The unit conversion of 300 cfm leads to approximately 510 m^3/hour. The room volume is 120 m^3 based on the room dimensions; so, the ventilation removal rate constant J is thus:

$$J = \frac{room\ ventilation\ rate}{room\ volume} = \frac{510\ m^3/hr}{120\ m^3} = 4.25\ hr^{-1}$$

C_0 is given in the premise of the equation as 2500 Bq/m^3; therefore, the application of Equation 9.4 yields the following:

$$C = C_0 e^{-Jt} = \left(2500 \frac{Bq}{m^3}\right) e^{-(4.25hr^{-1})(1\ hr)} = 36 \frac{Bq}{m^3}$$

As with radioactive decay, the concentration given by this equation theoretically will never reach zero but can approach it for all practical purposes.

9.2.1.3 Containment of Radioactive Material

One of the most effective means to reduce internal exposure is to contain the radioactive material so that workers are not exposed. Some methods that can be used involve the use of fume hoods such as one might see in a chemistry laboratory. When operating properly, the fume hoods are at negative pressure with respect to the surrounding room. If radioactive material is handled in the hood with the proper airflow, then room air is drawn into the hood at a velocity that prevents airborne radioactivity from exiting the hood into the room. The air is then drawn into the hood ventilation system and filtered before being discharged.

Another approach is to use a containment system known as a glovebox or a series of gloveboxes that are connected to each other. The interior of the glovebox is maintained at negative pressure, and the workers would insert their hands into the gloves to allow them to work with the material without the attendant risk of airborne radioactivity or skin contamination. Air from the glovebox is filtered before being discharged to the environment.

Figure 9-3 below shows a diagram of both a glovebox and a fume hood that can be used for handling radioactive materials to minimize the risk of the worker inhaling airborne radioactivity.

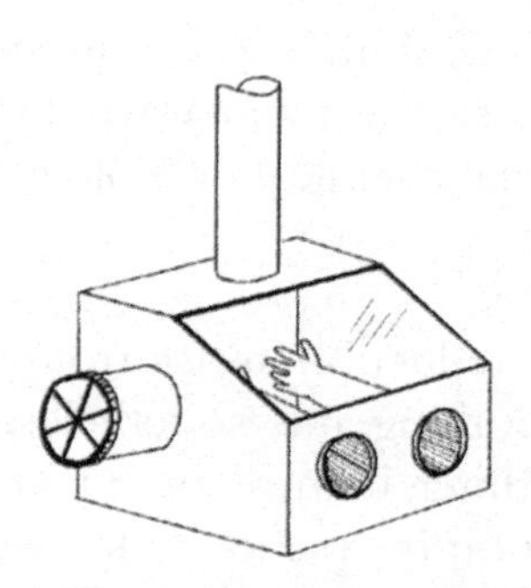
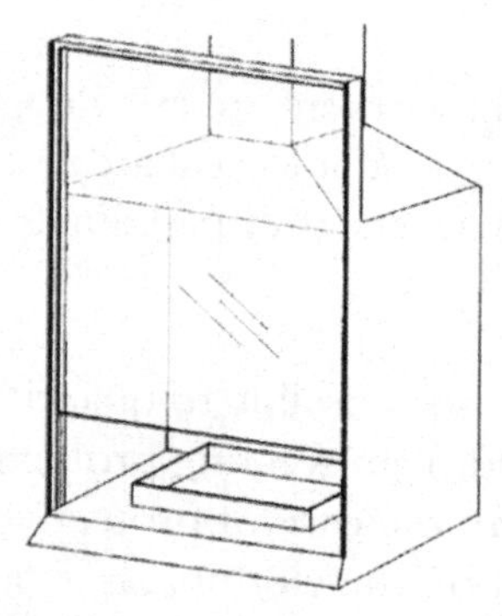

Figure 9-3. Illustration of a glovebox (left) and a fume hood (right) to reduce worker exposure to airborne radioactivity. (Image courtesy of IAEA from IAEA Practical Radiation Technical Manual PRTM-5, "Personal Protective Equipment," 2004).

9.2.2 Protective Clothing and Protective Actions

It is common to require workers to wear some form of protective clothing that is commensurate with the risk when the worker either enters a contaminated area or handles radioactive material. Protective clothing may include one or more of the following:

- Waterproof shoe covers worn over fabric covers
- Waterproof gloves worn over fabric gloves
- Coveralls
- Head covering (cap or full head covering)

The goal of the protective clothing is to reduce the risk of getting contamination on the skin where it could cause radiation dose or could become an unintentional ingestion hazard if it is transferred to the hands unwittingly.

While protective clothing is a viable method for reducing the risk of intake, there are other basic behaviors/actions that are used to reduce the risk of intake as well. In particular, the following are forbidden when entering radiological areas where radioactive material may be present:

- Eating
- Drinking
- Smoking
- Chewing tobacco products, gum, or another foodstuff

In addition, workers are encouraged to wash their hands immediately after exiting a radiological area. This is similar to what would be asked of workers after handling potential chemical or biological hazards.

While it is not possible to quantify the reduction in risk from these actions and from wearing protective clothing, it is nonetheless good practice. In the case of protective clothing, if any contamination should get on the clothing, it is (a) not on the person's skin where it may be difficult to remove and (b) separated from the person's skin to greatly reduce the possible dose from the beta particles that may be emitted from the contamination.

The type and extent of protective clothing will depend in large degree on the risk and type of facility. For some facilities, lab coats, gloves, and shoe covers are sufficient. In other facilities, full coveralls may be necessary. The material from which the protective clothing and gloves are made will vary as well. In some cases, the clothing is made of cotton and is laundered after each use so that it is reusable. Other facilities may choose to use disposable clothing. There are advantages and disadvantages to both, and it is a facility-specific decision to make. Table 9-1 below shows some of the considerations when selecting the type of protective clothing.

Table 9-1. Considerations of cotton versus disposable protective clothing.

Type of coverall	Advantages	Disadvantages
Cotton	• Breathable fabric • Relatively comfortable • Reusable • Minimal static electricity buildup	• Laundering cost • Returned laundered garments may still have slight contamination • May soak up liquids leading to accidental skin contamination
Disposable	• Lightweight and compact • One use and done	• Solid waste disposal cost • May not be breathable fabric and may tear easily • Cost of replacing

9.2.3 Respiratory Protection

For those situations where it is necessary to have a worker perform tasks in an area with airborne radioactivity, respiratory protection may be prescribed to reduce the intake. The use of respiratory protection is not trivial, however, and is not a matter of simply giving someone a mask to wear. There is a three-step process to follow before a person is provided with any respiratory protection:

1. The person must have a medical examination to ensure that the respiratory protection will not provide an undue stress on the person that could threaten the person's health. Each form of respiratory protection puts physical strain on the wearer and can exacerbate respiratory or cardiological issues that the person has.
2. The person must undergo a quantitative "fit test" for respiratory protection to ensure that the shape of the person's face is compatible with the device to provide a sufficient level of protection.
3. The person must be sufficiently trained in the proper wearing and use of respiratory protection.

The degree to which respiratory protection equipment is effective is quantified by its protection factor (PF). The protection factor is defined as shown below in Equation 9.5.

$$PF = \frac{concentration\ outside\ the\ respirator}{concentration\ inside\ the\ respirator} \quad (9.5)$$

From its definition, it is easy to see that it is desirable to have a protection factor that is as high as possible. The assigned value of protection factor depends on the type of respiratory equipment as well as the fit of the equipment on the individual.

There are a variety of respiratory protection types in use, with advances being made on a regular basis. However, respiratory protection typically falls into two categories: filtration-type respiratory protection and supplied-air respiratory protection.

Within the category of filtration-type respiratory protection, two of the common ones are half-face respirators and full-face respirators as shown below in Figure 9-4. Filtration-type respirators tend to have lower protection factors than supplied-air respiratory protection. This is due to several factors. First, filtration cartridges are not 100% effective in removing all airborne contaminants. Secondly, the seal of the respirator around the face is not always perfect, especially for the half-face respirator where movement of the cheeks of the person can cause a breach of the seal. Thirdly, because of the nature of filtration respirators, when a person inhales, there is a negative pressure inside the respirator compared to outside the respirator so that any leaks in the seal will cause airborne material to enter the respirator around the seal rather than through the filters.

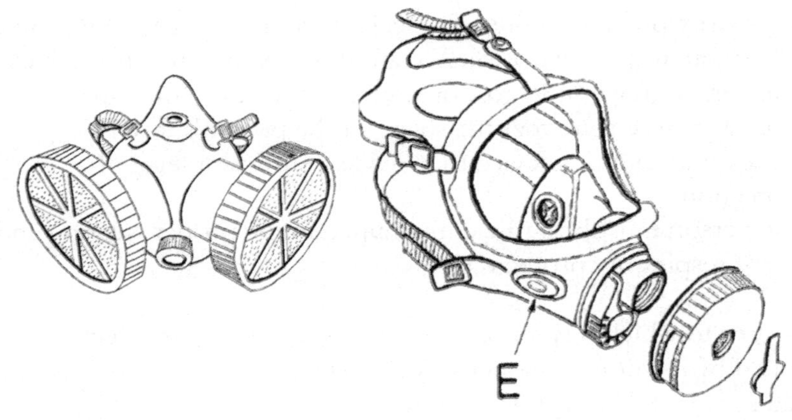

Figure 9-4. Half-face respirator (left) and full-face respirator (right) showing the exhalation valve (E). (Image courtesy of IAEA from IAEA Practical Radiation Technical Manual PRTM-5, "Personal Protective Equipment," 2004).

There are also versions of these filtered-air respirators that are powered, meaning that a fan/compressor draws in room air, filters it, and then supplies it to the breathing mask. This tends to make it easier for the person to breathe, keeps positive air pressure inside the mask, and thus results in higher protection factors.

Supplied-air respiratory protection uses air from outside the room to provide breathable clean air to the worker through a mask. In some cases, it is provided through a portable tank similar to what firefighters wear; this is referred to as a self-contained breathing apparatus (SCBA). In other situations, the air may be supplied via compressed air lines installed permanently in the facility (designated only for breathing air as opposed to compressed air for operating tools or equipment). For especially contaminated areas or where airborne tritium is an issue (tritium migrates through protective clothing easily and can be absorbed through the skin), a full suit that is supplied by external air may be used. The suit is made of vapor-proof material to prevent vapors and gases from migrating inward through the suit, but the entirety of the suit is under positive pressure and thus keeps contamination and airborne activity away from the worker when properly used. Figure 9-5 below shows examples of supplied-air respiratory equipment.

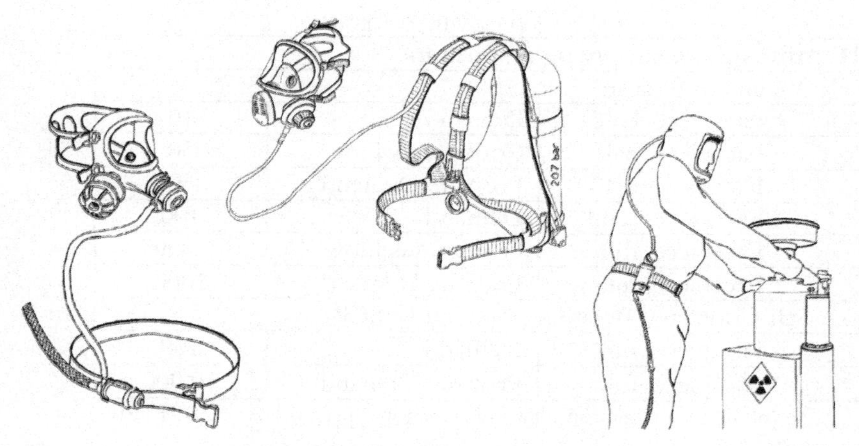

Figure 9-5. Examples of air-supplied respiratory protection, including air-line supplied mask (left), self-contained breathing apparatus (SCBA) (middle), and air-supplied full suit (right). (Image courtesy of IAEA from IAEA Practical Radiation Technical Manual PRTM-5, "Personal Protective Equipment," 2004).

Table 9-2 below shows some of the assigned protection factors allowed by the United States Nuclear Regulatory Commission (NRC) in the 10CFR20 regulations. Reviewing the values in the table bears out some of the generalities discussed above regarding half-face versus full face masks, filtration respirators compared to supplied-air respiratory equipment. For example, half-face masks generally have a lower protection factor. In addition, continuous flow or pressure demand mode for supplied-air respiratory protection provides a higher protection factor.

Table 9-2. Sample[a] of protection factors assigned for particulates stipulated in 10CFR20, as of 2023.

	Operating Mode	Assigned Protection Factor
I. Air Purifying Respirators		
Facepiece, half	Negative Pressure	10
Facepiece, full	Negative Pressure	100
Facepiece, half	Powered air-purifying respirators	50
Facepiece, full	Powered air-purifying respirators	1000
II. Atmosphere supplying respirators		
1. Air-line respirator:		
Facepiece, half	Demand	10
Facepiece, half	Continuous Flow	50
Facepiece, half	Pressure Demand	50
Facepiece, full	Demand	100
Facepiece, full	Continuous Flow	1000
Facepiece, full	Pressure Demand	1000
2. Self-Contained Breathing Apparatus (SCBA):		
Facepiece, full	Demand	100
Facepiece, full	Pressure Demand	10,000

a. The full list of protection factors is listed in Appendix A of 10CFR20.

One may wonder why one type of respiratory protection may be selected over another. An obvious consideration would be the airborne concentration in the room coupled with the anticipated work time; this is where the concept of DAC-hours discussed earlier is helpful in making decisions. A facility may decide to prescribe a particular type of respiratory protection if a worker is expected to be exposed to a certain number of DAC-hours.

However, the number of DAC-hours is not the only consideration. In some cases, a lower protection factor may be acceptable if the anticipated number of DAC-hours is relatively low and if a respirator with a larger protection factor would be unduly burdensome. For instance, the SCBA type apparatus can be restrictive in performing radiological work due to its physical size and weight and may not be necessary for potential low-dose situations. However, it may be crucial for certain circumstances where the anticipated number of

DAC-hours would be high without this type of respirator. Thus, decisions on respiratory protection must balance occupational safety and human factors as well as radiation protection.

Yet another factor in prescribing respiratory protection pertains to external dose. Often, the use of respiratory protection causes work to be performed at a slower rate. If the worker is being exposed to both airborne radioactivity and external radiation, wearing respiratory protection may reduce the internal dose but could cause a significant external dose to be received as the result of a longer work time.

One final note on respiratory protection pertains to the unique (and hopefully rare) topic of emergency egress in the event of an unexpected sudden release of radioactive material in the workplace that requires evacuation. In such a situation, workers should react quickly and not feel like they must acquire an official respirator prior to exiting the area. Any filtering medium can be used to try to reduce intakes even if it is not a perfect respirator. Layers of cloth, paper napkins, or other material placed over the nose and mouth while evacuating can significantly reduce the internal dose from such an event, even if the degree of reduction cannot be known in advance. The worker obviously should not delay exiting by any appreciable time, but quickly grabbing some filtering material and using it to reduce inhalation of particulates is wise. Bioassays conducted after the event can be used to assess any potential intake, but the worker should use any reasonable means to reduce the amount of radioactive material that is inhaled in emergency situations.

Free Resources for Chapter 9

- Questions and problems for this chapter can be found at www.hpfundamentals.com
- ICRP Publication 30, *Limits for Intakes of Radionuclides by Workers,* Ann. ICRP 2 (3-4), 1979. Available from www.icrp.org.
- ICRP Publication 60, *1990 Recommendations of the International Commission on Radiological Protection,* Ann. ICRP 21(1-3), 1991. Available from www.icrp.org.
- ICRP Publication 130, *Occupational Intakes of Radionuclides: Part 1.* Ann. ICRP 44(2), 2015. Available from www.icrp.org.
- International Atomic Energy Agency, *Personal Protective Equipment, Practical Radiation Technical Manual No. 5*, IAEA, Vienna, 2004. Available from www.iaea.org.
- NUREG/CR-5512, *Residual Radioactive Contamination from Decommissioning,* Pacific Northwest Laboratory, 1992. Available from https://www.nrc.gov/docs/ML0522/ML052220317.pdf
- The regulations of the US Nuclear Regulatory Commission and the US Department of Energy as pertains to radiation protection can be viewed in their entirety online at:
  - 10CFR20 (NRC) - https://www.nrc.gov/reading-rm/doc-collections/cfr/part020/index.html
  - 10CFR835 (DOE) - https://www.ecfr.gov/current/title-10/chapter-III/part-835?toc=1

Chapter 10 Principles of Radiation Protection in the Workplace

Having now covered the fundamentals of dosimetry, we are ready to discuss how radiation dose, whether from internal or external sources, serves as a driver for how radioactive material and radiation sources are handled in the workplace and in the environment.

The ICRP has consistently delineated three concepts that form the basis of radiation protection so that the benefits of using radiation can be realized without unnecessary harm as a result. These three concepts are justification, limitation, and optimization; the ICRP expounded on them in Publication 103 as follows:

- ***The principle of justification:*** *Any decision that alters the radiation exposure situation should do more good than harm.*
  This means that, by introducing a new radiation source, by reducing existing exposure, or by reducing the risk of potential exposure, one should achieve sufficient individual or societal benefit to offset the detriment it causes.
- ***The principle of application of dose limits:*** *The total dose to any individual from regulated sources in planned exposure situations other than medical exposure of patients should not exceed the appropriate limits recommended by the Commission.*
- ***The principle of optimisation of protection:*** *the likelihood of incurring exposures, the number of people exposed, and the magnitude of their individual doses should all be kept as low as reasonably achievable, taking into account economic and societal factors.*
  This means that the level of protection should be the best under the prevailing circumstances, maximising the margin of benefit over harm. In order to avoid severely inequitable outcomes of this optimisation procedure, there should be restrictions on the doses or

risks to individuals from a particular source (dose or risk constraints and reference levels).

(ICRP Publication 103)

The limits mentioned in the second point above are the ICRP limits; however, the concept applies equally well to dose limits that may be imposed by government entities or to administrative limits that a facility may impose upon itself. The most recent ICRP dose limits are shown below in Table 10-1.

Table 10-1. Summary of ICRP Dose Limits from Publication 103.

Dose Quantity to Limit	Occupational Limit	Public Limit
Effective dose	20 mSv per year averaged over any 5-year period and no dose exceeding 50 mSv in a single year	1 mSv per year, although it could be allowed to be higher in a single year so long as the five-year average is still less than 1 mSv per year
Equivalent dose to lens of eye	150 mSv per year	15 mSv per year
Equivalent dose to skin	500 mSv per year averaged over 1 cm^2	50 mSv per year
Equivalent dose to hands and feet	500 mSv per year	--

The United States has some variation from the limits shown above. For example, the equivalent dose limit for skin in the United States is 500 mSv per year as the ICRP specifies; however, the area of skin over which it is averaged in the US is 10 cm^2, not 1 cm^2. Also, the annual dose limit for workers is 50 mSv per year as specified by both the Nuclear Regulatory Commission in 10CFR20 and by the Department of Energy in 10CFR835. The exact value of the limit does not change the analysis that must be performed when evaluating radiological work but does impact what design features may be necessary to comply with the applicable dose limit.

While the concepts of justification, limitation, and optimization may seem abstract, there are nonetheless some practical aspects to them. For example, in some facilities, it is not uncommon to have a room with external dose rates above 10 rad/hour. How often should this area be surveyed, especially since the act of performing the survey to determine the dose rates could cause an appreciable dose to the person performing the survey? How does one balance the need for surveys against the need to minimize unnecessary dose?

There is some judgment involved in applying the principles of justification, limitation, and optimization; one approach to the question of surveying an area with a high external dose rate might be the following:

1. The area must be surveyed if it is known that someone will enter the area to do work. However, if more time is spent on an annual basis performing the surveys than it is anticipated that a worker would be present in the area to do radiological work on an annual basis, then further analysis must be performed to justify the number of surveys and how they are performed.
2. Depending on the amount of time it takes to perform the surveys, the principle of limitation comes into play. If it requires five minutes to perform the survey in an area that averages 10 rad/hour, and the surveys are performed once per quarter by the same individual each time, then a total dose of over 3.3 rads per year could be expected to be received just to perform the surveys apart from any radiological work. While not technically an overexposure exceeding the ICRP dose limits, it is uncomfortably close to the limits and warrants additional review.
3. If the surveys must be performed and the limits must not be exceeded, then there is opportunity for optimization to ensure that doses are as low as reasonably achievable (abbreviated ALARA). Some options to optimize the performance of surveys may be:
    a. Keep the area locked and inaccessible. Only perform surveys immediately prior to known work in the area, but do not perform surveys otherwise.

b. When performing the survey, use remote tools such as a detector on an extendible pole that can be inserted into the area while the radiation protection person stands at a distance further from the area.
c. Design a permanent detector system in the area that has a remote readout at a control panel away from the area. While this may seem like the obvious solution, the act of installing the equipment plus the annual recalibration of the detector may cause more dose than implementing one of the other options.
d. Put the detector on a robot that can enter the area to make the measurements without a person having to physically enter the area for surveys.

A similar approach could be made for other scenarios as well. Radiation protection is a safety-related discipline, and control of radiation in the workplace is achieved by analyzing exposure scenarios in advance. Therefore, it is important for the radiation protection professionals to anticipate what could happen that can cause radiation dose. Then they must be innovative in methods to reduce or eliminate radiation dose. At the same time, there is no single solution that may work for every situation; so, it is important to always review the radiological work being performed to find the optimum methods for controlling radiation dose.

Because the method for an individual to receive dose differs for internal exposure and external exposure, there are different techniques for optimizing the dose depending on the source. These have been discussed in Chapter 6 and Chapter 9. However, in this chapter, we will look at an integrated approach that will help form the basis for protective measures that are put into place.

10.1 Workplace Characterization and Surveys

It is impossible to make good decisions with regard to radiation protection without knowledge of the radiological conditions that exist in the workplace.

The amount and type of characterization will depend on the facility, but some types of surveys and monitoring are discussed in the following sections.

10.1.1 Surveys

Typically, the term survey is used to denote the action of collecting and documenting data on external dose rates and contamination levels. This would be comprised of the following:

- **External radiation survey** – Portable instruments are used to measure the external exposure rate or absorbed dose rate in various locations. The instruments used for this purpose are described in Chapter 5. The proper instrument must be chosen based on the anticipated radiation field so that any external dose rates due to beta, gamma, and neutron radiation are properly measured and documented.
- **Loose contamination survey** – A paper-type medium is used to wipe over areas that are possibly contaminated such that a portion of the contamination is transferred to the paper, which is then counted on an instrument (this is often called a smear survey). In the United States, loose contamination is reported in units of disintegrations per minute (dpm) for a sample size of 100 cm^2 of work area (the paper should be wiped over an area of approximately 100 cm^2 before being counted). Thus, the reported units are dpm/100 cm^2 to denote the activity detected in the sample size area.
- **Fixed contamination** – If no loose contamination is present, a portable instrument (typically a GM counter) can be passed over surfaces to detect the presence of fixed contamination that is not easily removed and thus poses no internal hazard.

To document and make the information more usable, the results are recorded on a survey map. A separate survey map is typically produced for each room/area so that sufficient spatial detail is possible for workers to see the radiological conditions. The map should be drawn with detail to show equipment and furnishings in the room so that it is clear where measurements were made. Figure 10-1 below shows how a survey map for a given laboratory room in a facility might be laid out. Work areas are clearly marked as to their relative size and placement; in addition, survey locations are labeled so that reported measurements can be correlated to those locations.

The choice of survey locations is largely a judgement call; however, they are typically chosen based on accessibility to workers and also to likelihood of changing radiological conditions. For instance, in Figure 10-1, the sink area, worktable, fume hood, and storage cabinet are all areas where contamination could be present as a result of handling radioactive material. The locations in the middle of the room are places that could be occupied for reasonable amounts of time.

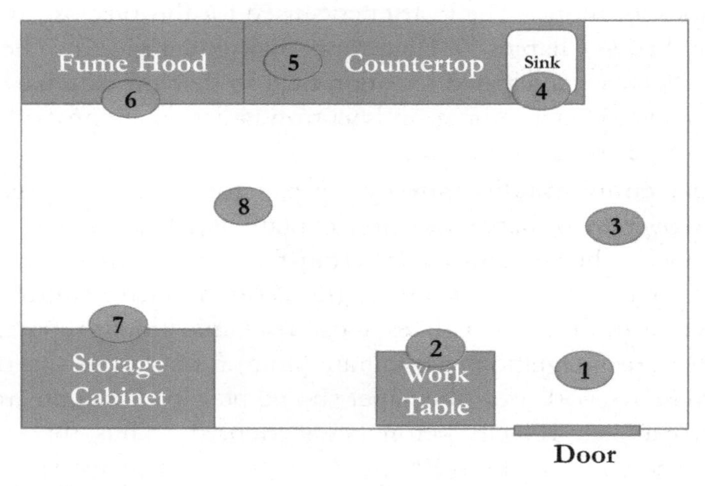

Figure 10-1. A survey map for a laboratory room denoting locations where survey measurements would be performed.

For each location identified in Figure 10-1, it would be expected to have results of the external radiation survey and contamination surveys be printed on the map so that workers can quickly assess the radiological conditions to which they may be exposed. If air monitoring results are available, then they should also be noted on the survey map. Besides being used by radiation protection personnel, one goal of the survey map is that all workers who enter the area should be familiar with the radiological conditions in the area so that they are informed and are able to work safely. Thus, the survey map of a given area should be accessible by all workers prior to entry into the area.

10.1.2 Airborne Radioactivity Sampling/Monitoring

Another part of characterizing the workplace involves measurement of any airborne radioactivity that may be present. This can be accomplished via a variety of methods depending on the degree of radiological hazard. Entire books and journals are devoted to air sampling and monitoring; as with all the topics in this book, it is more complex than can be covered in a single reference. Therefore, the discussion here is intended to be introductory in nature.

An air sampler uses an air pump to draw air through a thin filter of high collection efficiency and is often used and placed in the room to collect airborne dust particles on the filter, which is removeable. The filter would be taken from the sampler (maintaining care to retain all material on the filter) and counted on a detector system. Thus, there is a time delay in the collection of the sample (in this case, the filter paper) and the measurement of the activity on the filter in a laboratory setting.

In addition to the filter that collects particulate material, the air samplers may also have a charcoal cartridge placed behind the filter to collect isotopes of iodine, which adsorb onto the charcoal. As with the filter, the cartridge is removable and would be taken to the laboratory for analysis, typically gamma spectroscopy to identify the isotopes and the amount present.

Air samplers can be operated for several purposes:

- **Short-term sampling** – in this mode, the sampler may be brought into an area temporarily prior to work taking place to monitor the airborne radioactivity before and during work activity.
- **Long-term/average sampling** – in this mode, the sampler is operated continuously around the clock. The filter medium is removed and replaced on a set schedule so that the average concentration over the sampling time interval can be estimated.

Figure 10-2 shows examples of air samplers that might be used in the workplace. The one on the left is termed a high-volume air sampler because it is intended to be used for short timeframes at high flow rates. The sampler on the right is a low-volume sampler intended for longer periods of operation but can also be used for shorter periods of operation. Both samplers have an air pump that pulls air through the filter along with an indicator for the elapsed time, flow rate, and total air volume drawn through the filter. Usually, the high-volume sampler has a larger diameter filter size to facilitate the larger air flow without putting an unnecessary strain on the air pump.

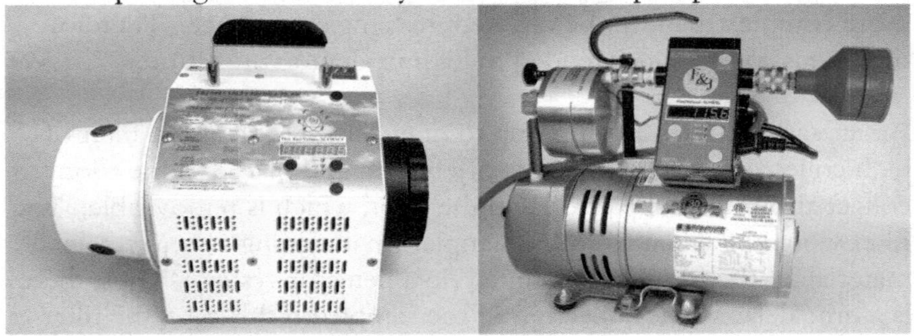

Figure 10-2. Examples of a high-volume air sampler (left) and a low-volume air sampler (right) (Photos courtesy of F&J Specialty Products, Inc. www.fjspecialty.com)

The radioactivity collects on the filter medium at a rate depending on the flow rate through the filter and the half-life of the radionuclide of interest because the radionuclide may be decaying in place on the filter as it is being collected. The usual equation for predicting the activity on the filter as a function of sampling time is given in Equation 10.1 below:

$$A(t_{sample}) = \frac{CF\varepsilon}{\lambda}(1 - e^{-\lambda t_{sample}}) \qquad (10.1)$$

Where

- $A(t_{sample})$ is the activity on the filter as a function of sample collection time
- C is the average concentration in the air (Bq/m^3)
- F is the flow rate of the air through the filter (volume per unit time)
- ε is the collection efficiency of the filter (typically, this value exceeds 99%)
- λ is the decay constant for the radionuclide
- t_{sample} is the sample collection time

Obviously, the time units included in t_{sample} , λ, F, and C must all be consistent with each other so that the units divide out properly in the arithmetic.

One practical point to consider is the issue of filter loading, whereby so much particulate matter collects on the filter that the flow rate is reduced instead of staying constant. A flow rate that decreases with time invalidates Equation 10.1; therefore, when operating an air sampler in a particularly dusty environment or when operating for long periods of time, the sampler should be checked to ensure that the flow rate is not impeded.

Astute readers may have noticed a similarity in the structure of Equation 10.1 with Equation 1.7 earlier in the text. Consequently, the same mathematical simplification can apply to Equation 10.1 that was discussed in Chapter 1. Specifically, if the half-life of the radionuclide of interest is very long compared to the sample collection time (i.e., the radionuclide is not decaying appreciably while being sampled), then the activity buildup on the filter can be predicted based on Equation 10.2 below using the same terms as in Equation 10.1.

$$A(t_{sample}) = CF\varepsilon t_{sample} \qquad (10.2)$$

This equation logically predicts a linear buildup of activity on the filter with time assuming that the flow rate, concentration, and collection efficiency stay constant. Understandably, the product of flow rate F and sample time t_{sample} is the total volume of air drawn through the filter. Since some air samplers have a readout that shows the total volume, it may be easier to use this value rather than the flow rate and sample time.

Of course, Equations 10.1 and 10.2 are used ultimately to determine the average concentration in the room when the activity on the filter is known based on measurement. The approach to solving for the concentration depends on the half-life of the radionuclide of interest by comparison to the involved timeframes.

If the radionuclide of interest is long-lived by comparison to the sample time, delay time before counting, and the count time, then the arithmetic is relatively straightforward as shown in the following example.

Example: A low-volume air sampler operating at 2 cfm (ft^3/minute) is operated for 4 hours in a room with airborne Cs-137. The air filter is assumed to have a 99.5% collection efficiency and is removed and counted in a laboratory 30 minutes later on a gamma spectroscopy detector which measures 20,000 counts of the 662 keV gamma ray during a count time of 20 minutes with a detection efficiency of 3.5%. What is the concentration of Cs-137 in the room?

Solution: First, we take note of the fact that the half-life of Cs-137 is 30 years and that it will be in secular equilibrium with Ba-137m. The half-life is much longer than the sample collection time, the delay time prior to counting, and the count time so that we do not need corrections for any radioactive decay. Also, we must distinguish between the collection efficiency of the filter and the detection efficiency of the detector. Often in the workplace, we use the word "efficiency" in a variety of contexts, and it is imperative to keep the concepts straight.

Logically, the activity measured on the filter should be equal to the following:

$$Activity = \frac{recorded\ counts}{(detection\ efficiency)(count\ time)(Yield)}$$

From the decay scheme of Cs-137, the yield for the 662 keV photon from Cs-137 is 0.85. Thus, the activity based on the foregoing information is:

$$Activity = \frac{20{,}000\ counts}{\left(0.035\ \frac{counts}{gamma}\right)(1200\ seconds)\left(0.85\ \frac{gamma}{decay}\right)}$$

This gives an activity on the filter of 560 Bq.

Equation 10.2 can then be rearranged to solve for C, since the half-life of the radionuclide is much longer than the sample collection time. There are some unit conversions that are necessary to give meaningful results.

Flow rate = 2 cfm = 944 cm^3 per second
Sample collection time = 4 hours = 14,400 seconds

Thus, the concentration is

$$\frac{A(t_{sample})}{F\varepsilon t_{sample}} = C = 4.14\times10^{-5}\ \text{Bq/cm}^3 \text{ or } 41.4\ \text{Bq/m}^3$$

If the radionuclide half-life is short so that decay occurs during the collection time, the delay time, or the count time, then the arithmetic gets more complex but is not insurmountable by any means. It just requires us to be diligent in applying the concepts. Figure 10-3 below shows a graphical representation of the activity on the filter as a function of time if the half-life of the radionuclide is 55 minutes, but the other parameters mentioned in the example above remain the same.

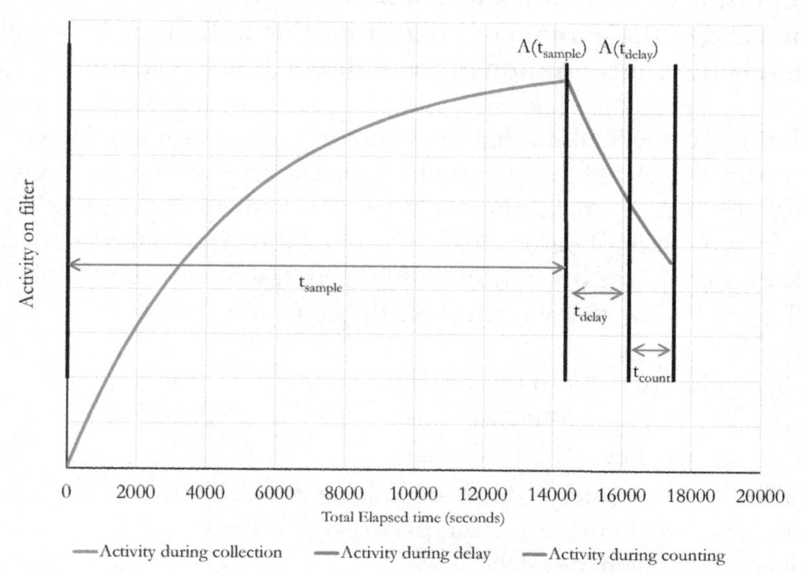

Figure 10-3. Illustration of the activity on an air filter as a function of time when the half-life is short compared to time intervals of interest.

As shown in the figure, the activity on the filter initially increases with time as predicted by Equation 10.1. Once the filter is removed from the sampler, normal radioactive decay takes over, and the activity on the filter continues to decay. The measurement of the number of counts in the laboratory can be used to determine the activity at the beginning of the count period. That information can then be used to calculate the activity on the filter at the end of the sampling time period.

The example below takes the previous example's information and applies it to a shorter-lived radionuclide.

Example: A low-volume air sampler operating at 2 cfm (ft^3/minute) is operated for 4 hours in a room with airborne activity for a fictitious radionuclide with a half-life of 55 minutes. The air filter is assumed to have a 99.5% collection efficiency and is removed and counted in a laboratory 30 minutes later on a gamma spectroscopy detector which measures 20,000 counts of the nuclide's gamma ray during a count time of 20 minutes with a detection efficiency of 3.5%. The yield of the gamma of interest for this radionuclide is 0.85. What is the concentration of the radionuclide in the room?

Solution: We will take a slightly different (but hopefully still logical) approach to solving this problem. With a short-lived sample, our detector system is measuring the integrated activity of the sample, or the total number of decays that occurred during the counting period. It is important to extract information from the premise and apply it to Figure 10-3 so that we can solve the equations correctly.

t_{sample} = 4 hours = 14,400 seconds
t_{delay} = 30 minutes = 1800 seconds
t_{count} = 20 minutes = 1200 seconds
half-life = 55 minutes = 3300 seconds $\rightarrow \lambda = 2.1 \times 10^{-4}$ sec^{-1}
Flow rate = 944 cm^3/sec (from previous example)
Collection efficiency = 0.995

First, we need the number of decays (not the decay rate) that are measured during the count interval. From the information in the premise, the number of decays is calculated logically by:

$$Number\ of\ decays = \frac{number\ of\ counts}{(detection\ efficiency)(yield)}$$

$$Number\ of\ decays = \frac{20{,}000\ counts}{(0.035\frac{counts}{\gamma})(0.85\frac{\gamma}{decay})} = 672{,}269\ \text{decays}$$

By definition, this is the integrated activity over the count time. Thus, we can use Equation 1.7 to solve for A_0, recognizing that A_0 in any activity equation is always the activity at the beginning of the time interval of interest. In this case, the time interval is the counting time.

$$\tilde{A}(t_{count}) = 672{,}269\ decays = \frac{A_0}{\lambda}(1 - e^{-\lambda t_{count}})$$

Solving for A_0 gives a value of 634 Bq. Looking again at Figure 10-3, this is also the activity at the end of the delay period. Therefore, we can use the simple radioactive decay equation to get the activity at the beginning of the delay period by applying Equation 1.4 as follows:

$$A(t_{delay}) = 634\ Bq = A_0 e^{-\lambda t_{delay}}$$

A_0 in this particular equation should not be confused with the value of A_0 that we just calculated for the beginning of the count time. This current A_0 is the activity at the beginning of the delay time, which is the activity at time t_{sample}. Solving for A_0 this time gives a value of 925 Bq. Inspection of Figure 10-3 shows that this is also the activity on the filter at the end of the sampling period, which is $A(t_{sample})$. Thus, we are ready to apply Equation 10.1 to solve for the average air concentration.

$$A(t_{sample}) = 925\ Bq = \frac{CF\varepsilon}{\lambda}(1 - e^{-\lambda t_{sample}})$$

Substituting in the other variables and solving for the concentration yields a concentration value of 2.17×10^{-4} Bq/cm^3.

In the workplace, these calculations lend themselves well to a preprogrammed spreadsheet or to a standalone computer program to avoid transcription errors or misapplied equations. The use of a computer means that the same solution can be applied regardless of the half-life of the radionuclide: that is, the solution for the short-lived radionuclide that we just found can be applied equally well to long-lived radionuclides.

When work is being performed that may cause localized changes in airborne concentrations (such as opening a piping system or ventilation system that may have contamination in it), stationary air samplers may not always be located such that they collect a sample that is representative of what a person would breathe in that location. In those cases, there is the option to have the worker wear a personal air sampler (PAS), which is battery operated and lightweight enough that it does not put a physical strain on the person. The filter head assembly can be clipped to the lapel of the person's clothes so that the filter is sampling air near the worker's breathing zone. Figure 10-4 below shows examples of PAS systems. The calculations and analysis of data are the same as for the stationary air samplers.

Figure 10-4. Examples of personal air samplers that can be used for representative air sampling in the breathing zone of the worker. (Photo courtesy of F&J Specialty Products, Inc. www.fjspecialty.com)

In addition to air samplers, a continuous air monitor (CAM) can be placed in the room to give a real-time indication of airborne radioactivity. A CAM has a filter medium and radiation detection system integrated into a single package so that the airborne concentration can be tracked over time and displayed. In addition, alarm setpoints can be entered into the system so that audible and visual alarms can alert workers to exit the area if airborne concentrations exceed specific values.

Unlike air samplers, CAMs are able to give information on changing airborne concentrations as a function of time. Thus, if the concentration has been observed to change, the timestamp from the CAM can be used to help investigate the cause of the change. A facility may have permanently installed CAMs or may choose to use portable instruments for work-specific monitoring or for emergency response assessments. Figure 10-5 below shows one example of a CAM.

Figure 10-5. Continuous air monitor (CAM) for measuring alpha and beta airborne activity. (Photo courtesy of Ludlum Measurements www.ludlums.com).

A common configuration in a CAM is to have a fixed filter where the particulates accumulate on the filter as in an air sampler. A radiation detector is placed such that the radiation from the filter is continually measured. However, there is still an integrating time over which the average concentration is determined since Equation 10.1 still applies to the buildup of activity on the filter. Thus, the algorithm in the detection system must account for this integrating time as it evaluates changes in the activity levels recorded by the detector.

Regardless of whether air samplers or CAMs are used, one potential interference is due to radon progeny. Even though Rn-222 is a noble gas, the decay products are particulate in nature and can attach to dust particles that are then captured by the filter medium in either the air sampler or the CAM, contributing to natural background counts that may obscure workplace radioactivity. In the case of air samplers, the delay time between sample collection and sample counting can be chosen to allow some of the progeny to decay. In the case of CAMs, statistical methods are necessary to separate the contribution of the radon progeny to the radiation measurements so that meaningful estimates of airborne concentrations of workplace radionuclides can be made.

10.2 Planning Radiological Work

Before work is performed in a radiological area, thought and planning are necessary to ensure that the ICRP principles of justification, limitation, and optimization are maintained. One of the first steps is to clearly designate areas where radiological work will be performed and ensure that only properly trained individuals are allowed access to the area. These areas are typically called radiologically controlled areas (RCAs) or some similar designation. Access may be restricted (controlled) by locked doors, mechanical turnstiles that prevent accidental access, or other types of physical barriers.

When evaluating radiological work, it is important to recognize that activities generally can be categorized as one of the following:

- Routine/common work: these are work activities that are performed as a matter of routine in an area on a frequent basis, possibly on a daily basis. Usually, activities in this category are those that are performed as part of the normal functioning of the organization.
- Special work: these are work activities that are performed infrequently and may require specialized knowledge. Some examples of this would be repair work on a piece of equipment, modification of a room, installation of new processes, or something similar.

Both routine and special work activities may be present in the same area. For example, referring again to Figure 10-1 earlier in this chapter, handling radioactive solutions in the fume hood might be considered a routine work activity. However, modifying or repairing the fume hood itself would constitute special work. Using the sink to handle chemical and radiological material might be considered routine work, but activities to repair the drain of the sink would be special work.

All radiological work must be planned, and it requires a thoughtful approach to assess the work in advance. Many questions must be answered before work proceeds, and the radiation protection professional should learn to expect the unexpected. Some questions to be addressed include the following:

- What is the identity and quantity of radioactive material that will be present?
- What is the exact work that will be performed? Will someone be directly handling a radioactive source or a contaminated object?
- What sorts of things could go wrong that might cause radioactivity to spread either as contamination or as airborne activity?
- What protective measures are in place to mitigate incidents?
- What external dose rates are present, and what amount of time will people be present? What is the total external dose that is projected to be received by each individual and also by all workers performing this work activity?
- Is there a potential for intake? What potential internal dose would occur without respiratory protection? Would respiratory protection slow the work so as to increase the external dose that is received?
- Are there occupational safety hazards?
- Who specifically will be entering the area, and are there additional training requirements that need to be satisfied?
- Should a radiation protection technician be present continually while the work is occurring to monitor the work and make decisions on aborting the activity in the event of changing radiological conditions (this is called "job coverage" or "health physics coverage")?

As these questions, and others like them, are answered, it becomes clearer how to proceed in planning so that the work is justified, that no limits are exceeded, and that the work is optimized to keep doses ALARA using techniques described in earlier chapters.

Documenting the planning that occurs and the results of such planning will vary by facility. In some facilities, a procedure manual or policy manual may be used to document how routine work will be performed, especially if the projected radiological consequences are relatively low. However, more formal planning documents may be generated to govern special work that is a one-time event or that may have higher potential dose consequences.

For larger facilities with lots of varied processes, a system involving Radiation Work Permits (sometimes called Radiological Work Permits, but both are abbreviated as RWP) may be employed. Typically, the person or work group needing to perform work would initiate the RWP process through the radiation protection (RP) organization. Based on collaboration between the work group and the RP organization, information about the work activity and radiological hazards leads to the creation of a document that has a unique identifier number that serves as the authorizing document for that particular work to be done. The finished RWP would specify items such as the following:

- The exact work to be performed and where it will be performed
- Persons authorized to perform the work, typically a particular group of individuals in a specific work group
- Dosimetry requirements (single dosimetry, multiple dosimetry, etc.)
- Protective clothing and respiratory protection requirements
- Expected radiological conditions
- Expected work duration for each entry and overall project duration
- Allowable dose to be received under the permit without further evaluation of the task
- Actions to take in the event of unexpected or changing radiological conditions

- The extent of health physics coverage

The RWP does not take the place of any specialized training or work procedures that may be necessary to perform the work. Rather, the RWP specifies in writing the understanding of the RP organization and the workers of what protocols will be in place to ensure radiological safety.

10.3 Implementing and Tracking Radiological Work

In facilities that do not use a RWP system, implementing and tracking of radiological work is limited mostly to monitoring the radiological conditions in the facilities and tracking the doses received by personnel. Changes in radiological conditions or changes in worker doses would indicate the need to identify what task may have been the root cause of such changes and, if necessary, implement modifications to work procedures.

In facilities with the RWP system, all work is linked specifically to a numerically identified permit that makes work tracking by task more manageable. Typically, prior to beginning work in a radiological area, a worker would:

1. Identify the proper RWP under which the work will be performed
2. Acquire secondary dosimetry (usually an electronic dosimeter) so that the worker has a real-time indication of accrued external dose
3. Read the RWP thoroughly to be familiar with all requirements and to ensure that no changes have occurred since the work was last performed
4. Sign in/log in on the RWP, which is usually tracked via computer database as to the person's identity and the date and time entered.
5. Enter the radiologically controlled area (RCA) through the appropriate access point.

After the work is completed, the person would:

1. Perform monitoring to ensure the worker is not contaminated (this is called "frisking out" even when using an automated contamination monitor)
2. Exit the RCA through the appropriate exit point, which may be a different point than the entrance point.

3. Sign out of the RWP; the worker's date and time of exit would be recorded along with the reading from the electronic dosimeter to track any external radiation dose that might have been received.
4. Return the electronic dosimeter to the appropriate location/rack

Because of the specificity of the RWP system and the ability to track all doses associated with work tasks via computer databases, the RWP system allows RP personnel to track doses over time associated with particular tasks and to identify any changes that might be used to reduce dose. Further, the dose tracking gives information to help in the future with estimating possible radiation doses with similar tasks that may not be identical to any existing work. Finally, the tracking allows for radiological investigations that may be necessary after the fact. In particular, the RWP information can be used to estimate a worker's dose in the event of a lost or malfunctioning dosimeter, especially if the worker was part of a cohort that was performing work at the same time.

A Special Note on Medical Emergencies in Radiological Areas

Despite all the careful work that goes into planning radiological work, one possibility cannot be planned in advance: a worker who has a medical emergency while in a radiological area. Some examples of medical episodes include:

- Fainting or heat stroke
- Mechanical injury (fall, broken bones or other serious injury)
- Heart attack, stroke, or seizure
- Wounds that lead to serious bleeding

In any medical emergency, radiological controls (taking off protective clothing, logging out of RWP, frisking out, etc.) become a secondary concern. The person's medical well-being **always** takes top priority. While it is certainly desirable to leave the workplace in a safe condition, radiation dose or contamination are usually minor risks compared to immediate life-threatening conditions. In the event of a medical emergency, the worker should be removed promptly from the RCA and given medical treatment without delay. Dealing with radiological controls and paperwork and possible contamination can be handled after the fact when the worker has been given proper medical care.

10.4 Some Final Considerations

Obviously, there is more to a radiation protection program than what is contained in this chapter. This chapter's intent is to give some of the technical bases behind the radiation protection program. There will be other considerations as well depending on the facility, such as:

- Receipt and transportation of radioactive material
- Waste management
- Training
- Radioactive materials license application
- Dose reporting

Much of this is regulation-specific, and thus no attempt is made to go into detail in this chapter. However, the material in this chapter has been presented in the hopes that it provides the technical basis to explore these other specific topics in the workplace.

Free Resources for Chapter 10

- Questions and problems for this chapter can be found at www.hpfundamentals.com
- ICRP Publication 103, *The 2007 Recommendations of the International Commission on Radiological Protection*, Annals of the ICRP 37(2-4), 2007. Available at www.icrp.org
- International Atomic Energy Agency, *Health Effects and Medical Surveillance, Practical Radiation Technical Manual No. 3 (Rev. 1)*, IAEA, Vienna 2004. Available from www.iaea.org.
- International Atomic Energy Agency, *Workplace Monitoring for Radiation and Contamination, Practical Radiation Technical Manual No. 1 (Rev. 1)*, IAEA, Vienna 2004. Available from www.iaea.org.
- The regulations of the US Nuclear Regulatory Commission and the US Department of Energy as pertains to radiation protection can be viewed in their entirety online at:
    - 10CFR20 (NRC) - https://www.nrc.gov/reading-rm/doc-collections/cfr/part020/index.html
    - 10CFR835 (DOE) - https://www.ecfr.gov/current/title-10/chapter-III/part-835?toc=1

Chapter 11 Environmental Health Physics Fate and Transport Modeling

Facilities that use radioactive material have the potential to release material to the environment where members of the public and biota can encounter it. This chapter and the next will deal with the area of health physics called environmental health physics. The current chapter deals with modeling calculations to estimate the impacts on the environment. The next chapter will deal with measurements of environmental media and the design of an environmental monitoring program.

It is important to understand that the same science that governs the transport and movement of radioactive material in the environment is also used when addressing nonradioactive constituents. Thus, the environmental health physicist is often tasked with dealing with nonradiological releases (e.g., air pollutants and liquid chemical releases) as well as radiological releases. This chapter and the next will focus on radiological emissions with the caveat that similar considerations would be followed for nonradiological emissions.

The first step in emission modeling and monitoring is to identify potential pathways for radionuclide movement from a facility to humans. Figure 11-1 below shows just a few examples of pathways that could be considered. Not every facility will have every pathway (and some facilities may have many more pathways than other facilities), but it is incumbent on the environmental health physicist to identify those pathways for which it is reasonable that nearby residents could encounter radiological emissions from the facility.

There are three main sources of radiation dose as follows:

- Direct radiation from the facility (gamma and/or neutron radiation), even if no releases occur
- Radionuclides in airborne effluents that are transported to the environment
- Radionuclides in liquid effluents that are transported to the environment

Airborne effluents can be in the form of gaseous radionuclides or in the form of particulate radionuclides suspended in the air. The particulate radionuclides may settle out of the atmosphere and deposit on the ground or on surrounding vegetation in addition to being an external radiation hazard and an inhalation radiation hazard. After depositing on the ground or vegetation, additional pathways can lead to ingestion of radionuclides by humans.

Liquid effluents are released to surface waters (a river, stream, lake, or ocean) where the radionuclides would be diluted by the volume of water. Once there, the radionuclides could be concentrated in aquatic vegetation or in aquatic animals that could be ingested by humans. In addition, the surface water could be used as drinking water by municipalities or could be used for irrigation purposes. Further, the surface waters and shoreline could be used for recreational purposes so that swimming, boating, and lounging on shore could provide radiation exposure.

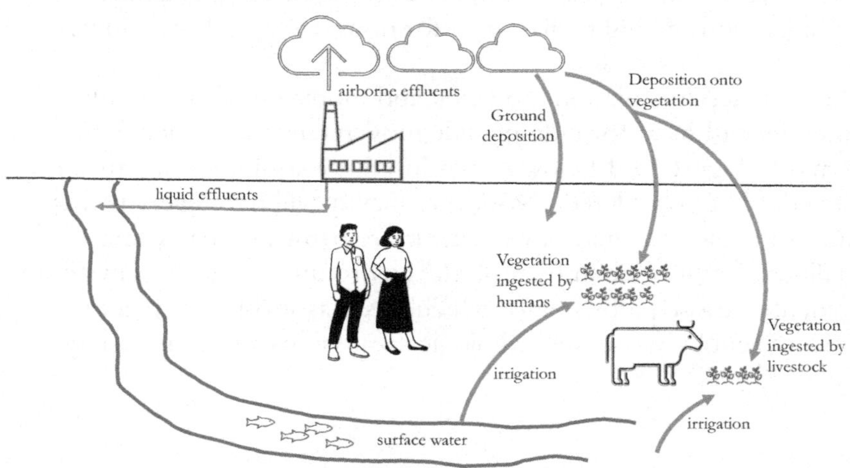

Figure 11-1. A small number of examples of how radionuclides from a facility are transported in the environment.

Each facility that has a license for radioactive materials must analyze potential pathways for how any releases of radioactive material may impact people and biota. The pathways may vary from facility to facility; for example, the area surrounding a given facility may not have nearby farms such that some of the vegetation and livestock pathways may not be applicable. However, it is always wise to consider as many pathways as possible and then eliminate from consideration those that are physically impractical or impossible.

In considering the diagram shown in Figure 11-1 above, some of the pathways to people that might be considered are as follows:

- Airborne release→ground deposition→direct radiation→people
- Airborne release→cloud submersion→direct radiation→people
- Airborne release→inhalation→people
- Airborne release→vegetation→people
- Airborne release→vegetation→animal→meat→people
- Airborne release→vegetation→animal→milk→people
- Liquid release→drinking water→people
- Liquid release→irrigation→vegetation→people
- Liquid release→irrigation→vegetation→animal→milk→people
- Liquid release→irrigation→vegetation→animal→meat→people
- Liquid release→shoreline→direct radiation→people
- Liquid release→shellfish→people
- Liquid release→fish→people
- Liquid release→swimming→people

The process for calculating the impacts can be tedious, and the doses from environmental releases are usually very small for routine operations owing to the small amounts of radioactivity that are released to the environment. For accident scenarios, however, larger quantities of radioactive material may be released in uncontrolled situations so that the doses could be appreciable enough to warrant actions to protect the public such as sheltering in place, evacuation, or other measures. In this chapter, we will focus on routine releases; however, accident analysis is also important and follows a similar analysis process.

The steps to perform an environmental dose assessment for a given pathway to people typically involve the following:

1. Determine the identity and amount of each radionuclide released (this is called the source term)
2. Use a model to calculate the dispersion of the material in the air or surface water, depending on the release mechanism. This will result in an estimate of the time-dependent concentration in the air or surface water at geographic points of interest.
3. Estimate the fraction of activity that is transferred from one medium to the next for each step of a given pathway. For instance, when surface water is used for irrigation, a transfer coefficient must be known for how much radioactivity in the water shows up in the vegetation. Then when an animal ingests the vegetation, a transfer coefficient must be known for how much of the radionuclide shows up in the milk or meat.
4. Use dosimetry models to calculate the dose to people depending on the pathway. The doses are typically calculated for three categories of receptors:
    a. A model of a representative individual who could theoretically live in the area and is positioned to receive a dose that would be considered a reasonable estimate.
    b. The total population surrounding the facility.
    c. A model of a fictitious individual whose theoretical lifestyle habits are often excessive and would create an unrealistic maximum dose from the environmental emissions. This model is sometimes called the Maximum Exposed Individual or MEI. For example, the MEI might be theorized to reside continuously at the site boundary to receive maximum doses due to airborne emissions but also simultaneously reside in a location to receive the maximum dose from drinking solely surface water. These assumptions (and others) may be excessive but often still result in a dose estimate that is very small, thus giving greater confidence that any reasonable scenario would result in extremely small doses.

In this chapter, we will take each of these four steps in turn. In the modern era, computer programs perform the computational work;

however, a fundamental understanding of the ideas and concepts behind the computer programs will help to make sense of results and provide a basis for questioning when calculations may be incomplete or in error.

The general methodology for how environmental dose calculations are performed is outlined in a variety of publications, textbooks, technical journals, and other references. In the 1970s, the Nuclear Regulatory Commission published a series of Regulatory Guides that has been definitive in their simplicity and the ease with which the concepts could be programmed into software. The basic concepts in these guides are still followed by many facilities even though the facilities may have modified some parameter values to adjust for their specific facility or to replace outdated values with newer ones. The Guides are freely downloadable from the internet and are:

- Regulatory Guide 1.111 *Methods for Estimating Atmospheric Transport and Dispersion of Gaseous Effluents in Routine Releases from Light-Water-Cooled Reactors*
- Regulatory Guide 1.113 *Estimating Aquatic Dispersion of Effluents from Accidental and Routine Reactor Releases for the Purpose of Implementing Appendix I*
- Regulatory Guide 1.109 *Calculation of Annual Doses to Man from Routine Releases of Reactor Effluents for the Purpose of Evaluating Compliance with 10 CFR Part 50, Appendix I*

As the titles imply, Regulatory Guides 1.111 and 1.113 describe how to calculate the concentration in air and water from releases, while Regulatory Guide 1.109 describes how to calculate the transport of radionuclides through the environmental pathways and then determine the dose to the public.

11.1 Source Term Estimation

The type and amounts of radionuclides released from a facility will vary depending on the facility itself and what operations are occurring in the facility. However, environmental health physicists are responsible for knowing the details of all emissions from the facility. Therefore, they must understand the operations and be able to either calculate or directly measure the identity and quantity of each radionuclide that is released.

It should be emphasized that facilities do <u>not</u> release material into the air or water as a means of wholesale disposal of radioactive material. Rather, the operation of a facility with radioactive material will naturally result in a small amount of radioactive material being distributed to the air and water as discussed in the following sections. However, emissions are minimized through engineering controls and are carefully monitored to ensure that impacts can be assessed.

There are generally three approaches to estimating the source term as follows:

- Direct measurement of the radionuclides as they exit the facility for cases where they are released in sufficient quantities to be detected
- Estimation of radionuclides that are hard to detect based on using a detectable nuclide as a tracer. For instance, it may be known that if Radionuclide X is detected, then Radionuclide Y must also be present. The amount of Radionuclide Y would then be estimated as some fraction or multiple of Radionuclide X based on knowledge of workplace conditions.
- Estimation of the released amount based solely on process knowledge of what work activities are occurring in the facility even if the released radioactivity is not detectable.

11.1.1 Air Emissions

Emissions of airborne radionuclides may come about through several means. One of the most common release modes is from fume hoods, gloveboxes, or similar type containment areas where radioactive material is being handled (these were illustrated in Figure 9-3 as means to reduce internal exposure). Some possible examples of facilities where this could occur are:

- a hospital that handles medical radionuclides
- a radiochemistry laboratory
- a facility that fabricates a product that includes radioactive material, such as a radiopharmaceutical factory

Another potential source of airborne emissions is from contaminated areas of facilities where the contamination could become airborne due to resuspension. For example, this could be an industrial facility with loose contamination; in addition, a hospital room with a patient who is excreting radioactive material would also be a possibility.

A third possible source of airborne emissions arises if the facility is producing radionuclides through nuclear reactions (e.g., a nuclear reactor or a particle accelerator). In this case, gases and particulate radionuclides created in the nuclear reactions may be collected and treated, but small amounts could leak into the immediate room(s) containing the device.

The ventilation systems for all of these scenarios are designed to draw air from the areas and then filter the air to remove as much particulate radioactive material as possible as discussed in Section 9.2.1.2. Usually, the filtration is through high-efficiency particulate air (HEPA) filters that remove in excess of 99.9% of airborne particles. Gaseous radionuclides are not removed by filtration, however, and other techniques may be employed to reduce the amount that is released. Such techniques are beyond the scope of this text but can be researched independently.

As discussed in Section 9.2.1.2, the filtered air is exhausted through one or more stacks outside the facility; the air from potential airborne areas is not recirculated inside the facility. These exhaust stacks may be mistaken by the general public for traditional smokestacks because of their height, but they are simply elevated release points for air from the facility with no smoke in the emissions (although in certain weather conditions, the exiting warm air from a facility could be fog-like in appearance and thus confused with smoke). Radiation detectors in the stack are used to identify and quantify the amount of radioactive material passing through the stack; if the amount of radioactivity exceeds a certain threshold amount, the detection system can be used to sound an alarm and shut down the ventilation system so that further material is not released from the stack.

The stack design is beyond the scope of this introductory text, but the height is selected to be much higher than surrounding buildings so that any material in the exhausted air can be suitably diluted by ambient air and wind before it moves away from the facility. In some cases, the stacks may be hundreds of feet high to ensure adequate dilution and minimize environmental impacts. An example of a ventilation stack is shown below in Figure 11-2.

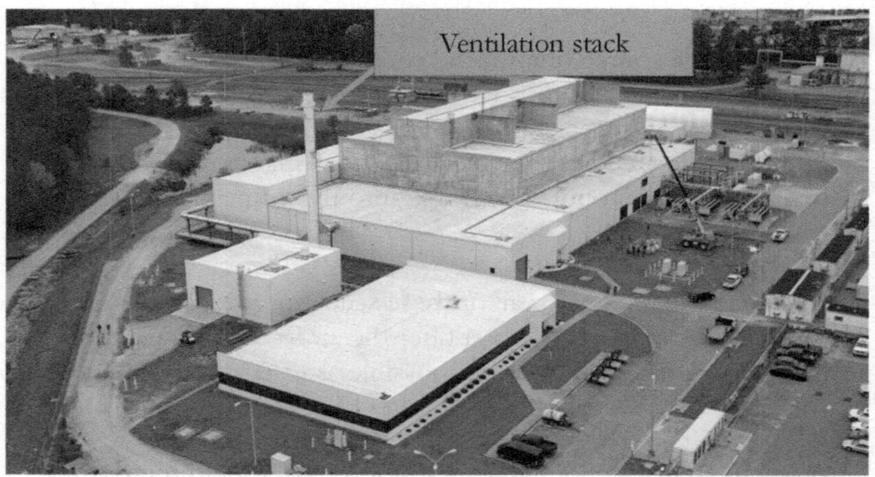

Figure 11-2. Ventilation stack at the Salt Waste Processing Facility at the Savannah River Site (Photo courtesy of the US Department of Energy)

For routine releases, it is typical to record and report the annual emission rate (termed "Q") in units of curies/year for each radionuclide. Understandably, the released amount may not be uniformly distributed in time throughout the year. For instance, emissions may be more prevalent during daylight hours due to facility operations compared to nighttime hours when some work activities may be reduced or shut down. Similarly, weekend emissions might be substantially lower than weekday emissions. However, the analytical methods employed for routine releases determine the annual impact based on the total releases for the year; therefore, Q becomes part of that calculation.

From the foregoing discussion, it should be apparent that routine airborne releases generally are through one or more ventilation stacks; the air is heavily filtered before being discharged to the environment to further minimize exposure of the public. This is not to say that other release scenarios are not possible; for example, suspension of contaminated soil that then blows downwind would also constitute an airborne release. However, for purposes of this chapter, we will look only at controlled releases through elevated stacks while recognizing that these principles can be applied to other scenarios.

11.1.2 Waterborne Emissions

Releases of radioactive material to surface waters could occur through several different methods. As with airborne emissions, the identity and amount of each radionuclide released on an annual basis must be known and documented by the facility.

One potential source for waterborne radioactivity is the use of cooling water to remove heat from equipment, since electrical components and mechanical components can generate an appreciable amount of heat. In a facility that handles radioactive material, some of the equipment may have the potential to be contaminated, but this equipment must still be cooled. For instance, in a nuclear generating facility, the nuclear reactor and spent fuel pool must be cooled. Water cooling is more efficient than air cooling; thus, water may be taken from surface water sources (river, stream, lake), circulated through a heat exchanger to remove heat from the desired system, and then discharged back to the environment.

A heat exchanger works on the principle of allowing heat to move between two isolated systems without mixing the contents of the systems. A schematic of a heat exchanger is shown below in Figure 11-3. While not shown, the plumbing may have cooling fins or a tortuous zigzag pattern within the component to maximize heat transfer from the component to the cooling water.

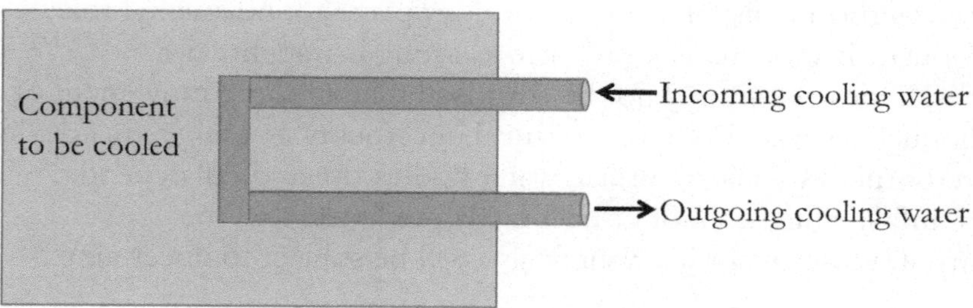

Figure 11-3. Schematic illustration of a heat exchanger where water is used to cool a component without mixing the cooling water with the contents of the component.

Ideally, there would be no ability for radioactive material to leak into the cooling water from the component. In practice, though, small amounts may transfer to the cooling water due to tiny leaks at fittings and other scenarios. As with air emissions, the amount of radioactive material that is eventually discharged is minimized and can be further reduced by various technologies, including passing the water through ion exchange resin or other media before the water is discharged to either a cooling lake or a stream/river.

Radioactive material can also enter the environment through the sanitary sewer system. Regulations do provide for the discharge of small concentrations of radioactivity via the sanitary sewer system, such as from a laboratory drain where tiny amounts of radioactivity in samples could be disposed. Following treatment and disinfection at a sewage treatment facility, the liquid part of the waste may be discharged to surface waters. Thus, this could be a pathway for radioactive material to enter the environment. Discharge to the sewer system does have several considerations as follows:

- The facility must be able to determine the identity and quantity of all radioactive material that enters the sewer system.
- Permission for such discharges to the sewer system relies on an assessment of the dilution provided by nonradioactive liquid streams as it exits the facility so that the concentration of radioactive material when it reaches the sewage treatment facility is minimized.

11.2 Dispersion Modeling

The dispersion of material into the environment is a highly complex process that cannot be relegated to a few simple equations. In the case of airborne dispersion, meteorology plays a fundamental role; obviously, if short-term weather forecasting is fraught with uncertainties, we should not be surprised that airborne movement of radionuclides will also have statistical uncertainty as well. Similarly, in waterborne dispersion, surface water bodies rise and fall depending on rainfall, which cannot be accurately predicted, and the concentrations in surface waters also will be subject to uncertainty.

This section will examine elementary methodologies that are applied for routine releases over the course of a calendar year. These methods are not necessarily realistic in every circumstance but do aid in providing screening values of concentrations in the environment. In cases where a facility may determine that these methods are overly conservative, adjustments and more robust models can be found in the literature and applied. However, we will use simpler models to illustrate the thought process when considering dispersion.

11.2.1 Dispersion of Airborne Contaminants

For a continual release of air from a stack, the material from the stack forms a plume that is moved away from the stack by the wind (advection). At the same time, the material in the plume diffuses in the vertical direction and also diffuses horizontally perpendicular to the wind. This movement of material is illustrated below in Figure 11-4.

By convention, the coordinate system in use for calculations assumes that the wind direction is always along the x-axis. Because the wind direction changes throughout the year, however, this means that the coordinate system shifts throughout the year, and this translation of coordinate systems must be addressed when attempting to determine the amount of dispersion throughout the year. This will be discussed later in this section.

The z-direction and y-direction of the coordinate system are illustrated in Figure 11-4. The y-direction is always horizontally perpendicular to the wind; thus, the y-direction must be translated for different meteorological conditions just like the x-direction. The z-direction is the vertical direction and thus does not depend on wind direction.

As illustrated in Figure 11-4, the anticipated concentration of material in the plume is highest in the middle of the plume, both vertically and horizontally. Whereas wind movement causes the material to move actively in the x-direction, material movement in the y-direction and z-direction is due to diffusion such that the concentration at the outer edges of the plume is less than near the center. At distances further from the stack, the amount of diffusion increases. Thus, the material spreads over a larger geographic area and is at lower concentrations at greater distances from the stack.

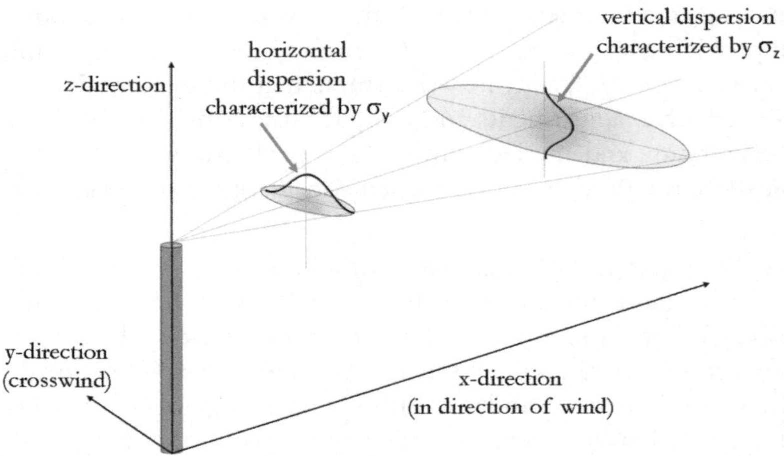

Figure 11-4. Idealized illustration of advection of material from an elevated stack by the wind in conjunction with diffusion in the horizontal and vertical directions perpendicular to the wind direction (drawing based on Turner 1970).

Obviously, the meteorology of the area around the release point will affect the amount of diffusion. We characterize this by a term called "atmospheric stability." Generally, a more stable atmosphere causes less spread in the y- and z-directions while a less stable atmosphere causes more spread.

The quantification of stability is often driven by the vertical temperature profile in the region. Neutral stability is achieved when the actual air temperature profile matches the adiabatic lapse rate, which is defined by the National Weather Service (NWS) as:

The rate of decrease of temperature experienced by a parcel of air when it is lifted in the atmosphere under the restriction that it cannot exchange heat with its environment. For parcels that remain unsaturated during lifting, the (dry adiabatic) lapse rate is 9.8°C per kilometer.
(https://forecast.weather.gov/glossary.php?word=lapse%20rate)

The Nuclear Regulatory Commission (NRC) in its Regulatory Guide 1.23 adopted a convention whereby the stability class can be categorized based on the vertical temperature difference measured between two points (ideally 10 meters from the ground and 60 meters from the ground). Table 11-1 below shows the criteria for identifying the stability class as a function of temperature gradient.

Table 11-1. NRC Criteria for designation of atmospheric stability categories based on vertical temperature gradient (from NRC Regulatory Guide 1.23)

Stability Description	Stability Category	Ambient temperature change with height (°C/100 m)[a]
Extremely unstable	A	$\Delta T \leq -1.9$
Moderately unstable	B	$-1.9 < \Delta T \leq -1.7$
Slightly unstable	C	$-1.7 < \Delta T \leq -1.5$
Neutral	D	$-1.5 < \Delta T \leq -0.5$
Slightly stable	E	$-0.5 < \Delta T \leq +1.5$
Moderately stable	F	$+1.5 < \Delta T \leq +4.0$
Extremely stable[b]	G	$\Delta T > +4.0$

a. The temperature difference is expressed as the temperature at an elevation relative to a point at a lower altitude. Thus, negative temperature changes indicate a temperature drop with increasing altitude.

b. Category G is reserved for extremely bounding calculations such as accident analysis. By assuming Category G, very little dispersion occurs, and maximum downwind concentrations can be calculated to provide a conservative overestimate of any consequences.

As discussed above, the adiabatic lapse rate that would correspond to neutral conditions is 9.8°C for every 1000 meters. This would indicate a decrease in temperature of 0.98°C for every 100 meters of elevation(-0.98°C/100 meters) and would thus fit in Stability Category D in Table 11-1. Categories A, B, and C correspond to more sharply cooling conditions with altitude and thus present greater instability in the atmosphere leading to more dispersion.

Conversely, Categories E, F, and G represent more stability and less vertical dispersion of pollutants because warmer air near the earth's surface will not migrate as quickly upward. If the value of ΔT is zero, then the air temperature is not changing with altitude, and we would say that an isothermal region exists. In addition, positive values of ΔT correspond to what are called temperature inversions, whereby the temperature actually gets warmer with altitude. Inversions have the practical effect of minimizing vertical movement of air (since warm air near the ground will not rise into warmer air above it) and keeping airborne pollutants below the altitude at which the temperature inversion occurs and can cause increased ground-level concentrations of pollutants. As a real-life example of this phenomenon, the author grew up in a geographic area approximately 10 miles from a paper mill. During the early morning hours on some days when temperature inversions were common, the hydrogen sulfide emissions from the mill were quite noticeable several miles away due to their pungent smell. As the sun rose and restored the normal temperature gradient of decreasing temperature with altitude, the hydrogen sulfide smell dissipated as the gas dispersed vertically.

The mathematics of calculating downwind concentration can be as complicated as one wishes it to be. Sophisticated equations have been derived using a variety of methodologies to determine the concentration at any given (x,y,z) location with respect to the release point; for acute releases during an incident, these robust equations may be quite appropriate. However, these equations are usually based on calculus principles that cannot account for changing environmental conditions throughout the year. Thus, simpler equations are usually sufficient for annual screening calculations as they tend to overestimate the actual concentrations.

For continuous releases throughout the year, it is common to calculate the concentration using a Gaussian plume model that is historically based on methods refined by Gifford. There are two common versions of the equation that are used as follows:

Version 1: Equation 11.1 (from Turner 1970) is used to calculate the concentration along the plume centerline at ground level (a different form of the equation is used for concentrations off-centerline or above ground-level). As illustrated in Figure 11-4, the concentration away from the centerline in the y-direction decreases and follows a Gaussian distribution. Therefore, the centerline concentration would be the highest (most conservative) at a given distance from the release point.

$$\chi = \frac{Q}{\pi \sigma_y \sigma_z \mu} e^{-(\frac{h^2}{2\sigma_z^2})} \qquad (11.1)$$

where

- χ = the airborne concentration (Ci/m^3) at a coordinate (x, y=0, z=0); in essence, x meters away along the centerline of the plume at ground level. Even though x does not appear explicitly in the equation, both σ_y and σ_z depend on x.

- Q = the release rate (Ci/sec) of the radionuclide (typically the annual release amount divided by the number of seconds in a year)
- σ_y = the horizontal dispersion coefficient (m) as illustrated in Figure 11-4
- σ_z = the vertical dispersion coefficient (m) as illustrated in Figure 11-4
- μ = the wind speed (m/sec)
- h = the height at which the material is released (m)

<u>Version 2</u>: Equation 11.2 is a modification of Equation 11.1 and is based on Sagendorf (1974). A time-averaged version of this equation is implemented in NRC Regulatory Guide 1.111 to calculate the average concentration in a 22.5° wide sector ($\pi/8$ radians wide). In effect, this equation assumes a constant concentration across the width of an entire sector without a peak concentration along the centerline. Thus, it can be applied to a group of people living in a sector around a facility.

$$\chi = \frac{Q\sqrt{\frac{2}{\pi}}}{(\frac{\pi}{8})\sigma_z \mu x} e^{-(\frac{h^2}{2\sigma_z^2})} \qquad (11.2)$$

The numerical constants in Equation 11.2 are often combined for simplification to give Equation 11.3.

$$\chi = \frac{2.032\,Q}{\sigma_z \mu x} e^{-(\frac{h^2}{2\sigma_z^2})} \qquad (11.3)$$

where

- χ = the sector-averaged airborne concentration (Ci/m^3) at a distance x from the release point
- Q = the release rate (Ci/sec)
- σ_z = the vertical dispersion coefficient (m) as illustrated in Figure 11-4
- μ = the wind speed (m/sec)
- x = the downwind distance (m)
- h = the height at which the material is released (m)

Figure 11-5 shows how the sectors typically are arranged around a facility. As shown in the figure, there are 16 sectors of angle 22.5° each ($\pi/8$ radians). At each major compass heading, the sector extends 11.25° on either side rather than having the cardinal compass points as the boundaries of each sector. As distance is increased from the facility, the sector width increases, even though the angle encompassed within the sector remains the same. Equation 11.2 (and 11.3, by implication) account for this phenomenon, since the product of $\pi/8$ and x in the denominator of Equation 11.2 is the arclength of the sector at the given distance x (remember that arclength $s=r\theta$, where θ must be in radians).

In addition, it is also common to calculate the concentration at certain radii around the facility up to a distance of approximately 80 km (50 miles). This allows the determination of dose to individuals and also to the entire population surrounding the facility within the given radius.

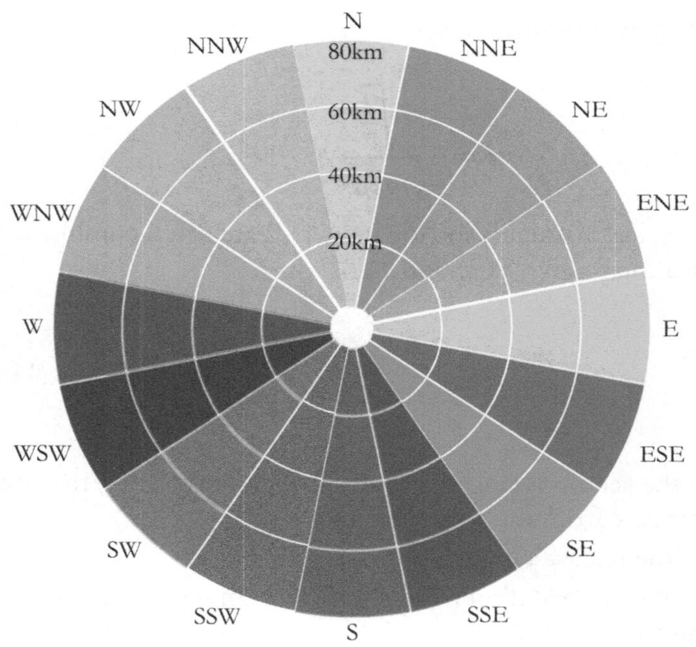

Figure 11-5. The area around a facility is commonly divided into sixteen sectors for purposes of calculating average airborne concentrations.

As indicated in Equations 11.1 through 11.3, the downwind concentration is directly proportional to the release rate Q; thus, it is common for a facility to use historical meteorological measurements to prospectively calculate values of χ/Q in units of sec/m^3 (i.e., calculate the downwind concentration assuming a release rate of 1 Ci/sec). These values can be stored in computer programs and then applied to specific release rates when the rates are measured or calculated. Thus, χ/Q (pronounced "Chi over Q") is the normalized concentration per unit release rate and can be applied to any pollutants (even nonradioactive ones) that are released from the facility with certain modifications as discussed in the next section.

11.2.1.1 Practical Use of the Gaussian Models to Calculate χ/Q

There are some practical points to consider when trying to apply Equations 11.1 through 11.3. Some involve selecting the proper parameters to use in the equations. Others involve interpreting the results and understanding the conditions under which the results can be applied.

11.2.1.1.1 Parameter Selection

In reviewing Equation 11.1, there are several parameter values that must be selected: h, σ_y, σ_z, and μ. For Equations 11.2 and 11.3, the distance from the facility is used directly in the equation, and σ_y is not used.

The release height h influences the downwind concentration because greater release heights result in lower downwind concentrations. As a first estimate, the release height should be taken to be the physical height of the stack. If the air from the stack is released with a high velocity, it is possible to calculate an effective release height that is greater than the physical stack height. However, this is less conservative, and some facilities unfortunately have underestimated their downwind concentrations by improperly calculating the effective release height. Therefore, this author encourages the use of the physical stack height for the value of h in calculating χ/Q.

The mean wind speed μ is experimentally determined using local meteorology. Some facilities have an onsite meteorological tower to obtain wind speed. Alternatively, the local wind speed can be obtained from nearby weather stations (such as airports or television stations) that report the wind speed and direction.

The atmospheric dispersion coefficients σ_y and σ_z are dependent upon the assigned stability category. If the facility has an onsite meteorological tower, then it most likely has temperature measurements at two different elevations so that the stability category can be quickly assessed using the definitions in Table 11-1. However, some facilities incorporate other considerations, including cloud cover, wind speed, and surface roughness (e.g., trees and hills cause different dispersion than open flat land) in identifying the stability category. As a first estimate, though, the temperature difference can be used to determine the stability category, and it then remains to identify the appropriate quantitative values for the dispersion coefficients.

A search of the literature reveals that there are many formulations for determining the numerical value of the dispersion coefficients. In many references, graphs are provided; in others, equations are provided to estimate the dispersion. At distances close to the facility (within 10 km or so), there is reasonable agreement between many of the methods for the horizontal dispersion coefficient (σ_y). For the vertical dispersion coefficient (σ_z), however, there may be significant differences between the estimation methods, especially for Categories A and B, where the greater instability causes rapid vertical dispersion at close distances to the release point. However, with the uncertainty surrounding any estimate of concentration, it is difficult to say that one method is superior to the others. It remains largely a judgement call on the part of the environmental health physicist as to the appropriate method to use.

One formulation that lends itself well to computer programming is incorporated by the Environmental Protection Agency (EPA) in its CAP88-PC computer program. The values for σ_y and σ_z are calculated based solely on distance as a function of stability category as shown in Table 11-2 below.

Table 11-2. CAP88-PC equations for calculating σ_y and σ_z as a function of distance "x" (in meters) from the release point[a]

Stability Category	Equation for σ_y (m)	Equation for σ_z (m)
A	$0.22x(1+0.0001x)^{-0.5}$	$0.20x$
B	$0.16x(1+0.0001x)^{-0.5}$	$0.12x$
C	$0.11x(1+0.0001x)^{-0.5}$	$0.08x(1+0.0002x)^{-0.5}$
D	$0.08x(1+0.0001x)^{-0.5}$	$0.06x(1+0.0015x)^{-0.5}$
E	$0.06x(1+0.0001x)^{-0.5}$	$0.03x(1+0.0003x)^{-1}$
F	$0.04x(1+0.0001x)^{-0.5}$	$0.016x(1+0.0003x)^{-1}$
G[a]	σ (Class F) – 0.5[σ(Class E) - σ(Class F)]	

a. For Category G, the respective dispersion coefficients are calculated by subtracting half the difference between the values for Categories E and F from the value for Category F.

Figures 11-6 and 11-7 below show the predicted dispersion coefficients from the CAP88-PC formulations in Table 11-2. As might be expected, categories designated as more stable have less dispersion compared to those categories that are designated as less stable. Also, the dispersion coefficients rightly increase as the distance from the release point increases, reflecting the expectation that the plume should widen both horizontally and vertically at greater distances from the release location.

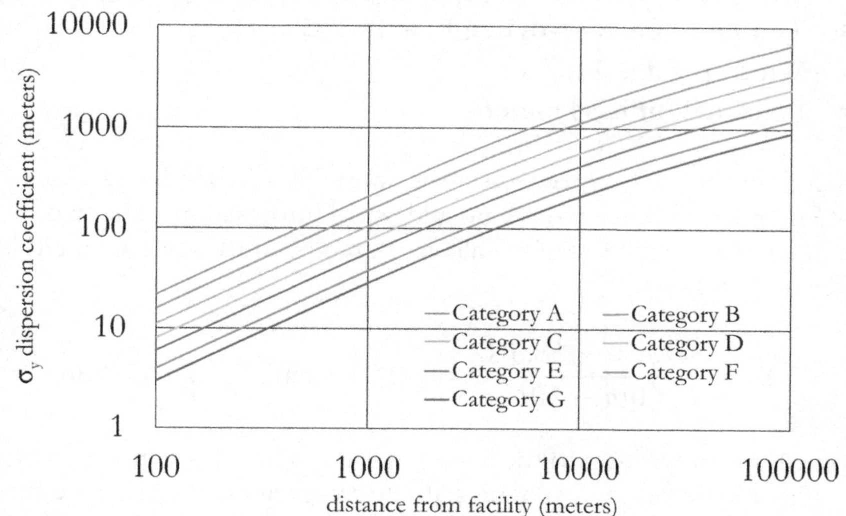

Figure 11-6. Horizontal dispersion coefficients σ_y (m) based on the CAP88-PC formulation.

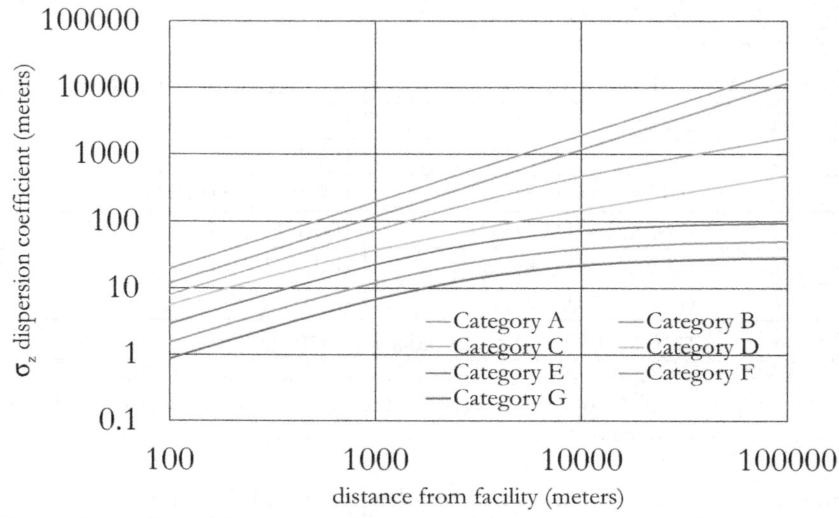

Figure 11-7. Vertical dispersion coefficients σ_z (m) based on the CAP88-PC formulation.

Example: Calculate χ/Q for both the centerline plume model and the sector-averaged model at a distance of 1000 meters given the following information:

- **Temperature at 10-m height is 18.3°C**
- **Temperature at 60-m height is 17.8°C**
- **Wind speed is 3 m/sec**
- **Stack height is 50 meters**

Solution: The wind speed, distance, and release height are given in the premise and thus require no additional information. The two temperatures can be used to calculate a temperature gradient per 100 meters as follows:

$$\Delta T = \frac{17.8°\text{C} - 18.3°\text{C}}{60m - 10m} = -0.01°\text{C/m} = -1°\text{C/100m}$$

Based on comparison to the values in Table 11-1, this temperature gradient indicates atmospheric stability category D. We then use the equations for category D in Table 11-2 to calculate the horizontal and vertical dispersion coefficients at 1000 meters distance from the release point.

$$\sigma_y = 0.08(1000\,)(1 + 0.0001(1000))^{-0.5} = 76.3\ m$$
$$\sigma_z = 0.06(1000\,)(1 + 0.0015(1000))^{-0.5} = 37.9\ m$$

We are now ready to calculate the concentration along the plume centerline using Equation 11.1 with both sides divided by Q as follows:

$$\frac{\chi}{Q}(plume\ centerline) = \frac{1}{\pi\sigma_y\sigma_z\mu} e^{-(\frac{h^2}{2\sigma_z^2})}$$

$$\frac{\chi}{Q} = \frac{1}{\pi(76.3m)(37.9m)(\frac{3\ m}{sec})} e^{-(\frac{(50m)^2}{2(37.9m)^2})}$$

$$= 1.54\times10^{-5}\ \text{sec/m}^3$$

For the sector-averaged concentration, we use Equation 11.3 with both sides divided by Q as follows:

$$\frac{\chi}{Q}(sector - averaged) = \frac{2.032}{\sigma_z\mu x} e^{-(\frac{h^2}{2\sigma_z^2})}$$

$$\frac{\chi}{Q} = \frac{2.032}{(37.9m)(3m/\sec)(1000m)} e^{-(\frac{(50m)^2}{2(37.9m)^2})}$$

$$= 7.49\times10^{-6}\ \text{sec/m}^3$$

As expected, the sector-averaged concentration would be less than the maximum plume centerline concentration calculated above, assuming the same release rate in both scenarios.

11.2.1.1.2 Corrections to χ/Q for Radioactive Decay and Changing Meteorology

The example above shows the ease with which an initial estimate of χ/Q can be made with a minimal amount of information. However, there are some obvious corrections that are generally made when considering real-world scenarios for routine releases over the course of a year.

For short-lived radionuclides (half-lives on the order of hours or shorter), it may be desirable to account for radioactive decay so as not to grossly overestimate the downwind concentration. It is relatively straightforward to correct for radioactive decay based on the travel time from the release point to the location for which the concentration is being calculated as shown in Equation 11.4.

$$\frac{\chi}{Q}(decay - corrected) = (\frac{\chi}{Q})_0 e^{-\lambda \frac{x}{\mu}} \quad (11.4)$$

Where

- $(\chi/Q)_0$ is the non-decay corrected value at the location of interest
- λ is the decay constant for the radionuclide of interest (sec^{-1})
- (x/μ) is the transit time (sec) from the release point to the location of interest, calculated as the distance (in meters) divided by the wind speed (meters/sec)

The other major correction is for the reality that the meteorological conditions are not the same throughout the year. Stability classes, wind speeds, and wind directions all change within a given day.

The usual approach to this is to calculate an average value of χ/Q for each segment around a facility for the entire year. For example, Figure 11-8 shows a subset of how the area around a facility could be divided into areas numbered 1-48 (the actual number for a facility is usually much higher because the number of annular rings may cover a span from 0 to 80 km as mentioned earlier).

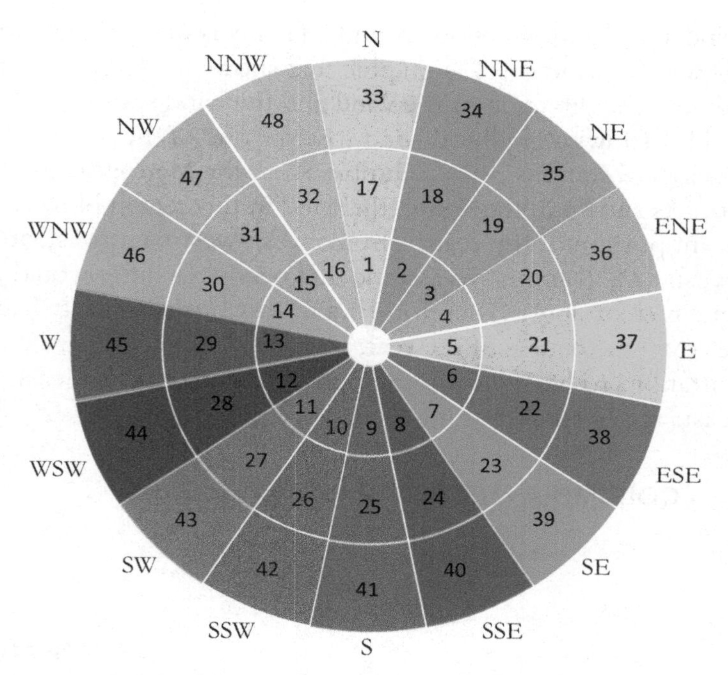

Figure 11-8. Segment designations around a fictitious facility to illustrate how average χ/Q values would be calculated for each segment.

If we look solely at Segment 19 (the second segment radially NE outward), for example, we might calculate χ/Q values for all seven stability categories and for five different windspeed ranges from calm to the maximum observed wind speed; in this example, that would represent 35 different χ/Q values for that one segment. Each individual value of χ/Q would then be weighted for the fraction of the year for which the wind was observed to be in the direction of Segment 19 at the given windspeed class and at the given atmospheric stability category. The sums of the weighted values would then give the average χ/Q value for Segment 19. This would be repeated for all the other segments around the facility such that each individual segment would have its own unique average χ/Q value. Thus, there could be literally a hundred or more average χ/Q values for a given facility, and the average annual air concentration in each segment would be calculated as the product of the release rate Q and the segment-specific average χ/Q value.

The wind speed and direction around a facility is often characterized visually with a "wind rose," a graphic that shows the fraction of the year of both the direction of the wind and the wind speed. Figure 11-9 shows a graphical wind rose for one of the meteorological stations near Columbia SC at the Metropolitan Airport. As shown in the figure, the wind in this region blows a significant portion of the year from the west, west-southwest, and southwest. It is common to assume that nearby facilities would have the same meteorology. Therefore, if a nearby nuclear facility had releases over the course of the year, we would expect higher concentrations to be 180-degrees opposite (i.e., to the east, east-northeast, and northeast of the facility's release point).

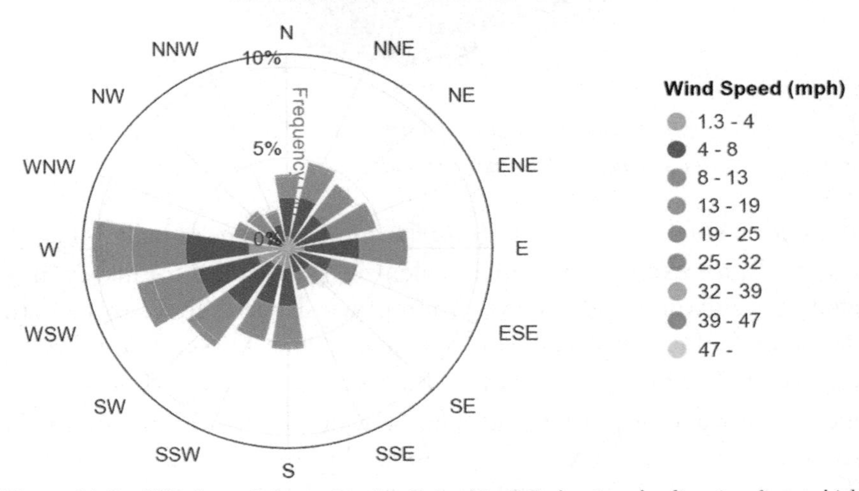

Figure 11-9. Wind rose information for Columbia SC showing the direction from which the wind blows (Image courtesy of cli-MATE, Midwestern Regional Climate Center. Purdue University. http://mrcc.purdue.edu/CLIMATE accessed on: 09/20/2023).

One important note about any wind rose is to properly interpret the direction of the wind. Historically, wind roses showed the direction from which the wind blows, in keeping with weather reports on television that give the direction from which the wind blows. However, some newer wind roses may show the direction toward which the wind blows, which can be used directly in obtaining the proper weighting factor for a given sector. A variety of online sources will provide wind rose data for airports and other meteorological stations around the country. Two sources that the author has found useful in allowing customization of location and periods of data are the following:

- Purdue University maintains the cli-MATE Data Portal with nationwide data at mrcc.purdue.edu. Registration is free, and the tools menu allows the generation of wind rose and other data for weather stations around the country.
- Iowa State University maintains the Iowa Environment Mesonet (https://mesonet.agron.iastate.edu/) that has a multitude of meteorological information. Under the "Apps" menu of the website is the option for wind roses. Selecting the proper network and station allows the user to obtain wind rose and other information for the given location.

The wind rose information is often tabulated into what is termed a joint frequency distribution file. This file has the weighted values for each stability class, for each wind speed, and for each wind direction. Obviously, this is a lot of data, and the reader would rightly assume that a computer program or spreadsheet would be quite useful for performing these types of calculations. Computer software for environmental calculations is discussed in Section 11.5.

11.2.1.1.3 Other Modifications to χ/Q

Other modifications to the calculation of χ/Q can be made for specific scenarios. For example, the non-gaseous components of the plume can be depleted as material deposits onto the ground surface from the release. This has the effect of reducing the concentration at further distances from the facility. Gaseous radionuclides, of course, would be unaffected by plume depletion.

Another modification can be made for the effect of buildings in the vicinity near the release point. Nearby buildings can cause wakes due to the wind movement and can influence the path and concentration of airborne material close to the release point.

Some models also account for the topography surrounding a facility, including ground cover, hills and valleys, and other features. As might be expected, drastic changes in topography over short distances affect the wind movement and cause turbulence as well as localized areas of high wind movement.

Most of these modifications are typically handled by software, as discussed in Section 11.5.

11.2.1.2 Deposition of Airborne Contaminants

In addition to calculating the amount of radioactivity in the plume at a given location, it is necessary to calculate the amount of material that is deposited in the various areas surrounding a facility. While the deposition amount may not significantly change the airborne concentration from plume depletion, deposition must still be addressed. Radionuclides that are deposited on the earth or on vegetation can be eventually ingested by local residents and thus present a pathway for additional radiation dose.

Deposition theoretically has two components: dry deposition and wet deposition. Dry deposition is the natural settling of particulates from the plume that pass through an area. Wet deposition is due to washout from rainfall; while this can be an efficient method for deposition, the number of hours of rainfall throughout the year is relatively small for most areas. Therefore, wet deposition is episodic while dry deposition is continuous. Calculations of deposition generally will focus on dry deposition because on an annual basis, the amount of wet deposition is generally not that significant. Models exist for wet deposition, however, and those calculations can be incorporated into software.

As with other quantities, the calculation of deposition can be achieved through several methodologies. The deposition amount must of necessity depend on the concentration of material in the plume; therefore, some formulations relate the deposition rate to the plume concentration via a factor called the deposition velocity v_d, which has units of velocity (m/sec) but in reality is not an actual velocity. Rather, the deposition velocity is the ratio between the deposition rate and the airborne concentration at a reference height above the ground. Mathematically, v_d is represented as shown below in Equation 11.5:

$$v_d\left(\frac{m}{sec}\right) = \frac{deposition\ rate\ \left(\frac{Ci}{m^2 sec}\right)}{airborne\ concentration\ \left(\frac{Ci}{m^3}\right)} \tag{11.5}$$

Values for v_d are based on experiment and the literature shows a wide range of values. For gases, v_d is zero (i.e., no deposition). For iodine compounds and other particulates, different values may be used. For example, CAP88-PC uses v_d values of 0.035 m/sec for iodines and 0.0018 m/sec for particulates.

Based on assumed values of v_d., Equation 11.5 would be rearranged to determine the deposition rate d_r as shown in Equation 11.6.

$$d_r\ \left(\frac{Ci}{m^2 sec}\right) = v_d\ \left(\frac{m}{sec}\right)\chi\left(\frac{Ci}{m^3}\right) \tag{11.6}$$

It is common to normalize this deposition rate relative to the emission rate Q (in curies per second) to give a parameter called D/Q, defined as shown in Equation 11.7. As with the X/Q calculation, values of D/Q can be calculated prospectively and stored in computer programs and then applied when the release rates of the radionuclides are known.

$$\frac{D}{Q}(m^{-2}) = \frac{d_r\ \left(\frac{Ci}{m^2 sec}\right)}{Q\ \left(\frac{Ci}{sec}\right)} = v_d\ \frac{\chi}{Q} \tag{11.7}$$

If we ignore weathering that could remove deposited activity and consider radioactive decay as the only removal mechanism, we can surmise that the deposited activity would build up on an exterior surface (soil or vegetation) as shown in Equation 11.8.

$$C_{surface} = \frac{d_r}{\lambda}(1 - e^{-\lambda t_b}) \tag{11.8}$$

Where

- $C_{surface}$ is the surface concentration (Ci/m^2)
- d_r is the deposition rate as defined above ($Ci/m^2/sec$)
- λ is the radiological decay constant in units of seconds to be consistent with the units for d_r.

- t_b is the buildup time, defined as the time over which the radioactivity is allowed to build up on the surface of interest in units of seconds to be consistent with λ.

For pasture grass, crops, or leafy vegetables, the buildup time would be set considering the growing season for the vegetation. Typical values would be on the order of 30-60 days. For soil, the buildup time is considered to be longer and cumulative; in Regulatory Guide 1.109, the NRC specifies using t_b=15 years for that scenario.

If weathering is included as another removal mechanism, then λ in Equation 11.8 would be replaced by an effective removal constant λ_{eff}, which would be the sum of the radiological decay constant and a weathering removal constant, but the structure of the equation would remain the same.

One additional consideration with soil is the potential for the radioactivity to be distributed in the top layer of soil where it can be taken up by the roots of plants that have been planted in it. By convention, the upper 15 cm of soil is considered the plow depth (the depth of soil that is disturbed by plows and mixed prior to planting crops). Thus, for calculations of plant uptake through the roots, the surface-deposited activity can be assumed to be homogeneously distributed in the upper 15 cm of soil as shown below in Equation 11.9.

$$C_{soil} = \frac{C_{surface}}{(\rho_{soil})(0.15\ m)} \qquad (11.9)$$

Where

- C_{soil} is the concentration in the soil (Ci/kg)
- $C_{surface}$ is the surface concentration (Ci/m^2)
- ρ_{soil} is the density of the soil (kg/m^3)
- 0.15 m is the assumed plow depth of the soil

11.2.2 Dispersion of Waterborne Contaminants

As might be expected, the calculation of radioactivity concentration in surface waters can be just as challenging as that for airborne releases. One major difference, however, is that smaller surface water bodies (e.g., rivers, streams, cooling lakes) usually have well-defined physical boundaries; this is different than airborne releases which can disperse throughout the atmosphere. For our discussion, we will focus on these smaller water bodies, since larger water bodies (e.g., oceans, the Great Lakes) require more extensive modeling.

The following sections will consider dispersion under two scenarios:

- Nontidal rivers
- Small lakes with inflow and outflow (such as cooling lakes)

One simplification that we will make is to assume that we calculate the concentration for a well-mixed zone. For a river, this means that the concentration is calculated downstream sufficiently far that the release has uniformly mixed within the water horizontally and vertically. For a lake, this means that the discharge has mixed uniformly in the volume of the lake.

Figure 11-10 below shows a conceptual image of dilution in a river from a discharge point. Similar to airborne dispersion, there is a source term Q expressed as a release rate per unit time. The volume of water in the river provides the necessary dilution to allow the calculation of concentration at a downriver point where the material is assumed to be well-mixed. In addition to the parameters shown in Figure 11-10, the flow velocity (in m/sec) must also be known. This can be measured directly at the river or may be approximated as the flow rate divided by the cross-sectional area of the river (i.e., divide the flow rate by the product of the depth and width of the river).

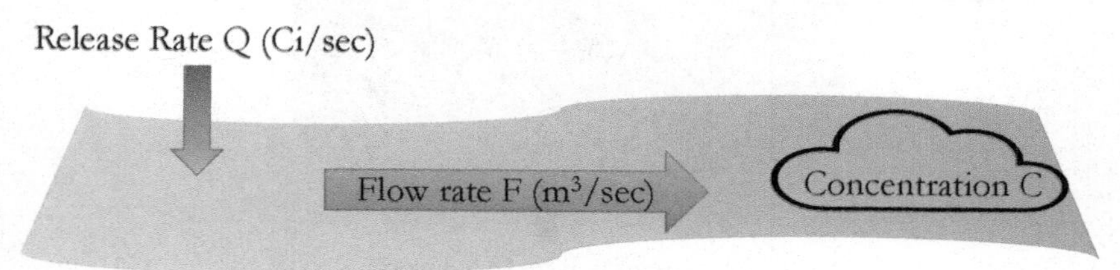

Figure 11-10. Illustration showing the release of radionuclides into a nontidal river that provides dilution.

Based on simple dilution, the concentration at a point downstream of the discharge point is given in Equation 11.10, which also accounts for radioactive decay between the discharge point and the point of interest for the calculated concentration. This equation is valid only for locations sufficiently downstream where complete mixing has occurred.

$$C = \frac{Q}{F} e^{-\lambda \frac{d}{v}} \tag{11.10}$$

Where

- C is the concentration in Ci/m^3 in the river
- Q is the release rate in Ci/sec
- F is the flow rate in m^3/sec
- λ is the radioactive decay constant (sec^{-1})
- d is the downstream distance (m)
- v is the stream velocity (m/sec) either measured directly or calculated as F/(river width × river depth)
- d/v is the decay time between the discharge point and the location of interest

Figure 11-11 below shows a conceptual drawing of a lake with an inlet and outlet. We assume that the average annual flow rate into the lake equals the flow rate from the lake (i.e., minimal evaporation). The release of effluents into the lake is at a constant rate and may involve the drawing of cooling water from the lake and returning it to the lake.

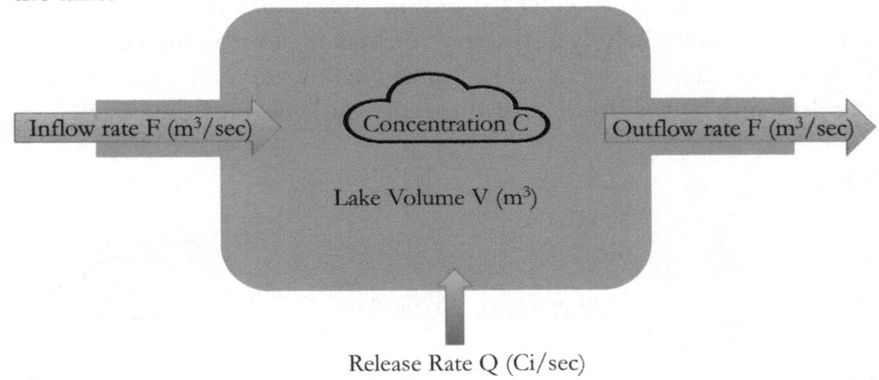

Figure 11-11. Schematic showing a lake with an inlet and outlet of constant flow rate that receives a source term Q from a nuclear facility.

Under these conditions, the source term Q is diluted in the volume of the lake. In Section 9.2.1.2, we discussed how airborne radioactivity could be diluted in a room that had inflow and outflow of air. We will use parallel concepts here as well for waterborne radioactivity. Similar to the ventilation removal rate, a removal rate constant for the lake can be calculated as F/V which will have units of inverse time (sec^{-1}). Higher values of F/V imply greater turnover of water in the lake and more rapid dilution.

Equation 11.11 (from Till 2008) describes the concentration C in the lake as a function of time based on the lake volume and the flow rate of the inlet and outlet.

$$C = \frac{\frac{Q}{V}}{(\frac{F}{V} + \lambda)}(1 - e^{-(\frac{F}{V}+\lambda)t}) \qquad (11.11)$$

Where

- C is the concentration in the lake (Ci/m^3)
- Q is the release rate (Ci/sec)
- F is the flow rate into or out of the lake (m^3/sec)
- V is the volume of the lake (m^3)
- F/V (sec^{-1}) is the removal constant considering only water flow into and out of the lake
- λ is the radioactive decay constant of the nuclide (sec^{-1})
- t is the time since the beginning of the release (sec)

The structure of Equation 11.11 is similar to that for Equation 11.8 where we considered the buildup of activity on the ground surface due to deposition from a plume. In Equation 11.8, the only removal mechanism was radioactive decay; therefore, only the radiological decay constant λ was included in that equation. The term ($F/V+\lambda$) in Equation 11.11 is an effective removal constant for the lake that includes radioactive decay as well as removal due to the inflow and outflow of water from the stream feeding the lake. Thus, that term shows up in the denominator of the equation as well as in the exponential piece of the equation. Similar to radiological half-life, the effective half-life for a given radionuclide in the lake is the natural logarithm of 2 divided by the effective removal constant. For radionuclides with a short effective half-life compared to the evaluation time, the concentration in the lake will reach a steady state concentration, which can be demonstrated mathematically.

The concentration calculated for the lake is also the concentration at the outlet of the lake where it discharges to a stream. Further downstream, the concentration will be decay-corrected for the travel time (d/v) between the outlet and the point of interest as shown in Equation 11.12.

$$C_{downstream} = C_{outlet} e^{-\lambda \frac{d}{v}} \quad (11.12)$$

Where

- $C_{downstream}$ is the concentration (Ci/m^3) at a downstream location
- C_{outlet} is the concentration (Ci/m^3) at the outlet of the lake
- λ is the radioactive decay constant (sec^{-1})
- d is the downstream distance (m)
- v is the stream velocity (m/sec) either measured directly or calculated as F/(river width × river depth)

If other tributaries enter the stream, Equation 11.12 can be adjusted for the reduced concentration using simple dilution calculations. Otherwise, using Equation 11.12 with no consideration of additional dilution would present a conservative estimate of the concentration.

11.3 Transfer of Activity Between Environmental Media

Once radioactive material has entered the environment, we must be able to account for its movement to each environmental medium until it causes a radiation dose to people. As shown earlier, Figure 11-1 indicates that there could be many steps for which the movement must be tracked.

NRC Regulatory Guide 1.109 takes an approach that relies on first-order kinetics and the use of transfer coefficients or bioaccumulation factors. In general, there are three general mathematical approaches that are employed to account for the time-dependent nature of the radioactivity in each medium.

1. If activity is being transferred from one medium to the next via physical means (water irrigation of crops, airborne deposition, etc.), the general approach is to use first-order kinetics that accounts for both the buildup of radioactivity and simultaneous radioactive decay to integrate the concentration rate over time much like we integrated the activity over time using Equation 1.7 back in Chapter 1. Thus, the equation will include a factor similar to that shown in Equation 11.13.

$$\frac{(1 - e^{-\lambda t_{acc}})}{\lambda} \qquad (11.13)$$

   Where

   - λ = radioactive decay constant **or** an effective removal constant that includes radioactive decay and another removal mechanism
   - t_{acc} = the time over which the activity accumulates in the medium
2. If the activity is no longer accumulating in the medium, then simple radioactive decay would cause the concentration to decrease in the media. This would be applicable in cases such as vegetables that are harvested and brought indoors from a garden or milk that has been stored after receiving it from a cow. The equation for the concentration in this case would include a radioactive decay factor as shown below in Equation 11.14.

$$e^{-\lambda t_{dec}} \qquad (11.14)$$

   Where

   - λ = radioactive decay constant
   - t_{dec} = the time over which the activity has decayed in the medium before moving to the next step of the transport process.
3. If an equilibrium can be established between two media, then a bioaccumulation factor or transfer coefficient may be used that relates the concentration in one medium to another medium. This is common for several scenarios such as the concentration

in plants that are grown on contaminated soil. In that case, a simple element-specific constant factor can be used to relate the concentration in the plant material to the soil in which it is grown. Bioaccumulation factors are also used in estimating the concentration of radionuclides in fish relative to the concentration of the radionuclides in the water in which the fish reside. In addition, there are bioaccumulation factors for the concentration in meat or milk based on the concentration in the feed or grass that is consumed.

As an example, Table 11-3 below shows some of the historical bioaccumulation factors from NRC Regulatory Guide 1.109 for transfer from soil to plants. Other references may provide differing values, however. For example, one of the newest expanded sets of data is contained in ICRP Publication 114 (*Environmental Protection: Transfer Parameters for Reference Animals and Plants*), available from www.icrp.org.

Table 11-3. Values of the bioaccumulation factor B_{iv} for selected elements as taken from NRC Regulatory Guide 1.109.

Element	B_{iv} (Veg/Soil) concentration in vegetation compared to soil (dimensionless)
H	4.8E 00
Na	5.2E-02
P	1.1E 00
Mn	2.9E-02
Fe	6.6E-04
Co	9.4E-03
Ni	1.9E-02
Cu	1.2E-01
Zn	4.0E-01
Sr	1.7E-02
I	2.0E-02
Cs	1.0E-02

While it is tedious to consider every possible pathway, a single example can help illustrate this process. Let us consider the pathway of an airborne release where the radioactivity is transferred to vegetation. This is presented as Equation C-5 in NRC Regulatory Guide 1.109 and is reproduced below as Equation 11.15.

There are three distinct parts of Equation 11.15 as follows:

- The product of d_i and the red part of Equation 11.15 estimates the concentration in the vegetation due to direct deposition on the vegetation from the air and the buildup of activity with time.
- The product of d_i and the green part of Equation 11.15 estimates the concentration in the vegetation due to uptake of radioactivity through the roots of the vegetation from the deposited activity distributed in the soil and the buildup of activity with time.
- The blue part of Equation 11.15 estimates the radioactive decay that occurs between the harvest and the actual ingestion of the vegetation; it corrects for decay while the material is being stored.

$$C_i^v = d_i \left\{ \frac{r(1-e^{-\lambda_{Ei} t_e})}{Y_v \lambda_{Ei}} + \frac{B_{iv}(1-e^{-\lambda_i t_b})}{P \lambda_i} \right\} e^{-\lambda_i t_h} \qquad (11.15)$$

(Equation C-5 from NRC Regulatory Guide 1.109)

Where

- C_i^v = the concentration of radionuclide "i" in vegetation
- d_i = the deposition rate for radionuclide "i" as defined in Equation 11.6 above (activity per unit area per unit time). Physically, this can be the deposition rate per unit area of exposed vegetation or the deposition rate per unit area of soil.
- r = the fraction of the deposited activity that remains on the plants
- λ_{Ei} = the effective removal constant of the activity from the vegetation due to radioactive decay and weathering
- t_e = the growing period for the plants prior to harvest
- Y_v = the agricultural productivity (kg of vegetation produced per unit area of land)
- B_{iv} = the bioaccumulation factor for radionuclide "i" for soil-to-vegetation transfer as discussed above with examples in Table 11-3
- λ_i = the radioactive decay constant for radionuclide "i"
- t_b = the assumed time period for long-term buildup of radionuclides in soil

- P = the product of the soil density and the assumed plow depth of the soil, as discussed in conjunction with Equation 11.9 above.
- t_h = the assumed time between the harvest of the vegetation and the ingestion of the vegetation by either animals or people.

As can be seen, both the red and green parts of Equation 11.15 include the factor from Equation 11.13 to estimate the buildup of radioactivity in the vegetation with time. The red part of the equation acknowledges two removal mechanisms: radioactive decay and physical weathering. Thus, the effective removal constant is used in that fraction. The green part of the equation has only radioactive decay as a removal mechanism so that only the radioactive decay constant is used in that fraction.

In addition, the blue part of Equation 11.15 accounts for the radioactive decay of the activity in the vegetation between harvest and ingestion using the concept from Equation 11.14.

Specific values for the parameters in Equation 11.15 can be found in Regulatory Guide 1.109 (mostly Appendix E) and other places in the literature dealing with environmental models.

The concentration in the vegetation calculated by Equation 11.15 can then be used to calculate concentrations for other pathways, such as milk produced from cows that eat the vegetation. Thus, additional parameters would be factored in, such as what portion of the cow's diet is comprised of contaminated vegetation, the radionuclide-specific transfer coefficient between vegetation and milk, and the amount of decay time between when the milk is gathered until it is consumed. While these calculations can be tedious, they follow the same thought process as discussed above, and the environmental dose assessor would thus track the movement of radioactivity along each step of the pathway with technically justifiable values for the numerical parameters.

11.4 Calculating Dose from Environmental Media

Once the concentrations in environmental media are known, it then remains to calculate the dose to people. Referring again to Figure 11-1, it is easy to see that there are a variety of pathways that depend on lifestyle habits of the receptor. For example, the amount of vegetables, milk, fish, or meat that a person ingests daily is highly variable. In addition, there must be an assumption of what fraction of those foods were grown locally and subject to emissions from the nuclear facility of interest.

For inhalation and external irradiation, the amount of time a person spends indoors compared to outdoors affects the dose calculation. Also, people may be away from their homes and at work for a fraction of the day, adding another layer of complexity. Again, these lifestyle habits can markedly affect the dose calculation.

In addition, the general public is comprised of all age groups; so, it is common to calculate doses for people at a variety of ages. This requires (a) lifestyle information on the various age groups and (b) dose coefficients for the age groups.

As discussed earlier, the usual approach is to assume a reasonable lifestyle for representative individuals for a representative estimate of the dose that may be received. A second calculation can then be performed assuming more conservative (and in some cases extremely conservative) lifestyle habits to place a bounding estimate on the dose estimate. The extremely conservative calculations are for a model known as the Maximum Exposed Individual (MEI) mentioned earlier.

Table 11-4 below shows the usage factors from Regulatory Guide 1.109 to show the variety of quantitative values that are based on lifestyle habits for both the average individual and the MEI.

In looking at the values in the table, any person could find deviations in their own behaviors (e.g., ingestion rates for certain foods, shoreline recreation, etc.) from those presented in the table and may question the accuracy of the resulting dose calculations. Similar to the concepts discussed in the chapters on internal dosimetry, environmental dose calculations are not meant to represent an actual dose to a real person; rather, the calculated doses are only meant to be somewhat representative to ensure that the public is adequately protected following the release of radionuclides into the environment.

Table 11-4. Usage factors for a representative child, teen, and adult as presented in Regulatory Guide 1.109. These are meant for an average individual (Avg) and for the Maximum Exposed Individual (MEI).

	Child		Teen		Adult	
Pathway	Avg	MEI	Avg	MEI	Avg	MEI
Fruits, vegetables, and grains (kg/yr)	200	520	240	630	190	520
Leafy vegetables (kg/yr)	-	26	-	42	-	64
Milk (liters/yr)	170	330	200	400	110	310
Meat and poultry (kg/yr)	37	41	59	65	95	110
Fish (kg/yr)	2.2	6.9	5.2	16	6.9	21
Seafood (kg/yr)	0.33	1.7	0.75	3.8	1.0	5
Drinking water (liters/yr)	260	510	260	510	370	730
Shoreline recreation (hours/yr)	9.5	14	47	67	8.3	12
Inhalation (m^3/yr)	3700	3700	8000	8000	8000	8000

Dose coefficients for given pathways and individuals are radionuclide-specific and are too numerous to include in this book. However, the foundation for the dose coefficients has been presented for both external exposure as well as intakes.

NRC Regulatory Guide 1.109 contains an extensive set of dose coefficients. However, these coefficients are based on outdated ICRP dosimetric models that harken back to the 1950s and 1960s. Based on the concepts discussed in the earlier chapters of this book, the ICRP has published the following newer dose coefficients, which are available freely at www.icrp.org:

- ICRP Publication 144 – *Dose Coefficients for External Exposures to Environmental Sources*, which addresses external exposure scenarios.
- ICRP Publications 56, 67, 69, 71, and 72 – *Age-Dependent Doses to the Members of the Public from Intakes of Radionuclides*, which addresses the internal exposure scenarios. A downloadable database of the dose coefficients is also available through the ICRP website. However, it should be noted that these publications and database do not include the latest models adopted by the ICRP in

Publication 130, which was discussed in Chapter 7 of this textbook. The ICRP is in process of updating the dose coefficients to the newest models as of the writing of this book.

In addition to the individual doses that are calculated, it is common to calculate population doses. This is achieved by using population information for the area around a facility, usually based on census data and population concentration within the given sectors around the facility. Each person in a given sector is assumed to receive the same dose; therefore, the process is to calculate an individual dose for each sector, multiply that value by the number of people in that sector, and then sum the results for all sectors around the facility.

11.5 Computer Programs for Environmental Health Physics Calculations

As mentioned before, the calculations for environmental health physics are tedious because of all the different factors that go into the calculations. The calculations themselves involve simple arithmetic, but the sheer number of numerical values that must be drawn from various data sources requires great attention to detail and thus invites the potential for calculational errors. While it is possible to perform some elementary hand calculations to illustrate the principles, it is not practical for nuclear facilities to carry out all the calculations manually. Therefore, it is common to use computer programs that can be used to perform the calculations; the programs can store all the necessary factors and population information that have been discussed in this chapter, and local meteorological information can be incorporated into the programs to allow site-specific results.

Some facilities will write their own computer programs to allow maximum flexibility in incorporating site-specific information and assumptions. These can be written as standalone programs but can also be designed as spreadsheets or databases. Of course, commercial software programs are available for purchase that can use site-specific data coupled with standard dispersion and dosimetric models to calculate the radiation dose to offsite receptors. The results of these annual dose calculations can then be presented in an annual environmental report that is accessible to stakeholders and the general public.

Facilities with radionuclide releases to the air in the United States must demonstrate compliance with the Clean Air Act National Emissions Standards for Hazardous Air Pollutants (NESHAPs) as specified in Title 40, Part 61 of the Code of Federal Regulations (40CFR61), Part H by the Environmental Protection Agency (EPA). While site-specific programs may be used for annual reporting to the public, the EPA requires a particular set of parameters and assumptions to ensure consistent calculations across all facilities. To aid facilities in demonstrating compliance, the EPA has made two computer programs available that can be freely downloaded as follows:

- COMPLY – available from https://www.epa.gov/radiation/comply
- CAP88-PC – available from https://www.epa.gov/radiation/cap88-pc

In addition, the Nuclear Regulatory Commission has made available several computer programs through its Radiation Protection Computer Code Analysis and Maintenance Program (RAMP) website, located at https://ramp.nrc-gateway.gov/. Academic users can register free of charge and then gain access to computer codes such as the following (available as of the writing of this book):

- GENII (includes both air and water pathways)
- NRCDose(a combination of three programs: GASPAR, XOQDOQ, and LADTAP II for calculating doses from air and water pathways)
- RESRAD (includes modules for calculating dose due to RESidual RADioactive material in the environment)

While the user interface and input deck are different for each software program listed above, the concepts described in this chapter are sufficient to give the user a good starting point for using the programs. Reading the User Manuals for each of the software packages and dissecting the technical basis for the calculations will help the user understand the calculations that are being performed and ensure that the program is being used correctly for the scenario under consideration.

11.6 Environmental Health Physics Calculations for Accident Scenarios

One final point on calculating doses from environmental releases pertains to unintended accidental releases. Such calculations are performed for one of three reasons as discussed below.

1. To aid in the design of a facility, technical personnel must analyze all potential accident scenarios that could occur. This analysis involves calculating the possible consequences of any releases based on knowledge of the source inventory in the facility and the possible mode of dispersion (fire, tornado, etc.). Assumptions are then made on the amount of material that would be released in the given scenario in order to calculate the offsite dose consequences. Additionally, extremely conservative meteorology is often assumed. Stability class G, as defined in Table 11-1, is the usual default assumption to ensure that calculations represent the worst-case scenario.
2. Facilities will perform emergency drills to ensure that they are prepared to handle real-life accidents that may occur. As part of the drill exercise, a scenario will be played out as though it had actually occurred, and personnel will react and respond to the event to identify any shortcomings in their protocols and to practice carrying out their responsibilities. Decisions will be made based on offsite dose calculations, and thus the environmental health physics calculations must be performed as the scenario is carried out in real time with the values being reported to management who can make decisions regarding mitigation measures. The facility may conduct its own drill several times per year; it is also common to perform a drill annually with regulatory authorities and local government emergency personnel to ensure that communications and coordination between these various entities are sufficient and that any scenario could be handled effectively.
3. In the unlikely event of an actual accident, qualified personnel must be capable of quickly assessing offsite dose consequences based on either measured, calculated, or assumed release amounts of radioactive material. Real-time meteorological information is incorporated to assess the projected travel direction of contaminants as well as the possible dose to receptors.

In any of these scenarios above, the value of X/Q and subsequent concentration calculations are based on short-term meteorological information rather than average meteorology throughout the year. In addition, more conservative assumptions are usually incorporated to provide a bounding case scenario to aid in making decisions regarding protective actions or interventions that may be necessary to reduce the exposure of the public.

As with routine releases, there are software packages that are used to calculate doses in these situations. One program that is available through the NRC RAMP website (https://ramp.nrc-gateway.gov/) is RASCAL. It has historically been used by NRC personnel to independently assess dose consequences of accidental releases. In addition, site-authored software programs or commercial software packages are available that can be used to evaluate the consequences of accidental releases.

Free Resources for Chapter 11

- Questions and problems for this chapter can be found at www.hpfundamentals.com
- CAP88-PC Version 4.1 User Guide, Environmental Protection Agency, 2023. Available from www.epa.gov.
- ICRP Publication 56, *Age-dependent Doses to Members of the Public from Intake of Radionuclides - Part 1,* Ann. ICRP 20 (2), 1990. Available from www.icrp.org.
- ICRP Publication 67, *Age-dependent Doses to Members of the Public from Intake of Radionuclides - Part 2 Ingestion Dose Coefficients*, Ann. ICRP 23 (3-4), 1993. Available from www.icrp.org.
- ICRP Publication 69, *Age-dependent Doses to Members of the Public from Intake of Radionuclides - Part 3 Ingestion Dose Coefficients,* Ann. ICRP 25 (1), 1995. Available from www.icrp.org.
- ICRP Publication 71, *Age-dependent Doses to Members of the Public from Intake of Radionuclides - Part 4 Inhalation Dose Coefficients*, Ann. ICRP 25 (3-4), 1995. Available from www.icrp.org.
- ICRP Publication 72, *Age-dependent Doses to the Members of the Public from Intake of Radionuclides - Part 5 Compilation of Ingestion and Inhalation Coefficients*, Ann. ICRP 26 (1), 1995. Available from www.icrp.org.
- ICRP Publication 114, *Environmental Protection: Transfer Parameters for Reference Animals and Plants*, Ann. ICRP 39(6), 2009. Available from www.icrp.org.
- ICRP Publication 144, *Dose Coefficients for External Exposures to Environmental Sources*, Ann. ICRP 49(2), 2020. Available from www.icrp.org.
- International Atomic Energy Agency, *Generic Models for Use in Assessing the Impact of Discharges of Radioactive Substances to the Environment*, Safety Reports Series No. 19, IAEA, Vienna 2001. Available from www.iaea.org
- NRC Regulatory Guide 1.23, *Meteorological Monitoring Programs for Nuclear Power Plants*, US Nuclear Regulatory Commission, 2007. Available from www.nrc.gov
- NRC Regulatory Guide 1.109 *Calculation of Annual Doses to Man from Routine Releases of Reactor Effluents for the Purpose of Evaluating Compliance with 10 CFR Part 50, Appendix I*, US Nuclear Regulatory Commission, 1977. Available from www.nrc.gov

- NRC Regulatory Guide 1.111 *Methods for Estimating Atmospheric Transport and Dispersion of Gaseous Effluents in Routine Releases from Light-Water-Cooled Reactors*, US Nuclear Regulatory Commission, 1977. Available from www.nrc.gov
- NRC Regulatory Guide 1.113 *Estimating Aquatic Dispersion of Effluents from Accidental and Routine Reactor Releases for the Purpose of Implementing Appendix I*, US Nuclear Regulatory Commission, 1977. Available from www.nrc.gov

Software

- EPA environmental pathways modeling software
  - COMPLY – available from https://www.epa.gov/radiation/comply
  - CAP88-PC – available from https://www.epa.gov/radiation/cap88-pc
- The following environmental pathways modeling software is available through the NRC Radiation Protection Computer Code Analysis and Maintenance Program (RAMP) website, located at https://ramp.nrc-gateway.gov/
  - GENII (includes both air and water pathways)
  - NRCDose(a combination of three programs: GASPAR, XOQDOQ, and LADTAP II for calculating doses from air and water pathways)
  - RESRAD
  - RASCAL (Analysis of environmental dispersion following an accident)

Additional References

- Turner, D. B., *Workbook of Atmospheric Dispersion Estimates*, Environmental Protection Agency, Research Triangle Park NC, 1970.
- Sagendorf, J. F., A Program for Evaluating Atmospheric Dispersion from a Nuclear PowerStation, NOAA Tech Memo ERL-ARL-42, 1974. Available at https://www.arl.noaa.gov/documents/reports/arl-42.pdf
- Till, J., E., and Grogan, H. E. (editors), *Radiological Risk Assessment and Environmental Analysis*, Oxford University Press, Ocford New York, 2008.

Chapter 12 Environmental Health Physics Monitoring

The previous chapter of our text dealt with the calculations surrounding releases of radioactive material to the environment. We now turn our attention to the design and implementation of monitoring programs for environmental health physics to measure any radioactivity that may be detectable in the environment.

It is impossible to sample the entirety of the environment around a nuclear facility. Therefore, environmental programs tend to focus on:

- monitoring or sampling of media that would be most likely to contain radioactivity that could be detected
- monitoring in locations where it is most probable that radioactivity could be detected
- monitoring for external radiation that may either be emitted directly from the facility or by gaseous plumes that are released from the facility

Many samples that are collected as part of environmental monitoring will have no detectable radioactivity that is attributable to the facility, which is beneficial to demonstrate to regulators as well as nearby residents that releases from the facility are minimal. However, the facility must still collect samples to document the continued efficacy of the engineering controls that are in place to minimize releases.

Environmental monitoring may be performed by one or more of the following entities:

- The nuclear facility itself performs monitoring so that it can demonstrate compliance with applicable regulations.
- Regulatory agencies (federal, state, or local) perform confirmatory monitoring as part of their oversight. In addition, monitoring by

the regulators may reveal trends or issues associated with releases from multiple nuclear facilities that may not be easily identified by individual facility monitoring.

- Private individuals or citizen organizations may perform monitoring due to their stakeholder interest and concern regarding environmental releases.

The following sections give an overview of some monitoring principles.

12.1 Air Monitoring

Air monitoring in the workplace was discussed previously in Section 10.1.2. For monitoring in the environment, a low-volume air sampler is typically used and is placed in a shelter assembly to protect it from adverse weather conditions. An example of such a shelter with the air sampler inside is shown below in Figure 12-1.

Figure 12-1. An air sampler mounted inside a cabinet shelter to protect it from adverse weather (Photo courtesy of F&J Specialty Products, Inc.).

Because of the need for constant electrical power, the placement of environmental air samplers must be planned so that electrical utilities can be run to the locations. In addition, the samplers must be relatively accessible by personnel while being unobtrusive to avoid accidental damage by the public or tampering through vandalism.

Air samplers may be placed around the perimeter of a site boundary and may also be placed in locations a few hundred meters away from the site to ensure that the air in the general vicinity is adequately sampled. The number of air samplers is somewhat subjective, but a half dozen or more samplers would be a common starting point for the measurements. Generally, the samplers would be placed to ensure a full 360-degree coverage around the site so that no matter which way the wind blows, an air sampler will be in the path of any releases.

The frequency with which the filters are collected from the air samplers is not a rigid fixed number in all situations. It is common to collect samples on a weekly basis; the longer that the air sampler operates, the more particulate matter (dust particles) will build up on the filter, potentially clogging the filter so that air is not adequately drawn through the filter medium. Therefore, in dusty environments, it may be necessary to change out the filter more often than weekly.

As discussed in Section 10.1.2, the filters (and charcoal cartridges, if equipped) are removed, placed in a plastic sleeve or bag to preserve all material that has collected on the filter, and then returned to a radiation detection laboratory for analysis, typically gamma spectroscopy. The average air concentration can then be calculated using the equations presented in Section 10.1.2 based on the amount of detected activity, the collection efficiency, and the volume of air drawn through the sampler.

12.2 Rainwater Collection

As discussed in Section 11.2.1.2, wet deposition is the process whereby rain falling through the air will cause the washout of particulate matter from the air. While wet deposition is not expected to be a significant source of deposition, it is nonetheless relatively easy to collect rainwater around a facility to determine if any detectable radioactivity is present. Therefore, it is common at the air sampler stations to place a flat pan on the roof of the cabinet assembly with tubing that leads to a container jug inside the cabinet. When the air sampler filters are collected, any rainwater can also be collected and analyzed at the laboratory to determine the presence of any radioactivity that may have washed out from the surrounding air.

12.3 Vegetation Sampling

Vegetation could contain radioactivity through a couple of processes that we discussed in the last chapter:

- Airborne radioactivity can settle out of the air and deposit on the exterior of the vegetation.
- Radioactivity in the soil (through airborne deposition or irrigating with water containing radioactivity) can be absorbed through the plant's roots into the plant.

As part of an environmental monitoring program, it is common to collect vegetation around a nuclear facility for analysis in the laboratory. In some cases, gamma spectroscopy may be employed directly to identify any gamma-emitting radionuclides. Radionuclides with insignificant gamma emission may require additional chemical processing and other radiochemical analysis to detect the radionuclide of interest. For the special case of tritium (which cannot be detected via gamma spectroscopy), chemical means may be employed to extract any water content from the vegetation followed by liquid scintillation counting of the extracted moisture.

The following sections discuss the rationale and approach to vegetation sampling.

12.3.1 Edible Produce

Residents around a nuclear facility may have their own gardens in which they grow produce to consume through the growing season. An example of a local garden is shown below in Figure 12-2. Likewise, stores and fresh markets may sell produce that is grown locally. Both scenarios provide a potential pathway for ingestion of radioactive material that may come from the nearby nuclear facility.

Figure 12-2. Local gardens provide produce for residents and should be monitored when in proximity to a nuclear facility.

While some edible produce can be grown during the colder months of the year, the majority is still grown during the warmer months of the year. Produce is generally divided into leafy (for example, lettuce, greens, etc.) and non-leafy (fruits, garden vegetables, etc.) because the models discussed in Chapter 11 use different calculational methodologies for the two types of vegetation.

Some nuclear facilities will arrange with local residents or farmers to purchase produce from their gardens or may purchase produce from local markets. One advantage of this method of sampling is that it represents real crops grown in the vicinity of the facility. As mentioned above, radioactivity may be present in the vegetation due to airborne deposition or due to irrigation. In some regions of the country, farmers may obtain permits to draw irrigation water from nearby rivers or waterways that are downstream from a nuclear facility. Thus, obtaining produce from these farmers may maximize the opportunity to detect any radioactivity that may be released from the facility.

One common trend is for the nuclear facility to grow its own garden on the plant site. There are several advantages to this approach. First, the garden can be populated with a wide array of produce and can be managed to ensure a good harvest to provide a suitable sample size for analysis. Second, airborne deposition will generally be greatest near the facility (based on principles discussed in Chapter 11); therefore, a garden grown in close proximity to the facility is likely to maximize any airborne deposition to provide an upper bound estimate of any radioactivity content in the food.

12.3.2 Non-edible Vegetation

While edible vegetation is of interest due to the ingestion pathway, facilities are also interested in potential airborne deposition that occurs throughout the year. In particular, leafy plants with large surface areas for deposition are desirable for analysis. Thus, it is common to collect leaves from deciduous trees as well as leaves from smaller woody plants (one caution that the author learned the hard way while collecting samples is to be able to properly identify and avoid poison oak or poison ivy!!).

In addition, grasses may be collected for analysis due to the potential for grasses to be ingested (either directly from the field or after being harvested for hay) by livestock. The milk or meat from the livestock would then serve as a potential pathway to humans.

During fall and winter months, many plants go dormant and lose their leaves. However, it is possible to use foliage from evergreen plants for analysis, since the foliage often has a large combined surface area. For example, pine needles have been shown to be useful in analysis for both radiological as well as nonradiological contaminants (see https://news.ncsu.edu/2022/02/pine-needles-story-of-nc-pfas/ for an example of the use of pine needles for sampling of nonradiological contaminants).

12.4 Soil Sampling

Since contaminated soil can serve as a means for radioactivity to enter the food chain for humans, it makes sense to collect soil samples from locations around a nuclear facility. The approach is relatively straightforward in that it involves collecting soil from the top 15 cm of soil, which we identified in Chapter 11 as being the traditional assumed plow depth of soil that is used for growing crops. Once collected, the soil would be mixed to ensure homogeneity prior to analysis.

12.5 Surface Water Sampling

Since some facilities will discharge liquid releases to surface waters, it is imperative to be able to sample the water to determine if any detectable radioactivity is present. However, surface water sampling requires some careful thought and planning to properly interpret any measurements that are made. As we did in Chapter 11, we will consider only nontidal rivers/streams and small lakes with inlets and outlets for our discussion. Sampling in other water bodies carries additional considerations beyond the scope of this text but which can be researched if needed.

12.5.1 Sampling Locations

It is possible that a particular river/stream could contain radioactivity from several different facilities. For example, a municipality may discharge its treated wastewater (which may contain small amounts of radioactivity from commercial customers) to the same river that is also used as cooling water for a nuclear facility. Each entity that discharges to the surface waters must be able to identify its own contribution to any contaminants that are present in the water.

For this reason, facilities will typically have a sampling location upstream of their facility where none of their releases could impact the surface waters. Measurement results from this location provide a baseline for radioactive material (natural or otherwise) that may already be present in the water.

The selection of the proper downstream location relies on modeling principles as discussed in Chapter 11. The sampling location should be where the surface waters have been well-mixed, which means it should be far enough from the release point to ensure homogeneity. At the same time, it should not be so far downstream as to have been diluted by other tributaries or significantly depleted due to settling or radioactive decay.

If the facility is using a river for discharges, then it is sufficient to pick an upstream sampling point several hundred yards upstream of the discharge location. Depending on the size of the river and its characteristics, other sampling locations might be at one or more of several locations:

- A sampling location at the discharge point to determine the maximum concentration of radionuclides prior to entering the surface water
- A sampling location further downstream to obtain a more representative sample after mixing has occurred
- A sampling location where it is known that surface water is removed for other uses, such as irrigation or municipal drinking water (discussed in more detail below)

If the facility is using a small lake (such as a cooling lake for a power reactor), the upstream sampling location could be in the creek or waterway that feeds into the cooling lake. Mixing would occur in the lake itself, and the downstream location could be at the point where the lake empties into the continuation of the creek/waterway or other downstream locations of interest similar to that discussed above for rivers.

Figure 12-3 below illustrates these sample location principles for a river and for a cooling lake scenario.

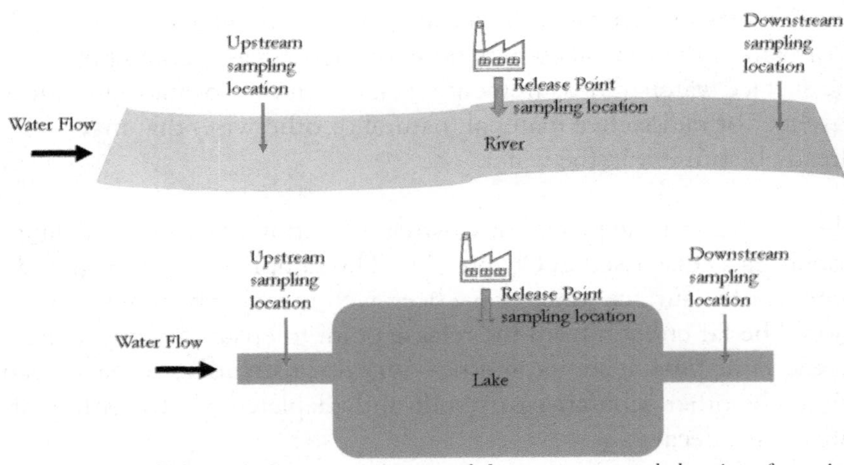

Figure 12-3. Illustration showing upstream and downstream sample locations for a river (top) and for a small lake (bottom).

12.5.2 Sampling Methods

The collection of a surface water sample can be as simple as taking a bucket with a rope attached to the handle and retrieving a sample of water from the appropriate location; this is referred to as a grab sample. The major limitation to a grab sample is that it represents a single snapshot in time and cannot take account for time variations in releases from a facility. While it may give a quick measurement result, interpretation of the values must always be made with the understanding that they do not necessarily represent conditions in the surface water throughout the year.

A more common mode of sampling is to use a composite sampling instrument. There are many designs for such instruments, but the basic premise is that a water sampling line is submerged in the waterway and is attached to a pump. The pump accurately draws a small amount of water on a predetermined time schedule over the course of a few days. The individual samples of water are mixed together, thus giving a composite such that the concentration would be more representative of an average value over time. The same considerations apply to the composite sampling instruments as apply to environmental air samplers: electrical power must be supplied to the instruments, and their locations should be selected to minimize accidental damage or vandalism.

Regardless of whether grab samples or composite sampling is employed, the sample should be drawn from a location where the water is well-mixed. Typically, this would be a location that is roughly halfway from the bottom of the waterway to the surface of the water. In the case of a river or stream, the location would also need to be roughly in the middle of the waterway.

Once the samples are returned to the laboratory, the water may be analyzed via gamma spectroscopy with no further processing; however, to increase sensitivity, some laboratories will use chemical separations or other means to try to concentrate any radioactivity into a smaller volume. Ion exchange resins can also be used to adsorb radionuclides from the water to concentrate them and make detection easier.

One additional consideration would apply when the surface waters are used as a source of drinking water by a municipality downstream of the facility. Treatment of surface water for drinking purposes does not generally remove radioactivity from the water; therefore, it would be incumbent to obtain drinking water samples from the water treatment facility for analysis.

12.6 Sediment Sampling

Sediment on the bottom of lakes or rivers can concentrate radioactivity due to adsorption of radionuclides. Thus, the concentration of radioactivity in sediments can be much greater than that in the surface water itself.

Collection of sediment can be obtained via a variety of methods, such as shovels or trowels or augers. For deeper water, a handheld mechanical clamshell grab dredge can be used. To operate the dredge, the clamshell mechanism is opened before being deployed from a location above the water (from a boat or bridge) via a rope or cable. When the dredge is lowered and strikes the bottom, the weight of the dredge operates the mechanism and causes the clamshell to close and scoop the sediment. The dredge can then be raised so that the sediment can be extracted and returned to the laboratory for analysis. Figure 12-4 below shows an example of a handheld dredge.

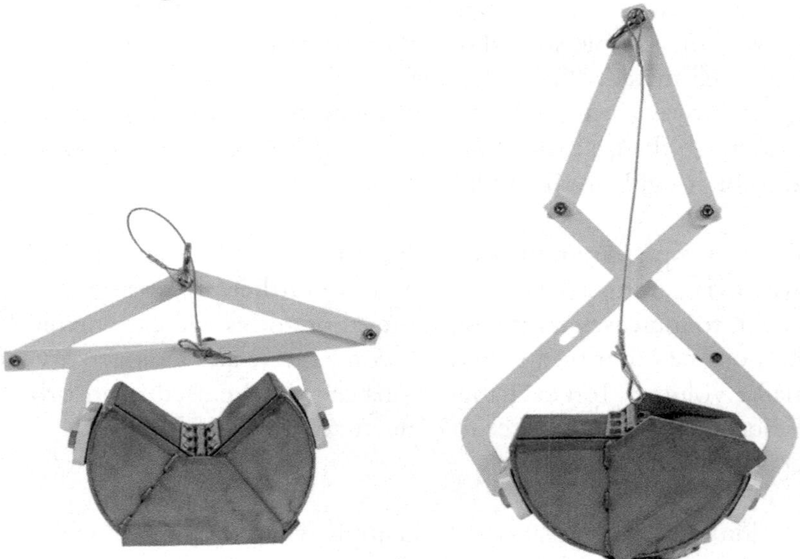

Figure 12-4. Example of a hand-held clamshell dredge used to gather sediment. The left image shows the opened dredge as it is deployed while the right image shows how the weight of the dredge causes it to close when it strikes the bottom of a river or lake (Photos courtesy of AMS, Inc. www.ams-samplers.com).

12.7 Aquatic Life Sampling

Section 11.3 discussed the concept of the bioaccumulation factor or the transfer factor. As it pertains to surface water, radioactivity in the water can transfer to aquatic life, thus making the monitoring of aquatic life a sensitive indicator of the water content. In addition, because of potential consumption of aquatic life by humans, this serves as a pathway for radiation dose and must be assessed. Obviously, there are other interests in collecting fish samples not related to radiological monitoring, such as monitoring fish populations and analyzing fish for chemical contaminants.

The collection of fish is often accomplished via a technique known as electrofishing or "fish shocking." This method of collection involves using a generator as a source of electricity in a boat and having electrodes placed in the water to create an electric field. The field temporarily immobilizes the fish and causes them to float to the surface where they may be scooped out as shown below in Figure 12-5. **It is important to note that electrofishing requires a government permit and is not legal for casual or sport fishing.**

Figure 12-5. Environmental technicians using electrofishing to gather fish from a river to document fish population information. (Photo courtesy of US Fish and Wildlife Service)

Aquatic animals known as filter feeders (e.g., clams, mussels, and other types of fish) often concentrate radioactivity due to the nature of their feeding action. Collecting them may involve more effort than that used for electrofishing; however, they can serve as a sensitive indicator of water contamination.

Obviously, aquatic animals may concentrate radionuclides in different parts of their anatomy based on the chemistry and metabolism of the animal. Only edible portions of the animals are of interest for dose assessment and are analyzed in the laboratory to assess potential radiation dose. Non-edible portions may be analyzed to address other concerns and for research on radionuclide movement, however.

12.8 Milk Sampling

Residents around a nuclear facility may have their own milk animals or may obtain milk through a local dairy. Commonly, the nuclear facility will purchase milk from a local dairy or farmer on a regular basis for analysis of various radionuclides.

While cows are the predominant milk animal, goats can also provide milk to local residents. Due to the difference in metabolism between goats and cows, goat milk can have higher concentrations of some radionuclides compared to dairy cows, as evidenced by the bioaccumulation factors for cows and goats shown below in Table 12-1. In particular, isotopes of strontium, cesium, and iodine are of interest since for nuclear reactors, these are common fission products that could be released to the environment. Therefore, if the environmental monitoring program identifies goat farmers in the vicinity of the facility, obtaining goat milk samples may be of interest.

Table 12-1. Selected stable element transfer factors for dairy cows and dairy goats as presented in NRC Regulatory Guide 1.109.

Element	Transfer factor for feed to cow milk	Transfer factor for feed to goat milk
Phosphorus	0.025	0.25
Iron	0.0012	0.00013
Copper	0.014	0.013
Strontium	0.0008	0.014
Iodine	0.006	0.06
Cesium	0.012	0.3

12.9 Groundwater Sampling

Residents around a nuclear facility may understandably question whether their groundwater contains radionuclides from the facility. While the study of how radionuclides migrate to groundwater is beyond the scope of this text, there are some scenarios that could lead to groundwater contamination. The scenarios generally begin with the contamination of soil; afterwards, rainwater causes the nuclides to migrate downward deeper into the soil where they can then leach into aquifers. The groundwater in the aquifers can then flow offsite to locations where homeowners or municipalities may drill wells into the aquifer to withdraw drinking water. Further, surface streams could cut into the shallowest aquifer at a seep, as illustrated below in Figure 12-6. The migration of groundwater into the stream then provides an opportunity for radionuclides to be transported into the stream as well.

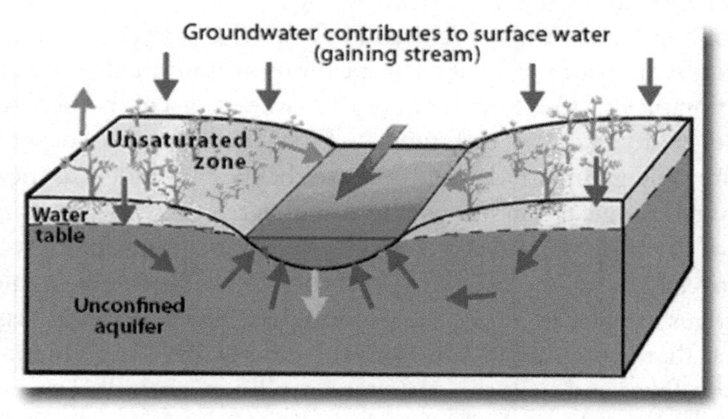

Figure 12-6. If a surface water stream cuts into an aquifer, the seep from the aquifer could enter the stream. (Image courtesy of the US Geological Survey, US Department of the Interior)

For areas where the risk of groundwater contamination is low, it may be sufficient to obtain groundwater from nearby residential wells that have already been drilled in the vicinity. For areas of known groundwater contamination, monitoring wells may be drilled at specific locations. Measurements of contaminants in the groundwater in the monitoring wells can then be used to map the migration of radionuclides in the aquifers and estimate impacts further downgradient from the source of the contamination.

Monitoring wells may be close to each other or may be spread over a large distance. One example of monitoring wells in close proximity to each other is shown below in Figure 12-7.

Figure 12-7. Example of multiple monitoring wells near each other (Photo courtesy of the NC Division of Water Resources).

12.10 External Radiation Monitoring

Environmental dosimeters (either TLD or OSLD as discussed in Chapter 5) are routinely placed at locations around nuclear facilities. The principles of measurement are the same as discussed before, but the badges do have two differences. First, the badges are designed to measure only gamma radiation because they are intended to measure general radiological conditions and not assign a dose to a person. Second, the badges are designed to be outdoors in inclement weather; therefore, they are enclosed in waterproof, sealed holders. Typically, the badges are placed above the ground on utility poles in accessible areas so that gamma radiation from the facility itself or gamma radiation from any airborne emissions can be measured. The number of badges depends on the facility, but it is common to have from tens to possibly a hundred environmental badges within several miles of the facility.

The badges are integrating dosimeters and thus must be swapped out on a regular basis. Typically, a quarterly exchange is followed, but some facilities may choose other frequencies. Most environmental badges will show only background radiation, although there may be some variation across the distance that separates the badges. Badges that are near the site perimeter of some facilities may show some elevated readings, however. In particular, boiling water reactors (BWRs) may produce gamma radiation that is detectable at the site perimeter, although the actual radiation dose is still quite low.

Despite the typically low readings on the badges, reviewing the results is still important for the environmental health physicist to identify any outliers and possible locations where new measurements should be made. In particular, unexpected changes in readings should be investigated. For example, the author is familiar with a situation involving an elevated reading for an environmental dosimeter that was on an electric utility pole on the edge of a rural farm. Bulk quantities of fertilizer had been placed near the utility pole on the ground and left for an extended period of time, and the gamma radiation from naturally occurring materials in the fertilizer had been detected by the dosimeter.

12.11 Sewage Monitoring

As discussed in Chapter 11, facilities can obtain permission to discharge small amounts of radionuclides to the sanitary sewer system. While the amount from any one facility may be trivial, the combined amount from multiple facilities within a given area should be addressed as well. Thus, regulators will typically go to sewage treatment facilities to obtain samples that can be analyzed. While the prospect of gathering a sewage sample is somewhat unpleasant, it is nonetheless necessary to gain a complete picture of radionuclide discharges in the area served by the treatment plant.

As part of the sewage treatment, liquids are extracted from the solids and then sanitized before being discharged to surface waters. Thus, it would be important to obtain a liquid sample after treatment in addition to the residual sludge, since different radionuclides will partition differently between the liquid and sludge phases.

12.12 Annual Environmental Reports

When the measurements have been made from monitoring, and the calculations have been made regarding modeling of annual releases, what becomes of all this information? Typically, it is used by the nuclear site to prepare an environmental report that is published on a yearly basis and includes information from the previous calendar year. In the past, the reports were submitted to the regulators and were available in print form for members of the public to request a hard copy.

In the modern era, regulatory agencies and nuclear sites routinely make annual environmental reports available on the internet for free download. These reports are quite detailed and explain the measurement process, measurement locations, and measurement results. In addition, the results of modeling for airborne and waterborne emissions are presented; the reports use language consistent with what has been discussed in both Chapters 11 and 12 of this text but will also include nonradiological impacts along with radiological impacts.

Because of the time required to analyze, assemble, and review the data and text, it can take several months to almost a full year to publish each year's report. Thus, the most recent annual environmental report available online will not be for the current calendar year and could be a couple years old due to production time.

Figure 12-8 below shows examples of the covers of some of the reports available for NRC licensees and DOE sites. As mentioned above, the full text can be downloaded as a pdf file and studied to understand the measurements and assumptions.

Figure 12-8. Sample covers of the Annual Environmental Reports for the Farley Nuclear Plant, Los Alamos National Laboratory, Oak Ridge Reservation, and Palo Verde Nuclear Station.

To find the annual environmental report for a given facility, it is necessary only to use an internet search engine and search for the site's name and the search terms *annual environmental report* (quotes are not usually necessary around the search terms). For NRC-licensed power plants, the reports will be accessible via the website of the Nuclear Regulatory Commission. Sites operated by the Department of Energy generally will have their reports on their respective sites.

In addition, some state agencies who perform confirmatory environmental monitoring will have their reports available as well. In that case, those reports can usually be located through the state regulatory agency responsible for radioactive materials and radiological protection.

12.13 Conclusion

We now come to the end of this chapter as well as the end of this textbook. As has been stated several times, the goal of this book was to provide the basic foundation of radiation protection at the professional level rather than the technician level. While there are many other areas in radiological protection, mastering the content in this text should prepare the professional-level health physicist to be able to adapt to the other areas.

It is the hope of the author that this book has been a help in the journey towards a professional career in health physics/radiological protection.

Free Resources for Chapter 12

- Questions and problems for this chapter can be found at www.hpfundamentals.com
- International Atomic Energy Agency, *Environmental and Source Monitoring for Purposes of Radiation Protection*, IAEA Safety Standards Series No. RS-G-1.8, IAEA, Vienna 2005. Available from www.iaea.org.
- NRC Regulatory Guide 1.109 *Calculation of Annual Doses to Man from Routine Releases of Reactor Effluents for the Purpose of Evaluating Compliance with 10 CFR Part 50, Appendix I*, US Nuclear Regulatory Commission, 1977. Available from www.nrc.gov
- NRC Regulatory Guide 4.1, *Radiological Environmental Monitoring for Nuclear Power Plants*, US Nuclear Regulatory Commission, 2009. Available from www.nrc.gov.

Additional References

- Moeller, D. W. *Environmental Health (4th edition)*, The President & Fellows of Harvard College, Cambridge, MA, 2011.

Index

Made in the USA
Las Vegas, NV
08 February 2024